THE CITY

URBAN GEOGRAPHY

URBAN GEOGRAPHY

URBAN GEOGRAPHY

A Study of Site, Evolution, Pattern and Classification in Villages, Towns and Cities

GRIFFITH TAYLOR

Routledge
Taylor & Francis Group
LONDON AND NEW YORK

First published in 1949

This edition published in 2007
Routledge
2 Park Square, Milton Park, Abingdon, Oxon, OX14 4RN

Routledge is an imprint of Taylor & Francis Group, an informa business

Transferred to Digital Printing 2009

British Library Cataloguing in Publication Data
A CIP catalogue record for this book
is available from the British Library

Urban Geography
ISBN10: 0-415-41748-1 (hbk)
ISBN10: 0-415-48958-X (pbk)

ISBN13: 978-0-415-41748-8 (hbk)
ISBN13: 978-0-415-48958-4 (pbk)

Routledge Library Editions: The City

Canberra, an air view of the southern suburbs looking south from near Parliament House (See Fig. 173). The Sports Ground at the left is Telopea Park, and in the foreground is the suburb of Barton. In the background is Red Hill Ridge, with the Bimberi Range (6,000 feet) in the far distance at the right. Notice young trees, round-points and radial streets.

URBAN GEOGRAPHY

A Study of Site, Evolution, Pattern and Classification in Villages, Towns and Cities

by

GRIFFITH TAYLOR

D.Sc. (Sydney), B.E. (Mining), B.A. (Research) Cambridge
Professor of Geography in the University of Toronto
Formerly Professor of Geography at the Universities of Sydney and Chicago
President of British Geographers (B.A.A.S.) 1938
President of American Geographers (A.A.G.) 1941

With a frontispiece and 300 plans and diagrams

SECOND EDITION, REVISED

LONDON: METHUEN & CO. LTD.
NEW YORK: E. P. DUTTON & CO. INC.

First published February 3rd, 1949
Second edition, revised 1951

CATALOGUE NO. 5118/U (METHUEN)

PRINTED IN GREAT BRITAIN

PREFACE

TWENTY years ago I published the first volume of a series covering the field of Cultural Geography. It dealt with the borderland between geography and anthropology, and stressed especially the relations between Environment and Race. In 1936 the second volume (linking geography and history) appeared, and in this the interactions of Environment and Nation were investigated. During the war I wrote a third volume describing geographical aspects of the path to World Peace. This book is entitled *Our Evolving Civilization*, and is an integrated study of the whole field of Cultural Geography. It has now been published by the University of Toronto Press. It was logically the last of the series, but it formed the text of the Messenger Lectures at Cornell, which I gave there in 1944, so that it was ready earlier than the present volume.

This study of Urban Geography covers a field which, so far as I am aware, has not been traversed in any English text book. However, the shift towards urban life, obvious in all progressive nations, has resulted in lecture courses on the subject in all the important University centres of geography. Sociologists and Town Planners have however produced a score of useful text books in complementary fields, and it is hoped that the present volume will be of interest and value in most classes concerned with those subjects also.

The writer has been collecting data on town evolution ever since he was engaged in the preliminary surveys for Canberra in 1910. It will be found that almost all of the 200 towns studied have been visited in his journeys since that date. There are indeed almost an infinite number of settlements available for study, but in his courses of lectures on Urban Geography the author has naturally stressed those with which he is more or less familiar.

North America is perhaps especially suitable as a field of research, since its northern regions show occidental settlements in every stage of early evolution—which is not the case in the older settled regions of the world.

I have borrowed freely from memoirs by American and British geographers, and such references will be found to be fully acknowledged. I owe special thanks to Lewis Mumford, who kindly gave me permission to quote extensively from the best sociological study of the subject, i.e., *The Culture of Cities*. A number of my graduate students have furnished me with data dealing with communities which they have especially studied. Almost all the maps are due to my own rather amateur pen.

In conclusion the author realizes that this tentative study poses many unanswered problems, but such is the common lot of the pioneer. However it is hoped that future investigators of this rapidly expanding field of geography will find this volume useful in their classes, as well as a foundation for their own research.

GRIFFITH TAYLOR

TORONTO UNIVERSITY
September, 1946

CONTENTS

PART I. GENERAL FEATURES

PART II. HISTORICAL

PART III. TOPOGRAPHIC AND OTHER CONTROLS

ILLUSTRATIONS

PART I

GENERAL FEATURES

"Urbanism has become so powerful an influence in human life that the very term which we use to express larger social loyalties is *Citizenship*."

L. F. THOMAS.

CHAPTER I

INTRODUCTION

Various types of human agglomeration

THE peculiar province of the geographer is to interpret the relation between man and his environment. For a generation there has been debate as to the closeness of this relationship, geographers being divided into two schools on the question. The Possibilist stresses the human aspect of the relation, the Determinist the environmental factor. The present writer belongs to the latter school, and has devoted much research to an evaluation of the relation in regard to the three main types of human agglomeration. These essential types are Race, Nation and City.

It is generally accepted that Man has differentiated into several varieties which we call Races. There is no complete unanimity as to the major races of man; but in his various researches the writer has adopted the following five classes:—Negrito, Negro, Australoid, Mediterranean and Alpine races. Readers who are interested are referred to the book *Environment, Race and Migration* (Toronto and Chicago, 3rd edn., 1946) for the data behind this classification. Since there are about 2200 million people on the earth, it is clear that the human agglomeration which we call a *race* contains on the average about 400 million people. A somewhat closer approximation is given in the following table.

It may be well to explain that I have no belief in the classes instituted by Cuvier and Blumenbach about the year 1820. It is one of the inexplicable features of orthodox anthropology that it clings to these outmoded classes of Caucasian, Mongolian, Amerind, Polynesian, etc., etc. As I have explained at length in *Environment, Race and Migration* it seems to me obvious that we have similar migrations from the cradleland (in south central Asia) into *each* of the other continents (excluding Australia). Hence there are Alpine Broad-heads (brakephs) among the Asiatic, American and Polynesian aborigines. The same thing is true of the narrow-headed (dokeph) Mediterraneans; they also have spread from Asia into Europe, Africa, Polynesia, North and South America. For the purposes of this generalized table I have incorporated the Nordics with their kin the Mediterraneans. The Australoids include the wavy-haired aborigines of Australia as well as the much more numerous folk of the same race in southern India.

RACES OF THE WORLD (figures represent millions)

	Europe	Asia	Africa	N. America	S. America	Pacific	Australia	Total
Negrito	0	.05	.02	0	0	.01	0	.08
Negro	0	0	70	10	5	1	0	86
Australoid	0	40	0	0	0	.1	.1	41
Mediterranean	250	560	70	80	55	?	7	1022
Alpine	250	700	0	80	15	.5	0	1045
Total	500	1300	140	170	75	1.6	7.1	2194

My book supports the thesis that man differentiated and migrated in response to environmental changes. The habitats of the five races (before the day of widespread emigration) were arranged in a series of zones centred about south central Asia. These zones considered in conjunction with the 'fossil strata' left by earlier migrations, enable us to decide upon the cradleland of man, somewhere in south central Asia, and upon the order of evolution of the races. The evolution is indeed indicated by the order of the five zones; the marginal Negritoes being the earliest, followed by the Negroes, Australoids, Mediterraneans and Alpines. The latter, occupying the centre of the racial zones, are (according to the laws of biological distribution) the latest evolved of the races of man.

In the second volume of the series somewhat similar considerations in regard to Nations were discussed. The Nation is a cultural unit and is not of the same type as the biological unit which we call a Race. A dozen complex factors go to produce a Nation. Many of these are found to be man-made, depending on social, political and military factors. However, the environmental factor persists throughout the development of a nation, and it is easy to show how vital is its influence in the development of a stable and lasting nation. One may mention the importance of the plateau pastures (*Plaiuri*) of the Carpathians in the survival of the Roumanian nation, in spite of repeated invasions by Asiatic peoples.

There are about 30 nations in the continent of Europe, and the book *Environment and Nation* (2nd Edition, Toronto, 1947) is concerned only with these. Since the total population of Europe is about 500 millions, it is clear that the average population of a European Nation is about 17 millions. This is a much smaller agglomeration than the Race; which averages about 400 millions. Of course some nations are much larger than this average figure, as the following table shows. Indeed only

Roumania and Czecho-Slovakia are near this figure, the great number being much above or below 17 millions.

Largest Nations			*Smallest Nations*		
Russia in Europe	130	millions	Albania	0.8	millions
Germany	69	„	Estonia	1.1	„
Britain	48	„	Latvia	1.8	„
France	42	„	Lithuania	2.2	„
Italy	43	„	Norway	2.6	„

Just as the first volume, dealing with distribution of folk over the whole world led to somewhat new ideas in regard to Race, and interested anthropologists, so also it is hoped that the discussion of European Nations throws new light on some of the general principles of history. We shall see that a study of towns and cities will lead to certain novel conclusions which should be of value to Sociologists and Town-Planners.

How does the agglomeration of human beings which forms the unit in our present study, compare with those considered in the preceding volumes? The largest cities of the world, London and New York, consist of groups aggregating about eight millions. About 1911 there were 16 cities in the world whose populations exceeded one million, three being in the United States, two in Russia, and two in Japan. To-day there are over one thousand towns exceeding 10,000 people in the United States, and in some countries such clusters would be called 'cities'. There is, however, no generally accepted figure which determines whether a cluster is a city, a town or a village. Perhaps we may call a settlement with less than 500 folk a village, while a town has a population between 500 and 10,000. There is also a great difference of opinion as to the *optimum* number of folk who should form one such aggregation, whether town or city. There is some evidence that a figure of about 40,000 is well suited to present urban conditions. Perhaps we may say, therefore, that a figure between 10,000 and 40,000 is the size of the cluster with which the term urban geography is usually associated. However, the giant city of to-day (which Mumford has christened 'Megalopolis') unwieldy and unesthetic as it is, has passed through the early stages of pioneer dwelling, village, town, city and metropolis, before it reached its present self-suffocating condition. Hence our study of the City necessitates a discussion of all the earlier forms of urban agglomeration, if we would understand its evolution. In other words this volume is concerned with all the stages of the third and smallest human agglomeration, though the 'climax condition' may be thought to be a city of about

40,000 people. In conclusion, therefore, we may say that *racial* clusters contain about 400 million people, *national* clusters about 17 million, and *urban* clusters about 20,000 people.

It is rather surprising that there is so little regular study of the characteristics of Urban Geography in the leading Departments of Geography to-day. When we realize that even in a new country like Australia no less than 64 per cent of the population is urban (Fig. 1), while in Britain it rises to about 80 per cent, it is clear that the geographical factors controlling *city* life are of the greatest importance to a large proportion of the people of the world. The historians and sociologists have given much more attention to the urban effects of the Industrial Revolution than have the geographers. It results, as in the case of Race and Nation, that the environmental factors in this interesting development have been to a large extent ignored, or at least submerged in the discussion of the sociological conditions.

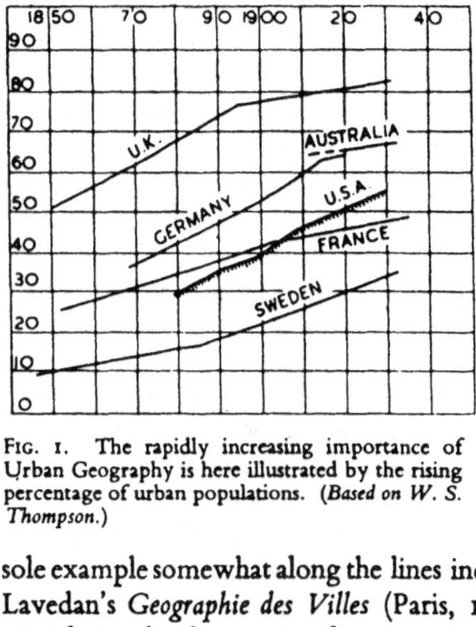

FIG. 1. The rapidly increasing importance of Urban Geography is here illustrated by the rising percentage of urban populations. (*Based on W. S. Thompson.*)

There are several valuable studies by geographers in closely allied fields; for instance, Geddes' *Cities in Evolution*, Van Cleef's book on Trading Centres, or Rodwell Jones' *Geography of London River* (London, 1931); but the sole example somewhat along the lines indicated above is in French, and is Lavedan's *Geographie des Villes* (Paris, 1936). However, this small but stimulating book contains few maps or diagrams, and it makes little attempt to analyse the factors which have led to the many different types of towns and cities. In effect it does not lay stress on the environment of the site, nor on the stage of evolution of a given town in a more or less inevitable cycle of development, nor on the relation of the major centre of a district to the smaller clusters; all of which seem to the present writer to be significant features in the geography of the cluster under consideration.

The most valuable studies in allied fields are a number of books dealing with Town Planning. Of these the most helpful is that by Lanchester (London, 1925). Several of them give useful accounts of the evolution of the town in classical and medieval times. Of course the outstanding work in this field is *The Culture of Cities* by Lewis Mumford (New York, 1938). It is not too much to say that this book has opened a new outlook on the evolution of cities. Especially useful is Mumford's concept of the Eotechnic and Paleotechnic types of city. I have used his classification to some extent in the sections dealing with the major features of town anatomy. There is no better study of the way in which social changes have brought about changes in the plans of cities; and he is the strongest protagonist of the Possibilist School because he is most interested in the *human* side of the problem. To many readers this is admittedly the most interesting side of urban geography; but the writer hopes to show that the complementary features, determined to a greater degree by the *environment*, must by no means be neglected if a full picture of the urban landscape—the supreme characteristic of our 20th century way of life—is to be appreciated.

Perhaps the main purposes of the present volume may best be understood by references to analogous work in other fields. Half a century ago there was a tremendous amount of data available regarding the various aspects of a normal topography. It was the great achievement of William Morris Davis to co-ordinate these data, and to produce his invaluable concept of a 'Cycle of Normal Erosion'. It is unlikely that we shall ever be able to analyse and synthesize the data of city development so adequately that we can produce as logical a concept of a 'Cycle of Town Evolution'. We can but try. For one thing we are dealing with the irrational actions of Man, rather than with the inevitable actions of Nature. Hence the City Cycle cannot be as clear as the Landscape Cycle.

I have in my previous volumes shown how helpful is the 'Zones and Strata concept', when applied to the differentiation and migration of races, languages, etc. The growth of a city starts at a nucleus, spreads in definite functional zones (as indicated by the types of buildings), and embodies a mass of data regarding outmoded or demolished buildings, which can be considered as equivalent to the *strata* in the 'Zones and Strata Concept'. I have worked out a number of lengthy examples of the use of this technique, primarily in the vicinity of Toronto and Chicago; but it is of course applicable to any other district. Turning to another aspect of the problem, Köppen has simplified the comparison of the innumerable climates of the world by his classifications and formulae. I have ventured to apply a new technique to urban evolution, which gives us formulae for any given town or city. These are quite manageable for towns with a population of 40,000 or less. Indeed it is easy to work backwards and to draw a plan *from the formula*, which is a close approximation to the real distribution.

THE SCOPE OF THE BOOK

Part I—General Considerations

In dealing with so varied a subject as villages and cities, it is very difficult to decide on the presentation of the vast body of material which is available in literature; or which has been collected by the writer ever since he made his first study in this field in 1910. In that year he became officially connected with the location and lay-out of the planned city of Canberra; the proposed capital of the Commonwealth of Australia. This city and its evolution will be referred to many times in the ensuing pages.

It will be of interest to many geographers to see what determined the writer to adopt the classification and order which is described in his book. In most cases the data in urban literature are collected under the heads of the respective countries; for instance, most advanced geographies of a given country give some description of the site and evolution of the chief towns. One could therefore have given a description of the various towns of Germany, with a view to finding what, if anything, determined the special hallmark of a German city. But since in these prosaic days most businesses and techniques are fairly standardised throughout the industrial world this did not seem a very hopeful procedure. After all it is not the national character of a town that the geographer is primarily anxious to ascertain. In the case of the present writer it is the site and the evolution which are of most interest. Hence it seems logical to make our classification and our order of procedure depend essentially on these two aspects of the problem.

It was soon evident that one would need to specialize somewhat in discussing such a large topic. In the case of my study of *nations*, I confined my attention mainly to the 30 *European* nations; for it seemed evident to me that the general principles evaluated in that study of 'Environment and Nation' in Europe, would be found to be in large measure applicable to nations developing in the other non-European centres of national development. But for these the material is not available, and in general is not described in the languages with which I am familiar. So also in the case of the siting and evolution of towns and cities, I soon came to the conclusion that the field must be narrowed to those towns which had developed according to the pattern of a European civilization in relatively modern times. The writer is familiar, through long residence, with many towns in America, Europe and Australia. However since the evolution of towns goes far back in the history of man, and therefore leads us to classical and eastern areas, these are described—usually from first-hand knowledge—in one or two of the earlier chapters. But in general the research deals with what Mumford would call Eotechnic (medieval) or Paleotechnic (industrial) towns; or with mixtures of the two, which are extremely abundant in the continent of Europe.

I have endeavoured to describe processes and stages in some detail in the case of a few outstanding examples in the earlier chapters. The later chapters take the *dominant feature of the environment* as the key to the city; and I have attempted to show how a town would develop say on a prairie location; or at the bend of a river around a hill-fort; along the line of a great mountain corridor; at the head of a lake; where roads or railways cross; etc., etc. It seems more logical to use the unaltering character of the site as our chief determiner rather than the varied functions of the final city. Most industrial cities of to-day pass through much the same stages of development (as I hope to show when introducing the concept of *infantile, juvenile*, and *mature* settlements) but the natural environment always greatly affects the actual pattern of the growing city. Hence New York and Chicago, which are both colossal industrial agglomerations, in somewhat the same stage of development have quite different patterns, due in large measure to their very different sites.

In many of my recent lengthy papers or addresses I have adopted the technique of the 'traverse'. For instance, in describing Eastern Algeria I describe the varying environments met on a traverse *across the isohyets* from the Mediterranean to Touggourt, some 180 miles within the Sahara. The determining factor in settlement was the rainfall, and my journey gave me type-samples of all the varying settlements in the country. In the present instance, where we are dealing (at first) with towns all over the world, it is clear that temperature is even more important a determiner than rainfall as regards the character of the village or town. Hence I describe in the chapter following the Introduction, seven sample settlements met with on a journey from the Pole to the Equator, at about ten degrees of latitude apart. There is little likelihood that there will be much overlap in the general conditions determining the siting of these towns, if they are chosen so far apart. As a matter of fact I was able to describe from personal knowledge a sample from each of the seven continents, while still adhering to the latitude condition.

Considerable attention is given to each of these seven samples, which are as follows. (1) The little settlement of Cape Evans, built by Captain Scott in the Antarctic in 1911. (2) The fishing centre and port of Bergen in Norway, the largest city within 20 degrees of the Pole. (3) The small pioneer fur-trading centre of McMurray in northern Alberta. (4) Peiping, the former Chinese capital, a gigantic but somewhat senile and oriental type of an Eotechnic City. (5) Biskra, a French military post in the northern Sahara. (6) Urandanji, a little township on the main cattle-route in the arid interior of Queensland, Australia. (7) Santa Marta, a typical Spanish town on the coast of Colombia in South America.

In each of these very varied examples I try to answer the five following questions:—

(1) What induced man to settle in this region?
(2) Why did he choose this particular site of all those available?
(3) What kind of a settlement developed on the chosen site?
(4) How far has the settlement progressed towards the climax stage, having in view the period since it was founded?
(5) Can we decide where the environmental factors and where the human factors have played the greater part in the present pattern?

At the conclusion of this section of the book, the reader will have a clearer picture of the principles of urban structure and function which the writer is trying to analyse and synthesize.

In the next sections (Chapters III, IV and V) an attempt is made to classify the "Seven Ages of Towns". The evolution of the little farming settlements along the Kankakee River (based on research by A. Meyer) give us a picture of the change from a wild environment, exploited by nomad Indians, to a drained and cleared region developed around many small farming towns and villages. The process is carried on by an investigation of the evolution of the large city of Toronto, from its beginning around 1790 to its present condition of a huge industrial city of 700,000 inhabitants.

In determining the critical stages in the growth of a large city the writer has come to the conclusion that the "Zones of Building" offer the best criteria. In the *infantile* stage there is no separation of the domestic and commercial areas, or of the richer and poorer quarters. As time goes on we see the separation of the shops from the houses, and then the development of a few scattered factories. These are *juvenile* conditions. Then the houses tend to segregate, the better-class on the border, the older houses changing into depressed areas. With the definite zoning of the houses and the separation of commercial and industrial districts we reach *maturity*. *Senility* is expressed by the decay of definite districts, and by the definite decline in the prosperity and growth of a town.

The next major section of the book (Chapters VI and VII) gives a brief description of the prehistoric and later classical settlements. It shows that

Aurignacian man in Europe built houses of the type still used by the aborigines in Australia. So also the pit-dwellings of later Paleolithic times (as the Quennells have pointed out) resemble those used by the Eskimo to-day, while the Neolithic settlements of Britain have much in common with those of aboriginal peoples in Colombia or New Guinea to-day.

Medieval towns have in a number of cases developed directly from those of classical times. Samples of towns in both these long periods of history are illustrated and described in Chapters VII to X, primarily with a view to showing how much modern towns owe to them. A number of interesting transition types in Britain are described in some detail. A special section deals with the growth of a typical medieval University, that of Cambridge, England.

In Chapter X a resumé is given of Mumford's classification of towns into Eotechnic, Baroque, Paleotechnic, Neotechnic and Biotechnic. Mumford, like the present writer, confines his attention very largely to European and American towns. Much the larger portion of his stimulating and lengthy book is given to a discussion of the haphazard development of the industrial European town. His chief aim is that of the town-planner, i.e., to replace the monstrous growth due to the land speculator and the factory owner, by a city which is capable of future æsthetic growth, and of which the citizen may well be proud.

Part II—Classification by Site

Before discussing the various classes of towns as controlled by their sites, it will be well to describe in some detail the way in which the topography has affected various typical towns. I have chosen Edinburgh, Quebec and Sydney for this purpose. Edinburgh dates back to the earliest historic times, and its evolution has been explained in one of the finest memoirs of urban geography (*Scot. Geog. Mag.* 1919). It has also an extremely attractive situation, although no large river, lake or sea in the vicinity lends the charm of water to the landscape. Quebec gives us the best example in North America of a transition city. Much of the eotechnic (i.e., walled, medieval) condition can still be made out. In no city of my acquaintance is the control by site more marked. Incidentally it is of great interest as a bilingual city. Sydney has one of the finest natural sites in the world, though as an old 'Sydney-sider' I may be permitted to extol the charms of Vancouver and Hong Kong—which have the benefit of higher elevations and view-points in the immediate vicinity of the city. Sydney is the second city of the Empire, and yet the first house only dates back to 1788. Hence its history is particularly easy to trace, though to date no real *historical* study has appeared in print.

It seems logical to the writer to discuss sites of towns in an order based essentially on varying importance of the topography. Thus some of the earliest settlements were based on a hill-fort or *Acropolis*; for instance, Athens has such a position, and so has Rome. There are dozens of examples in France and Spain, some of which are described in an effort to analyse their main characteristics. A second group is based on *Cuesta* structure. These towns are best studied in the Weald Basin or some allied structure. Mountain *Corridors* have towns of a special character, which the writer has studied in some detail in the north of Italy. Towns controlling *Passes*, or on a larger scale '*Gates*', are of great significance, especially in Europe. One thinks of Vienna, Belfort, etc., in this connection The distribution of towns on a *Plateau* such as the Harz, or on an '*Eroded Dome*' such as the Black Hills, or that classic example the English Weald, are arranged according to fairly constant patterns.

Ports have always been main factors in the growth of many important towns. They may develop in *Fiords* and deep narrow gulfs, as in Norway; or along *Rias* as at Sydney and in many north Spanish towns. River Estuaries are very common sites of important towns, though often the latter are placed so far inland that they seem to be river towns rather than ports. Open coasts and wide bays are not such satisfactory sites, but modern engineering can cope with almost any type of shore and produce some form of satisfactory roadstead. Various examples of such *Roadsteads* and coastal settlements are discussed.

River towns have always been important, and in the early days perhaps had a paramount importance, since modern supplies of water, brought from hundreds of miles away, in some cases, were not available. The presence of *Falls* was often the decisive factor in determining the site. Fine studies of the towns of Lowell and of Battle Creek are available for this type of town. In the case of *Meanders* the site of the town is conditioned by the curve of the river, as shown at Verona, Dresden, Basle, etc. River *Terraces* are often the main determiner in the plan of a town as in innumerable towns in British Columbia, e.g., Trail or Ashcroft. *Delta* sites, *Fan* sites and certain *Valley* sites all need consideration. Often enough in early days the presence of an *Island* in a river was a potent factor, as for instance at Paris. The distribution of towns around *Lakes* is often rather characteristic, and differs from that of river towns.

As the topography becomes less obvious it naturally exerts less control on a town site. Towns and villages which have developed on *Plains* are extremely common for obvious reasons. Unless there is a well-defined river the development of the town must necessarily depend to a much larger degree on purely man-made controls (i.e., railways) than in the case of a town where topography exerts a marked control. Partly owing to the abundance and prosperity of towns in rich agricultural areas in central United States, we find that some American geographers have paid little attention to topographical control in urban studies. To some extent a *uniform* environment is present at the dawn of many '*Clearing* Towns', cut out of vast forests. These also may be classed with towns where the environmental factor is somewhat remote, though the street plan usually shows some indication of the original small clearing. In *Deserts* underground water plays a large role, as for instance near Biskra in the Sahara. There are, however, many special features in Desert towns due to the unusual character of the climate; which is of course just as much an environmental control as is the topography.

Another section of the book deals with those towns which owe their existence primarily to special human needs or developments. Many *Mining* towns have developed in localities where no town would normally arise in such an unattractive position, but for the coal or metals found there. Many *Tourist* towns have been determined by factors which in general would deter settlement in other regions. Thus Davos has grown up at a height of 5,500 feet in a cold arid district in Switzerland, because of its value as a hospital site and for winter sports. Many centres of mountain-climbing are of this type, as are many mineral spas and resorts based primarily on wild scenery. In many parts of the world especially in plain country the crossing of *Railways* has led to the growth of a town. In a pioneer land, such as the Canadian Prairies, villages grow up at standard distances along the railway, say every eight or ten miles, quite independently of any other consideration whatever.

The subject of *Planned* towns deserves special consideration, and leads to a brief discussion of Garden Cities and Town Planning generally. *Boom* Towns and *Ghost* Towns merit some attention, and there are plenty of both types available for study in North America. Some rather miscellaneous aspects of urban geography are discussed in the latter part of the book. Among these are the Urban Outliers encountered on the approach to a huge city such as Chicago or London; and the Zones and Strata Concept applied to a city and its neighbouring towns, as illustrated by the relation of Toronto to Port Credit and Whitby. Other topics are the Use of Formulae for towns, and the conversion of such formulae into simplified diagrams which approximate to the original plan of the town; Conurbations and a general discussion of the density of towns in various given environments. This last technique can indirectly be used to ascertain the standards of living in a given region, as I show by contrasting the populations of southern Australia and Algeria. Some general rules which students should observe in making a detailed survey of a large town—primarily with a view to mapping the functional zones—appear in the last few pages.

CHAPTER II

SEVEN EXAMPLES OF SETTLEMENTS FROM TYPICAL LATITUDES

THE study of Urban Geography is yet in its infancy, and no generally accepted classification of towns has yet been put forward. There is something very interesting about an attempt to ascertain the essential features of what must, in the future, become a very notable and well-cultivated branch of geography. What are the essentials in this study? No answer that it is at all complete can yet be made. But the obvious procedure is to study a number of fairly typical settlements—in as great a variety of environments as possible—and then decide what are the most notable features that have come to light in such a catholic study.

No student of human geography will deny that in the broadest sense *latitude* is the variable which most controls human affairs. Hence in the following paragraphs I shall describe seven settlements on a traverse from polar to equatorial latitudes, which happen to have come under my own observation. Their positions will be about 10° of latitude apart. It should make the examples of more interest that each is situated in a different continent. In each sample an attempt will be made to decide the respective importance of the environment and the human factor in the development of the settlement. No accurate evaluation is possible, but it is obvious that the environment is usually of greater importance in the choice of the site, while the human factor becomes more significant as the settlement grows in size and importance.[1] It may be noted that geographers belong to two rival schools. The 'possibilist' believes that man can develop any country along one of *many possible* ways. The 'determinist' thinks that the environment is by far the dominant controlling factor in most parts of the world, and that there is usually only *one* best way, *decided by Nature*.

The following table shows some of the main characters of the seven samples chosen.

[1]This section is largely based on my Presidential Address to the Association of American Geographers, Annals A.G. Vol. XXXII, March 1942.

	Latitude	Continent	Place	Population	Type	Raison d'etre	Köppen
1	77	Antarctica	Cape Evans	25	Ice cap	(Science)	EF
2	60	Europe	Bergen	100,000	Cool temp. port	Fish, etc.	Cfb
3	57	N. America	McMurray	500	Taiga village	Fur, etc.	Dfc
4	40	Asia	Pekin	1,600,000	Oriental	Political	Dwa
5	36	Africa	Biskra	4,000	Desert	Dates; military	BWh
6	22	Australia	Urandanji	50	Semi-desert	Cattle	BShw
7	11	S. America	Santa Marta	30,000	Tropical port	Bananas	AW'i

Since the climatic data of these clusters are of importance in the picture, I have added a hythergraph chart (Fig. 1a). It shows that most of the chief

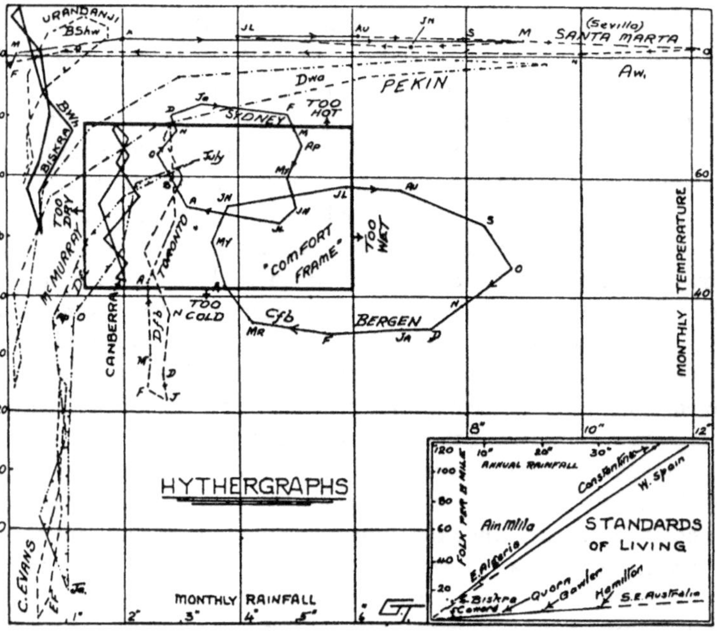

FIG. 1A. Hythergraphs of the places described in Chapter II, to show the widespread character of the discussion. Köppen formulae are added. The "Comfort Frame" includes the climates which are generally comfortable. *Inset* is a graph which compares the standards of living in three regions (Algeria, Spain, and South Australia), which all have the same environments.

environments of the world are represented. I have also inserted a 'Comfort Frame', which tentatively encloses the climates which most Anglo-Americans would call 'comfortable'. I do not however propose to dwell in any detail on climatic controls.

1. Cape Evans, a settlement on the Antarctic ice cap

We may learn a good deal as to the role of an adverse environment in the development of a settlement, by considering the first attempts to live in Antarctica. No doubt one of the little coal towns in Spitsbergen (such as Longyear City) would be more informative, but the major experience of the writer has been gained on the west coast of Ross Island in latitude 77° 30′ South. (Arctic conditions will be described later.)

The first question one should ask about any settlement is 'Why did man desire to establish himself there?' In the majority of cases the answer is 'Because he hoped to exploit Nature's resources in the vicinity.' In a very remote degree this is true of the small settlements on Ross Island, of which there are three (Fig. 2). Coal or metals may yet be found in rather inaccessible areas in Antarctica, though there is little prospect of immediate use.

The major attraction was that hereabouts open water is found nearest the South Pole. Hence Ross Island in the Ross Sea offered the best site for exploration, and for extending in this unique environment our knowledge of meteorology, magnetism, glacial erosion, geology, plant and animal ecology, etc.

Secondly, why was Cape Evans chosen rather than any other site? Here we have one of the best examples of environment control known to the writer. Every explorer wished to land his stores at the most southern spot available, as *early* in the summer season as possible. Hence the huts were invariably erected at the edge of the open sea, when the boats reached Ross Island, as usual, early in January. Thus in 1902 Scott placed his hut at Hut Point; the best of all sites, since he could here easily reach the southern Ice Shelf. In 1908 Shackleton's ship could not sail further south than Cape Royds, where he was isolated from the mainland by open water, or by the crevasses of Erebus during the best sledging months.

On Scott's second expedition, in which the writer was senior geologist, we found thick bay-ice south of Cape Evans on the 4th of January. This fixed the site of the hut at Cape Evans. The southern parties were able to reach the Ross Ice Shelf as late as the 24th of January, but in a few days open water separated the hut from the Ice Shelf. It is of interest that our sledging parties returning from the south were unable to return to our own hut at Cape Evans until the 13th of April. In the meantime we lived in the communal settlement at Hut Point. Two years later two men were drowned on this short crossing of 14 miles, because they tried to traverse the sea ice to Cape Evans before the new ice was thick enough. In the map (Fig. 2) I have added actual dates in 1911 when the sea froze near Hut Point; and when the ice broke away near Cape Evans.

Such then was the environment of Cape Evans. To the east extended uninterrupted ice slopes to the summit of Erebus at 13,000 feet. In places fields of crevasses prevented any crossing of the land ice; and of course the level sea-ice was far preferable for sledging, except in the season when it was liable to break away under one (in the summer).

The third question to answer is 'What kind of a settlement developed on the chosen site?' I have ventured to talk of Cape Evans as a 'communal hamlet'; since it reminds one of the similar settlements in Jugoslavia in

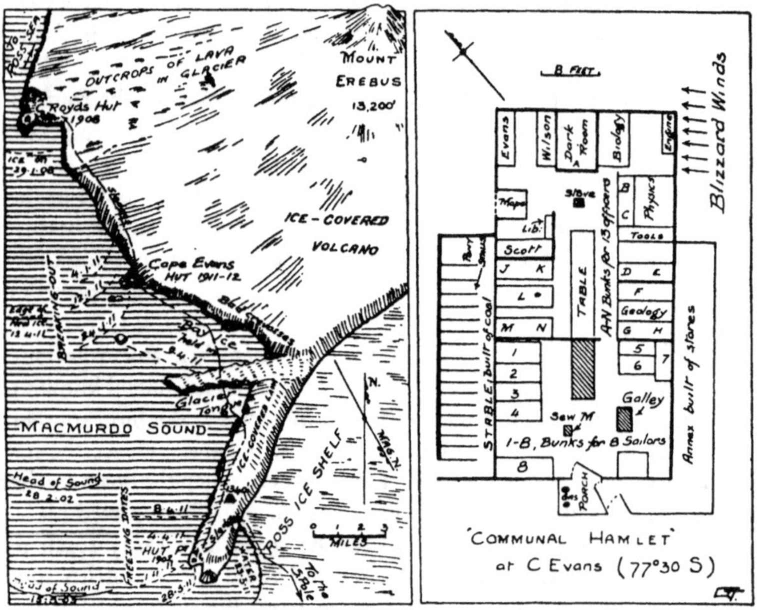

FIG. 2. (*Left*) A bird's eye view of the west coast of Ross Island in the Antarctic to show the settlements made by the three expeditions (1902, 1907, and 1910) where the environment permitted. Dates of the ice freezing and breaking away are added. (*Right*) Plan of the settlement at Cape Evans, Ross Island.

the late middle ages. (Still better would the term apply to the more complex hamlet at Little America.) The main hut (Fig. 2) had a floor area of 50 feet by 25 feet, and a simple roof with a central ridge. Three small windows, permanently shut and with doubled panes, admitted sufficient light in the summer, while in the long winter an acetylene plant in the porch gave us light.

Of considerable interest were the precautions to keep out the cold. In addition to five layers of wood and packing forming the walls, the hut was protected by side-sheds. These two annexes are clearly shown in the plan (Fig. 2). On the west the stables were built of compressed coal, while on the east an enclosed veranda was built of boxes of food, etc. To serve the varied needs of the twenty-five 'settlers' there were smaller buildings in the vicinity; such as the magnetic hut and the ice grottoes for storing food (and for pendulum work). There was also a great store of compressed chaff for the support of the ponies; while the dog kennels formed a lively 'faubourg' to the south.

At the end of the second winter the south-east blizzards had covered everything, so that the hut was almost invisible. But life inside the hut, with the temperature held at 45°, might be summarized in the phrase 'dark and dirty, but comfortable'. Outside the average temperature for July was around − 15°F.

If we use the analogy of living organisms in connection with this study of town development, then the hamlet just considered may be likened to an extremely simple animal consisting of only one cell. All functions are carried on in this cell, and no 'Zones' are present. It represents the simplest of human clusters, and is clearly in the very earliest 'infantile' stage of town development.

A few further notes on the environment may be added. The normal sledging season extends from about 20th September to 20th March. In these latitudes (77½° S.) the sun vanished about April 22nd and returned on August 22nd. However, the great cold and the almost constant blizzards made March and September very unpleasant for such sledging as had to be done. No vegetation (except rare moss and lichen) is found in the whole continent, so that no animals can live on the land. Practically all the food (except seal meat) had to be brought from civilized lands.

The density of human settlement under these conditions is too low to be charted. It may be noted that the nearest similar temporary settlement was the hut 400 miles north at Cape Adare; while Framheim (Little America) was about the same distance to the east. As far as the Antarctic continent is concerned, the geographer can say that the environment is supreme. Man (in the form of explorers and scientists) may *choose* to inhabit certain minute dots therein temporarily and at great expense, but the writer sees no argument in favour of almost complete human control (the dogma of the 'Possibilist') in our first example.

2. Bergen, a large town in the cool temperate portion of Europe

As noted earlier, our examples are to be taken at average intervals of 10 degrees of latitude. Hence our second example, as shown in the preceding table, is in a cool temperate climate; but is not typical thereof, since the coast of Norway has the warmest winter for its latitude in the world. However it is an extremely interesting town, for a variety of reasons. Owing to its position on the north-west of Europe it experiences almost constant, warm, moist, south-west winds. These produce a typical marine climate (Cfb, as shown in Fig. 1a), with a heavy rainfall and a very mild winter.

In addition to the unusual climate, Bergen has an unusual topography. Owing to the epeirogenic (i.e., *en masse*) uplift of the Caledonian Folds in Norway during late Tertiary times, the hinterland of Bergen is a plateau of about 4,000 feet elevation. The Caledonian folding (of 200

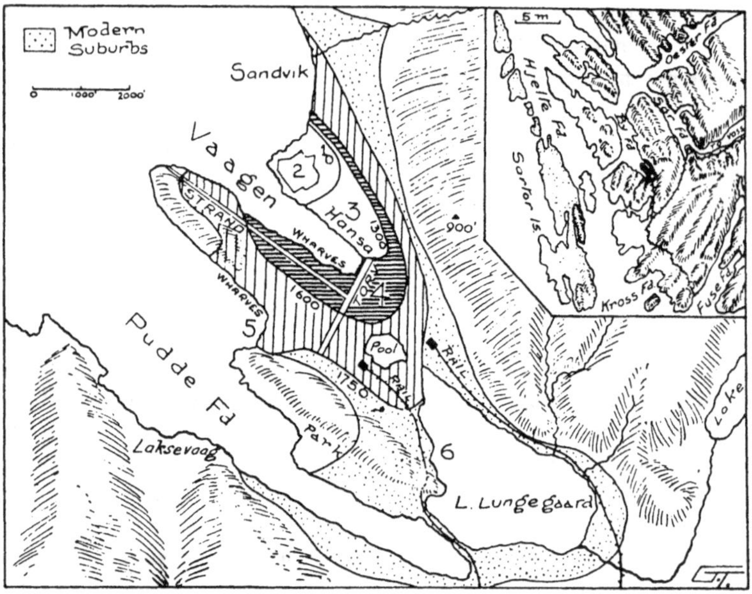

FIG. 3. Plan of the town of Bergen (Lat. 60° N.) in Norway, to show the control of the growth of the city by the topography. (*Based on H. Slanar.*) In the *inset* the site of Bergen (black) at the centre of the fold-arcs and fiords is charted.

million years ago) took the form of a series of parallel arcs, whose anatomy is clearly visible in the topography to-day, for the uplifted area consisted of alternating strips of hard granitic material and of softer Silurian schists. Erosion since the uplift, whether by water or ice, has removed the softer schists much more quickly than the harder eruptive rocks, and these 'Bergen Arcs' (crossed by what seem to be tension cracks) account for the regular valleys and gulfs which have in large part determined the site of Bergen (See the inset in Fig. 3). Thus By, Sor, Kross, and Fuse Fiords are hollows eroded in the arcs of soft schist; while Oester and Voss Fiords are perhaps transverse graben (depressions) due to tension.

Let us now answer our standard questions. 'Why did folk settle at Bergen in the first place?' We may safely assume that it was but one of the numerous sites used by the early Viking seamen. In the soft silts at the south side of the By Fiord, they dragged their boats ashore near the Sandvik of to-day (Fig. 3). In the early days Lungegaard Lake was connected to Vaagen Bay; while two morainic ridges separated Vaagen from Pudde Fiord. The town was founded near the present Bergenhus (2 in the plan) about 1075; and in later times several forts were built hereabouts to protect the harbour.

The Danish King Canute controlled Britain, Denmark, and Norway; and in 1070 Olaf founded Bergen; which lay midway between Edinburgh, Opslo (Oslo) and Nidaros (Trondheim). The name Bergen is derived from Berg-vin, i.e., mountain meadows, and refers to the abundant grass on the slopes of the By Fiord. Throughout the 12th century English traders were gradually replaced by Germans belonging to the Hanseatic League. During the next few centuries they built large warehouses all along the north side of the Vaagen, which is called Tydskebrygge (i.e., Deutsch Quay) to this day. About 1600 the power of the Hansa declined, and trade had all passed into Norwegian hands by 1750.

Before examining further the growth of Bergen it may be well to point out that it is the most northern city containing 100,000 people. This is the more remarkable since it is not a capital city, having been replaced by Oslo several centuries ago. (Helsinki, Stockholm, and Leningrad are all further south.) Nor has it any fertile hinterland, as is obvious from the inset map. We may safely say that its advantageous position in respect to the fishing grounds, and to the trade of the North Sea and the Atlantic Ocean, led to its very considerable expansion. As the map shows, it is protected from storms by the 'skerry guard' of the low Sartor Isles. Areas of farm-land on the floors of the numerous glacier-cut troughs are relatively large to the south-east of Bergen. Especially after the Bergen-Oslo Railway was built in 1900 did the city grow vigorously. Fish at first, and later ship-building and an entrepot trade in grain, wool, and petrol, account for Bergen's prosperity. It is still the chief trading port for Norway.

Reference to the main map in Figure 3 shows that Bergen was confined to the east side of Vaagen until 1300. This was the Hanseatic centre, perhaps at its maximum about 1500. Then the swampy land at the head of Vaagen was gradually covered with streets and houses. The present shopping and office centre is along the Torv (market) and Strand streets. Here also is one of the largest fish markets in the world. By 1750 the area between Vaagen and Lungegaard was all occupied by buildings.

Bergen has been the scene of many fires, one of the worst occurring in the district around the Pool in 1916, but this area has all been covered with large stone buildings since that date. Modern suburbs have spread round Pudde Fiord and Lungegaard. Various factories, mostly connected with ships or fish, have sprung up in outlying areas like Laksevaag or Sandvik. The steep sides of the hills, Floi-Fell (900 feet) on the east, and Lovstaken (1500 feet) on the south-west, have prevented the city from spreading far from the original site around Vaagen.

We can clearly see the part played by topography and climate in the growth of a town as large as Bergen in this high latitude; the same as that of south Greenland, north Labrador and the Bering Sea. Yet many cultural factors are of course present. Bergen was an early centre of religious and political power, and these are man-made factors. Water-power, fisheries, tourist trade, and railways depend partly on physical, partly on cultural conditions. A determinist will stress the former and say that water-power depends primarily on topography and rainfall, fisheries depend on the temperature and food conditions in the sea. Tourism flourishes where Nature has displayed her choicest scenery. Railways must go where the natural corridors are easiest. Man can *choose* one of an infinite number of ways of exploiting a country; but it seems to the writer that our 'chooser' rather resembles the so-called 'yes-man'. If he is to prosper, he does well to choose what Dame Nature the Dictator indicates is the best for him!

My visit to Bergen in 1931 was too short for me to make a functional survey of the town; and so far as I know this is not available. Many of the historic data in my main map are based on H. Slanar (*Proc. Geog. Soc.*, Vienna, 1918). It is obvious that the city exhibits the usual plan; with the business near the Torv, and with more or less concentric zones of offices, apartments, and villas to the south. These zones are somewhat horse-shoe shaped (rather than circular), in accord with the unusually hilly topography of the site.

3. McMurray, a fur-trading river port in North America

In 1778 Peter Pond explored the region north of Lake Athabasca, and about 1790 the North-west Company established a fur-trading station at McMurray on the Athabasca River (Fig. 4 at A). It was situated on the main corridor of canoe-travel from the Great Lakes to the Mackenzie River. Canoes reached Lake Winnipeg via the Lake of the Woods. Thence they paddled north to Cumberland House on the Saskatchewan, where they entered the Sturgeon River. At the head of this stream the Frog Portage led to the Churchill River which rises in La Loche Lake.

At the north end of the latter was the famous La Loche Portage where Methye Fort was built (Fig. 4). A rough wagon road of 12½ miles was

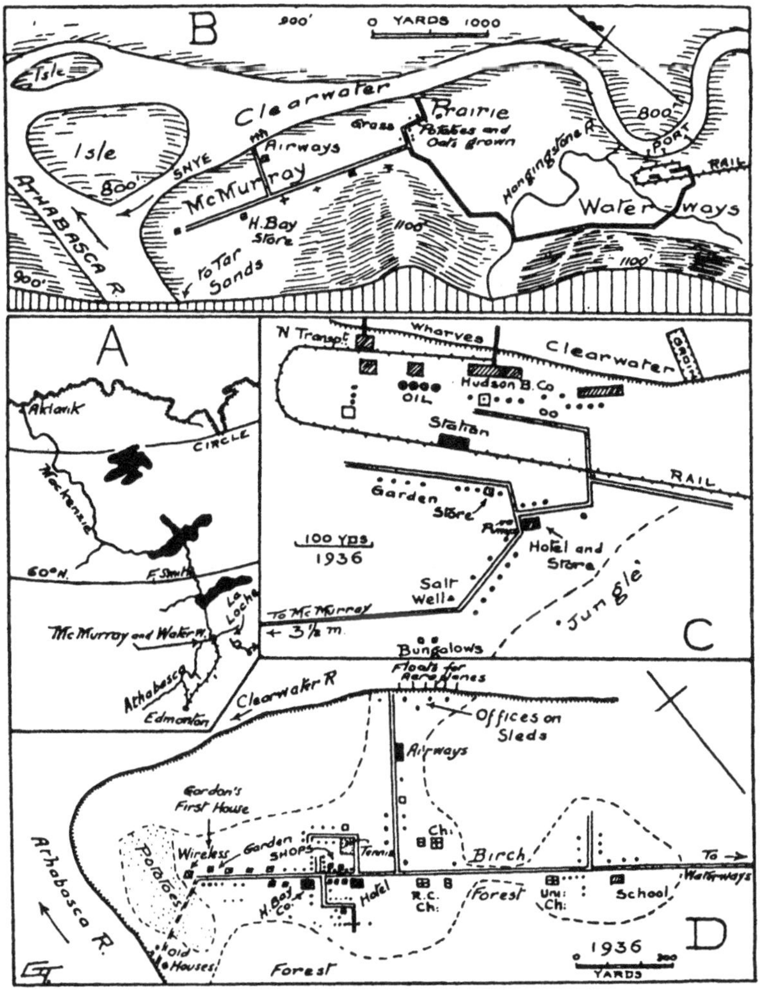

FIG. 4. Settlement conditions in the Taiga of north Canada. At *A* is a map of the Mackenzie Basin, showing McMurray and Waterways as well as the La Loche Portage. At *B* is a block diagram of the twin towns of McMurray and Waterways. At *C* is a detailed plan of the 'infantile' settlement of Waterways. At *D* the same for McMurray.

constructed here in 1875, so that heavy goods could be placed on the Clearwater River.

McMurray, the settlement under discussion, grew up forty miles west of the portage at the junction of the Clearwater with the mighty Athabasca

River. The waters of the latter, after passing through Lake Athabasca and Great Slave Lake, reach the Mackenzie River. The first steamer traversed the Mackenzie from Fort Smith Rapids in 1887, while in 1920 the railway from Edmonton reached Waterways on the Clearwater.

To-day the train traverses the Taiga north of Edmonton once a week, and I made my first survey in July, 1936.[1] At Lac La Biche (132 miles north) is the last of the farming country, though small patches of crops are grown right to the Arctic Circle. The chief inhabitants north of Lac La Biche are Cree Indians, engaged in trapping and fishing. The little towns, marked hereabouts on the map, usually contain fewer than half a dozen houses. It is therefore somewhat of a surprise to find that there are *two* towns at the end of the railway. There are Waterways and McMurray (Fig. 4 at B). Possibly it was intended to carry the railway beyond Waterways to the older settlement of McMurray. But deep water is available in the Clearwater; while rather high prices were demanded for the necessary lands at McMurray, so that there seems no likelihood of the railway being extended at present over the intervening three miles. Waterways controls the railway traffic, and here the loading of the steamers and scows for the northern country takes place. McMurray is still the centre for the local fur-trade, and for the growing air-borne traffic to the mines of the north. The new airport is 10 miles southeast.

There is little doubt in the writer's mind that this 'twin settlement' is destined to grow rapidly in the next few decades. Hence it seems of importance to learn just how the town has progressed in the years since the railway arrived. One must confess that Waterways shows no evidence of any design whatever (Fig. 4 at C). The railway curves round the station in a loop which serves the huge storage sheds of the Hudson Bay and Northern Transport Companies. The offices of the Companies are naturally close to these sheds. A large oil depot is between the two. Five small steamer-scow units link Waterways to the portage at Fort Smith. The river is open from mid-May to mid-October.

There are a dozen wooden houses, mostly little more than frame shacks, scattered in the vicinity. A few hundred yards from the station is the combined hotel and store—a more elaborate two-storey building—which is the natural centre for Waterways. Water for the hotel is obtained from the village pump across the street. One other smaller store is nearby. There was (in 1936) no church or school in Waterways; indeed about a score of wooden bungalows and shacks were all the other buildings in the place.

[1]For a description of these two sites in 1944 see my paper "Mackenzie Domesday"; *Can. Jnl. Economics*, Toronto, May, 1945.

Perhaps one reason for the lack of plan is the possibility of floods. In the spring and autumn the northern parts of the rivers are frozen, while the southern waters are still pouring their supply northwards. Hence widespread floods occur at times, which drown the houses of Waterways in several feet of water. Back of the settlement (Fig. 4) are rounded hills rising 300 feet above the river. These are clothed in a regular jungle of conifers and aspens. Natural clearings with considerable areas of pasture occur in places, and one such grassy meadow was grazed by the cattle of the early fur traders (about one mile east of McMurray) in the district now called Prairie (Fig. 4 at B).

For a settlement 140 years old McMurray is not very imposing. There is the straight main street about a mile long, which runs west to the junction of the two rivers (Fig. 4 at D). All the houses are shown in this large-scale map; and they are almost all frame houses and one storey high, though the shops often have a false two-storey front. However the Hudson Bay Store and the chief hotel are more imposing buildings, and most of them were neat and freshly painted. Indeed the school and churches are quite attractive buildings.

In addition to the big Company store there were three other stores, three cafés, two butchers, two barbers, and a drug store. The Bank, Post Office, and Dance Hall are in the shopping block. At the west end near the big river is the Wireless Station, while a few fur-dealers indicated the main local product. In the detailed map an area of several acres is shown as growing potatoes. These belonged to a Japanese farmer, and seemed to be flourishing.

The air-transport offices are on a branch of the Clearwater called the Snye (Fig. 4 at B). I was much interested to find that their haphazard arrangement along the bank of the river was due to each office being mounted on runners like a sled, so that during the flood periods in spring and autumn, they could be drawn to higher land to the south. This is a form of transhumance new in my experience. At the time of my visit the three air companies were shuttling ten planes to and from the northern mining fields, notably those near Yellowknife. Finally, about a mile from McMurray in the high shale cliffs of the Athabasca are to be found thick layers of Tar-sands, which in the future will be used as sources of petroleum. [See also the writer's book on Canada (1947)].

McMurray (perhaps excluding Fort Smith and Aklavik) is I believe the sole town north of the Alberta-Saskatchewan farm lands. But Churchill and some of the Yukon towns are doubtless larger. Howbeit it has few rivals as a high-latitude town in North America. It was founded about the same year as Toronto, but while the latter has grown to a city of 800,000, McMurray has a population of about 600 in summer and 300 in winter. Its satellite Waterways supports about 300 folk in summer, but the population falls to about 30 in winter. Man could 'choose' to go there in his thousands, but for environment reasons he prefers Toronto and Montreal!

McMurray however is large enough to show the beginnings of functional zones. The oldest houses are at the river end of the main street, but the business centre is the block between the hotel and the Hudson Bay Store. The churches and schools have been built in clearings cut out of the birch forest to the east. There is not yet any differentiation into first and second class residence zones, so that it may be classed in what later is called the *juvenile* stage (page 76). Waterways is too immature to show any plan, and is in the *infantile* stage.

The site of McMurray is clearly determined by the junction of the two streams of canoe traffic via the Athabasca and La Loche. The steamer port was placed here because of the rapids in the Athabasca just above McMurray. The next similar steamer port is at Fort Smith Rapids, about 250 miles to the north. But the railway is unlikely to be extended to Fort Smith since air traffic is already reducing the importance of steamers and railways in the empty northlands. Apart from its junction-position below the rapids, McMurray seems to offer no advantage over a dozen other sites in northern central Canada. But this definite environmental factor has given it a small lead, enough to make it the railway terminus and air-port. Cannot we agree again that man has 'chosen', out of many 'possible' sites, that which Nature had already marked out as the best?

4. Pekin (Peiping), a delta city in the temperate zone of Asia

Pekin offers a fair example of a large city which has developed under temperate conditions in a gigantic deltaic plain. It is larger, older, and much more complex in its evolution than any of the other examples, so that it represents the other extreme in our study from the settlement at Cape Evans. Yet we must try to answer our fundamental questions just the same. It is difficult to point to any environmental factor which has led to Pekin's dominance over most of the towns in the deltaic deposits of the Hoang-Ho. These cover an area of about 80,000 square miles, for though the plain extends south to Shanghai the southern portion (containing this city and Nanking) does not derive from the Hoang-Ho.

One would have expected the chief northern city to have developed either in the centre of the plain, or near the main river, or possibly at a good harbour on the coast. Pekin fulfils none of these conditions. In early historic days the capital of China was undoubtedly in Shensi. About the time of Christ it was moved east to Honan, and about A.D. 317 to the coastal province of Kiang-su. For a time around 960 it was placed at Kaifeng, which has a central position on the great river. However in the 12th century the Kin tribes of the north made Chung-tu their capital; and this town occupies the site later known as the Chinese section of Pekin (Fig. 5). The Mongol conquests resulted in Kublai Khan, about 1280, building a new capital (Khan-baliq) immediately to the north of the older Kin city of Chung-tu. This was the capital described by Marco Polo. Towards the end of the 14th century Nanking was made the capital, but from 1403 to 1911 Pekin continued to be the capital of China.

It seems clear that a considerable 'human' element is involved in the choice of the site of Pekin. In early days necromancers ascribed to the site a peculiarly fortunate character. No doubt the chief factor was political, since here the Chinese ruler could keep an eye on the marauding Mongols; while later the Mongol conquerors preferred to rule the Chinese from a city not too far away from their own native land.

The traveller reaches Pekin across a vast sandy plain; and no river or canal links Pekin to the outer world. The port, before the railway era, was a little town on the Pei Ho river to the east. To-day Tientsin is the chief entrepôt for the capital, and the writer reached it by this route in November, 1926. From the distance, the walls of Pekin give it the appearance of a gigantic box, with sides about five miles square—which has been flung down upon the plain. The present population is about 1,600,000, but about half the Chinese City of to-day is only sparsely occupied by houses, the southern portion containing many temple gardens and large open spaces. Ancient moats surround both portions of the city, while from the walls the numerous trees in the palace gardens and temples make a picturesque and unusual scene in this bare and level part of north China.

Turning to the actual plan of this Oriental city, to-day its arrangement is unique. The Chinese city consists of the usual narrow crowded streets of an Oriental settlement, flanked by innumerable small shops which cater largely to the tourist trade (as some of the street names suggest). The walls round this quarter were built in 1453, and are not quite so high as those around the Tatar city, which are 50 feet high and 40 feet thick. In the latter the streets are broader and the buildings larger and much more imposing. The Foreign Legations occupy the south-east portion of the Tatar city between the Hata Men and Chien Men Gates. They are enclosed in high private walls which enabled them to withstand the Boxer rebels in 1900.

A stream formerly crossed the Tatar city from north to south; and this has given rise to a number of ornamental lakes, which beautify the palace grounds of the Imperial City. The latter is enclosed in the centre of the Tatar city, and in turn it protects the Forbidden City where the Emperors spent most of their lives. In the northern part of the Imperial City is a small hill 200 feet high called Coal Hill. It is crowned by several small temples, and contrasts with the surrounding plains; but this 'eminence' is undoubtedly artificial in origin.

The Tatar City is covered with buildings, streets and palaces; and though many Manchus still live here, the majority of the people are now Chinese. It is difficult to chart functional zones for this city, for the ancient and modern plans are somewhat at variance. Originally the emperor's palaces and household occupied the Forbidden City. The government officials lived in the Imperial City, the ruling caste of Manchus in the Tatar City, while the subject groups of Chinese were in the adjacent but separate Chinese City. To-day the palaces are largely museums, while the government offices are mostly scattered through the southern half of the Tatar City, though some are still in the Imperial City. Probably the huge walls will be levelled in the near future, and the space converted into wide boulevards and gardens. It may be noted that though large areas of the Chinese City are empty, yet there are a few small 'suburbs' outside the walls, especially to the east of the Tatar City.

A glance at the map of China shows that there is no other large city in North China near to Pekin, except Tientsin. The magical and political factors (briefly referred to earlier) led to the city's birth, perhaps as far back as 723 B.C. At this time 'Chi' was the capital of the Yen kingdom, according to the Encyclopedia Sinica (Fig. 5). Given this start, no other city seems to have arisen to compete with it, though other sites such as Kaifeng would seem to have more to recommend them. In process of time the main roads inevitably led to Pekin, and six of these Imperial Roads (A to F) are shown in Fig. 5. Later the four main railways also radiated from Pekin, respectively to Mukden, Paotu, Canton and Shanghai.

In latitude and origin Pekin reminds one of Madrid, but the climate is rather different. There is a hot, wet summer and a very cold, dry winter. Owing to its eastern position in the huge land-mass of Asia Pekin has a rather extreme climate, as is clearly shown by the hythergraph in Fig 1a. The long extension of the graph to the right (in summer) indicates a marked rainy season, so that the graph for Pekin is fairly typical of Monsoon conditions.

The present plan of Pekin is essentially oriental and medieval. But an enclave of occidental culture is being incorporated in it, especially in the southern part of the Tatar City. We may perhaps compare this enclave to the neotechnic sections developing in certain parts of an occidental city. (I return to this concept later when discussing the plans and growth of Trento.) The palaces clearly form the nucleus of Pekin, even though to-day no emperor lives there. Around them is an administrative zone, and beyond that the upper-class town of the Tatar City. Smaller shops and slum areas are rather characteristic of the separate southern quarter called the Chinese City. I shall not try later on to give a formula for Pekin; but may remark that we are here confronted with an example of

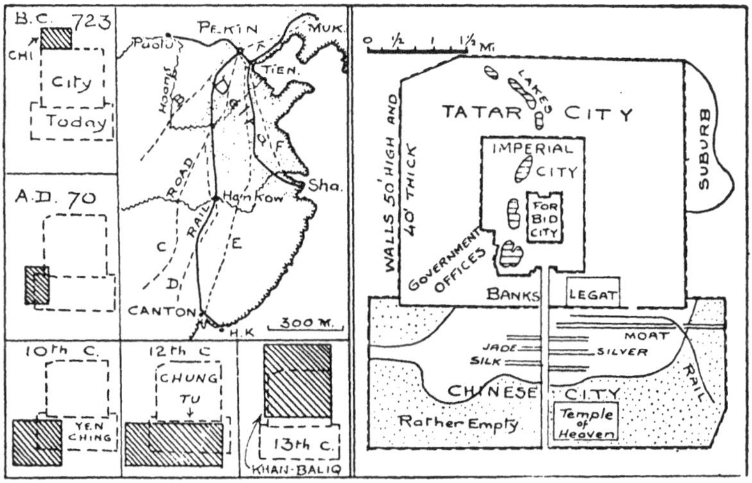

FIG 5. Various maps showing the evolution of the city of Pekin in the plains of north China. At the *left* is shown the evolution of the city (*from the Encyclopedia Sinica*). A small map is inserted to show the deltaic plain, and the roads and railways to Pekin. A generalized map of the 'concentric cities' of Pekin is given on the *right*.

city structure as complex as some of those composite minerals, whose chemistry almost defies the formula-maker. Luckily few towns are so complex.

It must be admitted that the choice of the site of Pekin is not due to any marked environmental factors. Obviously in such a rich agricultural region as the Hoang-Ho delta, some large distributing centres were bound to develop, but they are less characteristic of a primitive oriental culture than of a modern industrial occidental culture. We may perhaps grant that Pekin—situated in a vast region of *uniform* environment—is a good example to suggest that the possibilist theory does, under such conditions, explain the facts of geographical distribution.

5. Biskra, a desert town in the Mediterranean region

In 1938 I was able to carry out a life-long plan to compare the Sahara Desert with the arid areas of Australia. While the environments are alike, there is naturally very little resemblance in the human settlements. This may, at first sight, be taken as a triumph for the opponents of environment control. I am of the opinion, however, that the dissimilarities in the two types of settlement are merely due to different stages of development obtaining in Algeria and Australia.

Let us compare Coward Springs in Australia (450 miles north-west of Adelaide) with Biskra. Both lie on the edge of the desert, and have a

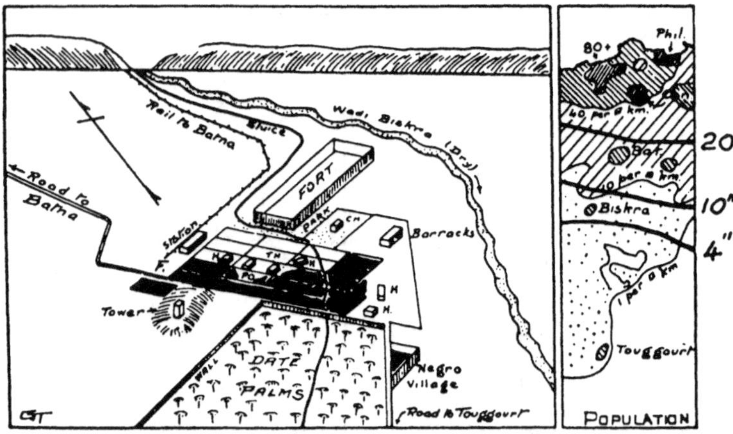

Fig. 6. (*Left*) Site-view of Biskra looking to the north-east. The town extends half a mile south-east of the station. The church, chief hotels, post office and town hall are indicated by initials. The chief Arab quarters are black. *F* is a factory site. (*Right*) Sketch map showing population density per square kilometre. Heavy lines are isohyets. (*Both maps from Geog. Rev.: April*, 1939; *isohyets added*.)

winter rainfall of four inches. The covering of sparse but regular vegetation in each case is similar in pattern though not in genera. Both, curiously enough, are situated on the edges of artesian basins. Both have railways, which have been running for many years, in the case of Coward Springs since 1890. When I visited Coward Springs in 1919 it was the chief settlement on the railway for fifty miles in either direction. A small date oasis had been planted here many years before, yet the town contained only four houses, two of which were empty. (Given an *excellent* environment, such as at Chicago, a settlement might add a million to its population in the same period.)

Let us now see what has happened in the Sahara with a far lower standard of living and a far greater population pressure. In spite of an equally unfavourable environment, Biskra is an apparently flourishing town with about 10,000 inhabitants, of whom 2,000 are French (Fig 6.) It is supported primarily by the date oasis, but also by the French military station, and to a lesser degree it acts as a caravan terminus and (in winter) as a tourist resort. My own visit was made in midsummer; when, it can readily be understood, our appearance was hailed as manna in the wilderness by the hordes of touts and beggars.

Biskra lies twenty-five miles south of the striking gateway through the Aures Mountains called El Kantara. Even in Roman times there were garrisons at El Kantara and to the south at Veskara (i.e., Biskra). Hence the town has had a long and eventful history.

The little river Tilatou runs south from the interior plateau to the marginal range, and thence through the gorge at El Kantara to the Sahara. At this settlement wheat is grown on little hill-fans just north of the gorge; while south of it in or near the stream-bed is a large date oasis which supports 6,000 Arabs. From El Kantara the water (in winter) flows south for miles into the veritable desert and reaches Biskra.[1] Thence it flows south, and, at times, reaches the salt playa called Shott Melrir, which is below sea level. Small mesas diversify the desert just north of Biskra. Oleanders grow along the dry bed of the stream with occasional patches of date palms. The vegetation is in tussocks with usually low tamarisk clumps or masses of reeds.

There is little of the pre-French town remaining at Biskra. A rectangular arrangement of streets, with the main axis running N.W.-S.E., extends for half a mile between the station and the group of large hotels in the south-east. The centre of the town contains the Town Hall and Post Office, while barracks are to be found near the wadi to the east. But the most noteworthy feature of the town is the large Fort (400 by 200 yards) which defends the town on the north. This is surrounded by stout brick walls about twenty feet high, and guarded by towers with slots for rifles. Inside this wall are the military offices, hospitals, as well as many small houses for officers, etc. At the extreme south-east corner of Biskra is the small 'negro village', though the dwellers seemed to be of Arab rather than negro origin. A large and roughly circular area of date plantations extends for nearly two miles to the south of the town.

The houses are built of brick or of irregular stones covered with stucco and painted white. In the town both the French and Arab quarters contain two-storey houses, but in the lower-class houses in the quarter south of the main town, the characteristic high Arab walls enclosed small yards on to which various flat-roofed rooms opened. A crowded open

[1] "Sea to Sahara," Griffith Taylor, *Geog. Rev.* (April, 1939).

market, as well as a mosque, are typical buildings in the Arab portion of the town.

Biskra is the terminus of a light railway which runs south to Touggourt, and collects the date harvest for export from the Sahara (Fig. 6). Possibly foreigners imagine that many such railways may run south into the rainless desert in the future. But it is the artesian water which makes possible the string of oases to the south of Biskra; and so far as I know there is no other locality in the French Sahara so well endowed. In prehistoric times the rains on the Ahaggar Plateau (500 miles to the south) gave rise to the large river Igharghar, which flowed north to Touggourt. Probably to-day some of the scanty waters flow far to the north along relatively deep-seated permeable beds, and replenish the artesian basin under Touggourt. No doubt waters from the Atlas Mountains also reach this basin.

The city-plan of Biskra shows us a fort in the north with a park belt and a zone of offices just to the south. The latter lies in the centre of a belt of better-class French houses. Still farther south is another zone of second-class houses, while yet south again is a belt of smaller oriental Arab houses.

How does Biskra fit into our determinist-possibilist debate? When we think of the enormous length of the northern border of the desert, which extends about 2,500 miles from the Wadi Draa to Suez in the east, it is clear that only a few districts are as favoured as Biskra. There is only one bygone Igharghar River, and probably no equal to the Touggourt artesian area. Figuig and Laghouat are similar desert outposts, but only the former is served by a desert railway. Hence I should make use of our Biskra example as follows:

Given desert conditions such as obtain in the Sahara, there are only a very few sites where the ameliorating factors have justified modern enterprise in developing the district as the French have developed Biskra. Even the military site was strictly determined by the water and the traffic conditions. We may note that these were also in operation in Roman times, centuries before the French took over Algeria. As usual I see Nature (i.e., the environment) acting as the dictator far more than man in the settlement problem under these difficult conditions.

The possibilist, it seems to me, puts the cart before the horse. He would say: 'Ah, but the vital trade in dates owes much to the artesian bores and to the desert railway; and surely these are due to human energy?' The determinist replies: 'Man can put down bores and build railways anywhere in the Sahara; but in the vast majority of cases he takes very good care to do so only where Nature has provided the conditions to make such expenditure worth while.'

In such exploitation Nature determines the route of development, man only determines its rate. To return to the parallel with Coward Springs in Australia. I see no sign of man utilizing this southern region yet. But when population pressure has increased to something like that present in Algeria, then we shall find man in Australia using the methods of development which Nature has driven man to use in the same environment at Biskra.

We may illustrate this phenomenon of population-pressure by comparing the population densities in three regions, each of which has much the same environment. They are western Spain, Algeria and southern Australia. All have a Mediterranean climate with similar and pleasant temperature conditions on the whole. In all three regions the rainfall diminishes in the interior. It is obvious that crop-densities (and therefore populations) depend primarily on rainfall.

In the graph inset in Fig. 1a the relation is expressed approximately by a straight line for each of these regions. Spain and Algeria exhibit very similar conditions; and, to the writer, this indicates that the population-pressure is about the same in each. Southern Australia shows a much lower density of population for the same rainfall. At a rough approximation, this seems to indicate that the population pressure in Spain and Algeria is about eight times as great as in southern Australia. The natural result is that standards of living are far higher in Australia. But it is entirely probable that population pressure in Australia will increase considerably as time goes on. Later on, the graph-line for Australia will climb to a position much closer to those for Spain and Algeria. The difference is not really due to human choice, but to the *immature* character of Australian settlement.

6. Urandanji, a semi-desert hamlet in the east of tropical Australia

For my sixth sample I have chosen a tiny village on the pioneer fringe of Australia, of which I made a sketch survey when I was investigating the rival trans-continental railways in 1922. Urandanji, tiny as it is, appears on most maps of Australia for it is the sole settlement in that part of Queensland. Camooweal is 120 miles north, while Boulia is the same distance to the south. Somewhat nearer are the mines at Dajarra and Duchess on the railway to Townsville. The striking feature about Urandanji's position however is that there is no town to the *westward for 1,200 miles*, until one reaches Marble Bar not far from the Indian Ocean. It is true that there are a few large cattle ranches near Barrow Creek (Fig. 7); but hardly a score of white folk inhabit these areas of sparse pastoral occupation. They are indicated as blank patches in the un-

4

inhabited portions of tropical Australia, which are shown by close ruling.

What sort of a settlement has grown up in this pioneer region of Australia during the fifty years or so of its pastoral development? With a total rainfall of only 11.4 inches, and an average temperature of 74° F., one could not expect much of a human agglomeration, especially as the evaporation is not far from 120 inches a year.

One day in August, 1922, we motored west across the rather rough landscape between the railway at Duchess and Urandanji. We passed two ranches in which the usual bare iron buildings (raised on wooden piles

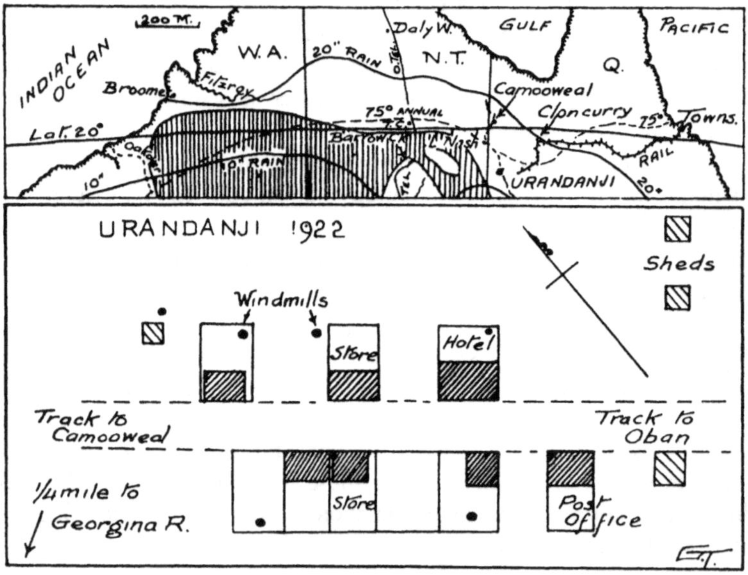

FIG. 7. (*Top*) A map of tropical Australia to show the position of Urandanji, just to the east of the widest expanse of empty land in Australia. Ruled areas are desert, and isotherms and isohycts are inserted. (*Bottom*) A sketch survey of the little settlement of Urandanji on the main stock-route from Camooweal to Marree.

to combat white ants) rather repel the traveller who comes from more favoured districts. Some twenty miles from Duchess we climbed up a stony pass amid giant ant-nests and clumps of desert spinifex grass (*Triodia*). West of this range the rock outcrops changed from sandstone to shale, with an immediate improvement in the grazing capacity. In these plains the soil is often a red gravelly clay, and carries sparse Mitchell Grass. Wild bustards and large kangaroos were occasionally seen.

The little township lies 150 miles on the hot side of the Tropic of Capricorn. It consisted (in 1922) of an hotel, two stores, the post office and a few other houses disposed on each side of the broad main street.

(Fig. 7). In all cases the walls are made of 'weather-board', i.e., planks cut from hard eucalypts (with a wedge section) and nailed in a horizontal position. Iron roofs and verandas are universal. The most prominent features were half a dozen iron windmills, which pump up river soakage from wells some fifty feet deep.

Though no large trees grow in the township, yet there are some fine hedges of acacia; and two or three of the gardens contained beautiful rose-trees. The Georgina River rises north of Camooweal, and after a course of 200 miles passes close to Urandanji. In times of flood the river may be several miles wide, but on my visit hardly any rain had fallen for the last six months. However there was a fine pool in the river, about half a mile long, near the township, which indeed owed its position partly to this vital factor, the river.

Further up the river in Northern Territory is Lake Nash (Fig. 7). It is the only permanent water available to the wild blacks in a dry season, and accordingly they tend to congregate here each November. A large cattle station has been established at Lake Nash, to the west of which is the Arunta Desert.

We may sum up by stating that under the environment conditions cited, with cattle ranches about twenty miles apart, a little town such as Urandanji has developed about half way between Camooweal and Boulia, i.e., 120 miles from either. The hamlet is on the great stock-route of cattle travelling from Northern Territory and Camooweal to the South Australian Railway at Marree. Some day Urandanji may be on or close to the logical transcontinental railway, which will link Darwin to the populous country in the south-east of the continent.

It is worth noting that the Urandanji district has much the same climate as the Darfur, Chad, and Middle Niger regions in Africa. These receive about eleven inches of summer rain, and this is sufficient to enable hundreds of small towns to develop in Africa. The average population density in the African belt is about six per square mile. In Australia it is about one-eighth of a person per square mile for similar environments. The writer does not agree that this illustrates human choice, i.e., that fifty folk choose to live in Darfur for one who chooses to live in Western Queensland. It is merely another example of the stages of development in man's occupation of a specific environment. In a pioneer country, with no population pressure and with high standards of living, we find an early stage with a density of only one-eighth.

Under the competitive conditions obtaining in the Sudan, six folk per square mile can be supported at a very low standard of living. Though it is highly unlikely that Urandanji will ever reach this high density; its population may advance considerably as pressure increases in the rest of Australia. However, a local resident (Mr. V. Bye) writes me in 1941 that only two more houses have been built in Urandanji in the last twenty years, while one original house has been burnt down and one store closed. Hence man has not exercised his powers of choice very noticeably as regards this possible site.

7. Santa Marta, a tropical monsoon town in South America

For my last example I have chosen a sea-level town only eleven degrees from the Equator. Here with an average monthly temperature of 81.5° F. we find a fairly typical tropical settlement, even though the rainfall of forty-three inches is rather small for such a low latitude. Santa Marta in Colombia is the most northern large town in South America. Moreover it lies south-west of the first settlement made in 1493 by Columbus in Santo Domingo. Since the constant trade-winds would drive the early Spanish ships southward to Santa Marta, it is not surprising that Ojeda built a fortress here in 1502, possibly before the foundations of Nombre de Dios in Panama. Thus our seventh example is in a sense the oldest settlement in North or South America. The town however was not properly established until 1525. My survey was made in December, 1930, when I made several short journeys into the Andes, which here reach their most northern position.[1]

A rocky cape projects like a hook to the south-west, and so affords ample protection from the north-east trade winds. Sixty feet of water is to be found in this bay (Fig. 8), and it is now one of the largest banana ports in the world. The coast is rocky and rises fairly rapidly to 9,000 feet in the San Lorenzo Range. Within forty miles are the ice-clad peaks of the Sierra Nevada, which rise to 18,000 feet or more. My chief aim in visiting Santa Marta was to find how the environment varied with elevation in a region so near the Equator.

Thus the settlement originated near the fine natural harbour as a Spanish outpost to control the Spanish Main. According to E. R. Tirado, towards the end of the 16th century the town already contained six of the present rectangular streets and a church in what is now the south-east of the town. It was however almost moribund by 1882, when the railway to the interior was begun. To-day its prosperity depends almost wholly on the banana trade, which is most efficiently directed by the American colony connected with the United Fruit Company of Boston.

As regards the town itself it is built on the flat plain north of the little

[1] "Settlement Zones in the Sierra Nevada de Santa Marta, Colombia," Griffith Taylor, *Geog. Rev. 21* (Oct., 1931).

Manzanares River. The town, sheltered from the trade winds, is muggy and hot. The natural vegetation on the coastal plain is largely xerophytic.

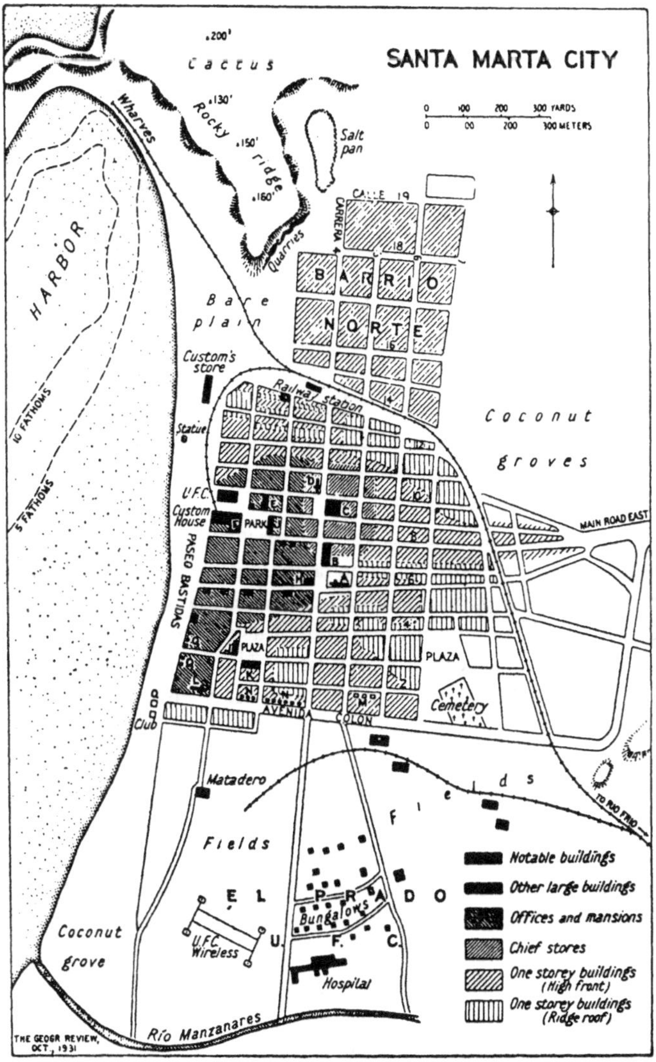

FIG. 8. Santa Marta City in Colombia, S. America, founded in latitude 11° N. in 1502.

It is very dusty in the frequent dry spells, yet no roads have been built into the surrounding hills, not even to the coffee haciendas some eighteen miles to the east, which are still reached only by mule tracks.

It is not difficult to distinguish that part of the town which is purely Latin-American from that which has developed mainly as the result of the recent commercial growth of Santa Marta. There are four types of buildings, which together constitute practically the whole of the town. These are arranged in definite zones, and are plotted in Fig. 8. The foreign business is chiefly in the west, and the local business in the centre. The poorer people live chiefly in the east.

One-storey buildings prevail save in the west of the town, and only a few of the larger public or private buildings have three storeys. Even the poorest Spanish houses have the peculiar projecting window-seats shut in by iron bars to protect the señoritas. Santa Marta is the capital of the large Department of Magdalena, and some of its official buildings are quite pleasing in appearance. In the south-west of the town are a number of villas in gardens, and they more closely resemble the houses of northern peoples.

North of the town has arisen a barrack-like suburb largely inhabited by labourers on the wharves. Very different is the suburb south of the city where under beautiful poinciana and acacia trees are the bungalows of the United Fruit Company.

How does Santa Marta illustrate our discussion on Possibilism? The fine harbour was no doubt a major factor determining the original site, while the lack of any easily exploited resources led to the long stagnation of the town. It will no doubt be stated that its present prosperity is largely due to the initiative of certain Bostonians. This is no doubt true; but the Bostonians were led to Santa Marta by the unique environment; in which numerous ice-fed streams flow down to the Caribbean plains. Irrigation is the key to the expansion of the town in the last forty years, and this is surely a gift from nature; developed of course by human initiative. The United Fruit Company could build a railway anywhere along the Spanish Main. They wisely 'chose' the one district where large-scale irrigation was possible. The present writer credits the expansion of a town based on a 'Hobson's Choice' of this kind to the account of the determinists.

Conclusions from the Study of Seven Settlements

What clues as to the kind of data needed in a study of Urban Geography can we gain from the descriptions in the preceding paragraphs? Cape Evans was unique since it was almost the sole collection of houses in a whole continent. Clearly the first need in such a settlement was *ease of access* from the surrounding world. It is by no means easy to land on Antarctica, and Captain J. K. Davis in 1911 skirted hundreds of miles of the coast fronting Australia before he found a place where he could land men and their equipment. The outstanding feature of the environment was the *low temperature*, and this led to the special type of architecture developed in the houses. Water supply was not difficult, since we melted ice; but all the *food* was of course brought from outside, except seal meat. This was eaten with a view to combatting scurvy in the hut, and was a main staple of food on our sledging journeys along the coasts.

In the second example dealing with Bergen, the importance of the *underlying structure*—which determined the central position of the town—is worth emphasizing. The *remarkable* climate led to the development of this large town not far from the Arctic Circle. The effect of the moraine ridges on the town's growth is notable. With the high plateau behind, tourism is of some importance. In dealing with McMurray and Waterways we learn valuable facts with regard to the growth of *twin settlements* where rival interests (e.g., fur and railways) arise. Here also the question of the clash of an advanced and a *primitive culture* is posed, though our description does not dwell much on the site of the Indian huts. Peculiar climatic conditions, such as the spring and autumn floods, are shown to have a great bearing on the amenities of Waterways.

Pekin and Biskra deal with Oriental towns with not much resemblance to the growth of the cities of our type of civilization. However the former city shows how difficult it is to ascertain exactly what caused the siting of Pekin in that particular spot in a *vast alluvial plain*. It seems to have been due in part to the deductions of a necromancer, rather than to any special advantage in the environment, and this may be taken by the possibilists as an argument in their favour! The segregation by nationalities, and the emphasis on *defensive city walls* are features which we find to some degree in the medieval cities of our civilization. The development of modern stores and apartments in one *new section* embedded in the medieval city is also to be seen in many of the older cities of Europe.

In Biskra we have a good example of the effect of *military* occupation on the growth of a city. Here also the *desert* conditions stress the importance of water-supply from the distant hills. Present also are *cultural divisions*, as seen in the French, Arab and Negro quarters. The relative densities of population in this part of Algeria, and in similar environments in southern Australia, are topics of considerable interest.

Our representative of settlement in Australia is Urandanji. Its chief interest depends on the sparseness of population in a new land where there are somewhat unattractive temperatures and precipitation. Under these circumstances the population, in the chief settlement for a hundred miles along a main stock route, has only grown to thirty, though it has been in existence for three quarters of a century. The last example is also in the tropics, in this case Santa Marta in South America. Here we see a city largely dependent on the energy of a *foreign business community*, the United Fruit Company of Boston. The port facilities, railway, and suburb of El Prado as well as the irrigation channels are due mainly to American enterprise. But the lack of all fair roads into the adjacent highlands is a peculiar feature considering the muggy tropical conditions in Santa Marta. The power of the *Spanish tradition* is however shown by some features of the architecture of even the lowliest dwelling.

CHAPTER III

THE EVOLUTION OF A CITY. EARLY STAGES

A.—The change from a hunting to a farm-village environment

AN unusual study by Alfred Meyer, *The Kankakee Marsh*,[1] gives us a valuable picture of a region as it changes from occupation by nomad Indian hunters to a community of farmers surrounding various small towns and villages. To-day very little of the wild landscape remains. The general environment formerly consisted of a broad crescentic area of marsh, about eighty miles long and ten miles wide. This lies to the

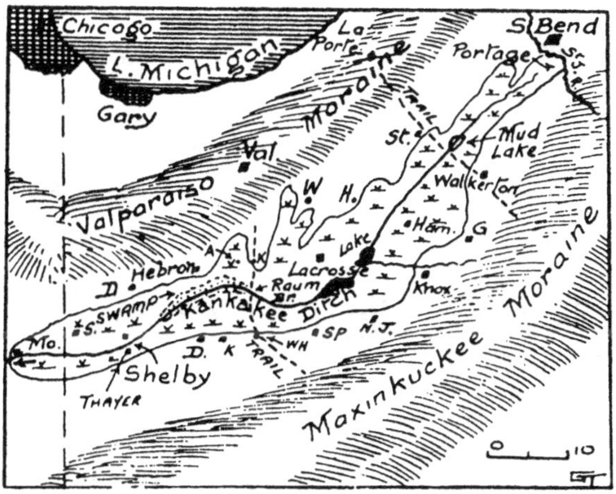

FIG. 9. The former Kankakee Swamp (Indiana), whose evolution has been described by A. Meyer. Mo. = Momence.

south of the great Valparaiso Moraine, and north of the parallel Maxinkuckee (or Marseilles) moraine (Fig. 9). It follows the curve of the terminal moraines in front of the former ice-lobe occupying the hollow of Lake Michigan. The marsh was drained by the Kankakee River, which flowed to the south-west, and then to the north-west, to join the Illinois River near Morris (Illinois). Its course through the marsh, from near South Bend at 720 feet, was marked by many meanders, swamps and shallow lakes, some of which are shown in Fig. 9. Indeed for much of the year the whole marsh was flooded, the waters of the Kankakee being held back by a natural dam of Silurian rock near Momence, at an elevation of 615 feet above sea level.

[1]Michigan Academy of Science, Vol. XXI, 1935, Chicago.

The original native cover consisted of three formations. First the swamp timber with large ash, elm, maple and oak trees was more widespread in the lower *western* marsh (to the east of Momence). The second type was the marsh proper, filled with sedges, wild hay, and wild rice. The third type consisted of sandy uplands, largely dunes which carried some oak. These dunes sometimes rose thirty feet above the swamp, and were rich with all kinds of edible berries.

Meyer describes four stages in the evolution of the cultural landscape in this interesting area. Before 1840 the wild marsh environment was the

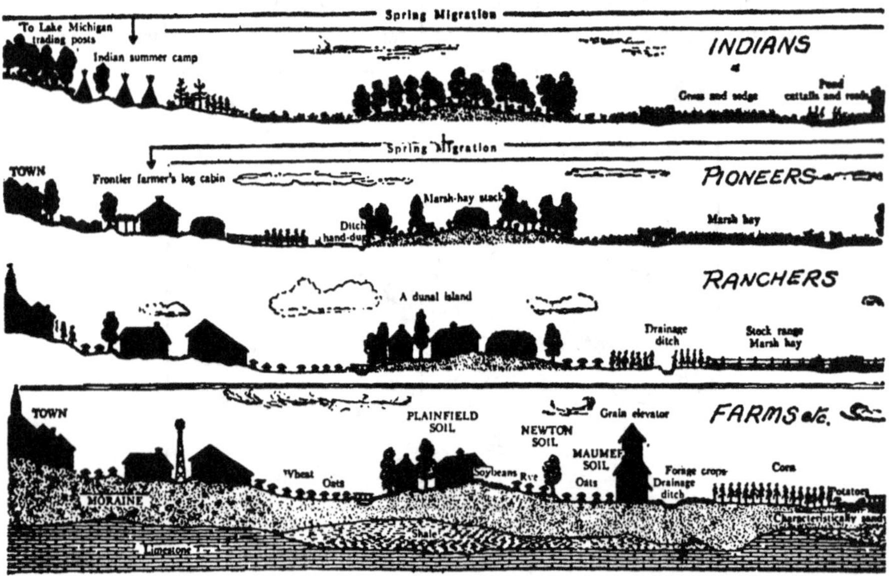

FIG. 10. Changes in occupation in Kankakee

refuge of the Pottawatomie Indians from their fierce Iroquois enemies. They raised their summer tepees on the drier moraine ridges, on each side of the marsh, where their ponies fed on the marsh hay. The head of the meandering Kankakee River reached almost to the St. Joseph River at South Bend. Here accordingly was a portage much used by Indians, and later by French fur-traders between Lake Michigan and the Mississippi basins, by way of the Kankakee and Illinois Rivers. In the fall the Indians migrated to the centre of the marsh, and lived on small sandy islets among the scattered oaks which grew thereon. Fish were plentiful in the lakes and streams, while the wild rice attracted vast flocks of birds. In these early days there were only two or three trails across the wide marsh, one leading to Valparaiso, and the other to La Porte. Meyer illustrates this stage of settlement by a cross section of the marsh, and this is reproduced from his paper in Fig. 10. The section is somewhat generalized, but represents a traverse about eight miles long.

The Indians abandoned most of this area about 1836, and their place was taken by pioneer settlers of English, Danish, or German origin. The environment did not change much in this second stage. Fur, fowl and fin remained the chief attractions to whites as to Indians. The settler built a log cabin on the drier margin, and often owned a shanty on one of the

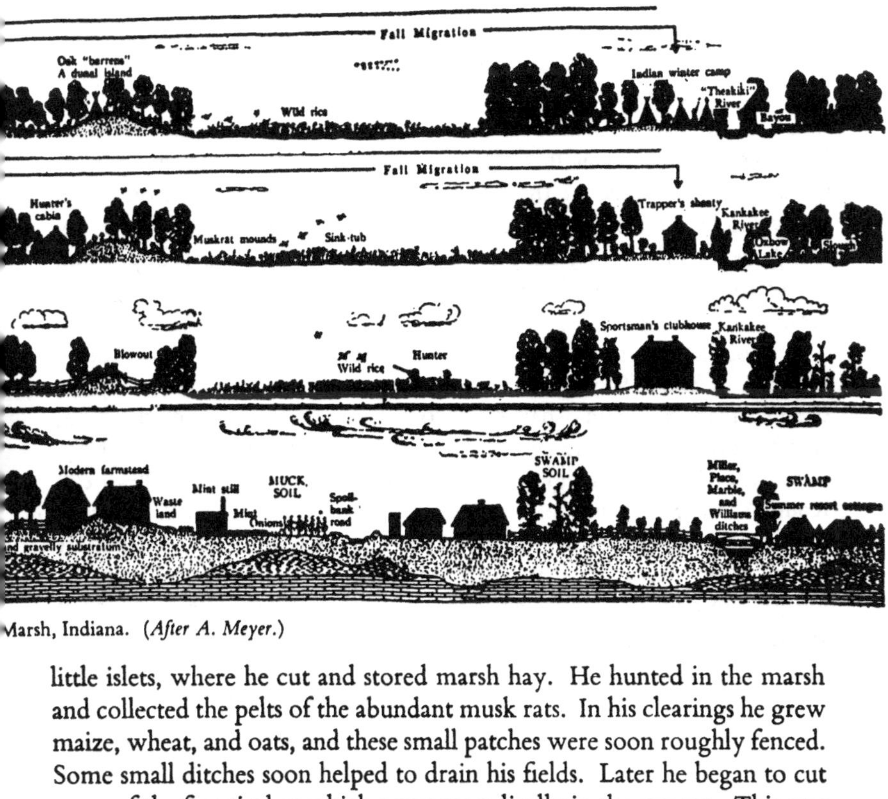

Marsh, Indiana. (*After A. Meyer.*)

little islets, where he cut and stored marsh hay. He hunted in the marsh and collected the pelts of the abundant musk rats. In his clearings he grew maize, wheat, and oats, and these small patches were soon roughly fenced. Some small ditches soon helped to drain his fields. Later he began to cut some of the fine timber which grew sporadically in the swamp. This was a great railway age; and there was a constant demand for railway 'ties', often cut from white oak. This stage occurred between 1840 and 1880 and is illustrated in the second traverse in Fig. 10.

Railways converging on Chicago soon crossed the marsh in several places. In the third stage the rectangular roads of the higher margins were extended right across the swamps, now becoming much drier. Cattle were grazed on the drained areas, especially after many of the trees had been cleared. Fences spread in many directions. A new development began when wealthy sportsmen from all the neighbouring large cities realized that wild life, especially birds, still remained in the many undrained areas. Club-houses sprang up along the river—as sketched in the third traverse (Fig. 10), replacing the trappers' shanties, which in turn had replaced the winter tepees of the Indian. About 1900 ditching on a large scale began, and in a decade or so the whole of the meandering Kankakee River was confined between the straight banks of the major drainage ditch. (Fig. 9).

The present stage is illustrated in the lowest of the four sections (Fig. 10), which shows that there is very little left of the marsh of Indian times. Draining has done away with almost all the fish, fur and fowl. The club-houses are replaced by small farms, and the earlier farms of the margins have become villages or small towns. Only along the actual margins of the straightened river are there some houses for summer tourists. The sandy knolls are now large farm wood-lots. Soy beans and rye grow alongside, and the loamy swamp soil is a vast maize field. It is hoped that reclamation officials will obtain control of the few scattered areas of wasted and useless swamp, and restore them as sanctuaries of wild life.

It is of interest to see how many small settlements have grown up in this area of about 800 square miles in the century since 1840. Meyer's, detailed map (two-and-a-half miles to the inch) and others of the kind, show about a score of settlements *within* the marsh area, such as Shelby, Lacrosse or Hamlet; or right on its boundary such as Hebron, Wheatfield. Kouts, Wanatah, St. Pierre, North Judson, Hanna, Knox and Walkerton, Of these Kouts (south of Valparaiso) has about 900 people, but almost all the others have less than 500. Along the edges of the swamp the villages are about six miles apart, while within the former marsh they are about twelve miles from each other.

B. Stages in Pioneer Settlement on the Mackenzie River

In a study of the development of towns in our European civilization, the two continents best suited for the purpose would seem to be Europe and North America. In other continents, such as Australia and South America, the conditions are in general much further removed from those which obtained in the original cradle of our civilization than in the two specified earlier. Furthermore it is only in North America, perhaps, that pioneer settlements over large areas are still to be observed, and these are best studied in the northern margin of Canada.

During the summer of 1944 it was the good fortune of the author to make an extensive study of the settlements all along the Athabasca-Mackenzie route, from the dense settlements of the Prairies in Alberta to the frozen seas north of the Mackenzie Delta. This great route of corridor rivals that of the Saint Lawrence in the south-east of the Dominion. Its present stage of development, with small settlements more than fifty miles apart, resembles conditions along the Saint Lawrence corridor about the middle of the 18th century; and though we may easily push the analogy too far, we may be sure that a hundred years hence, the Mackenzie route will be marked by small towns where to-day are only the tiniest of settlements.

In the following table the essential features of sixteen of these pioneer settlements are summarized. Data as to population, stores, churches, and houses are given, while there are other amenities of civilization, such as airports, wireless and police which show that our modern pioneer has notable advantages over the settler of the 18th century.

PIONEER SETTLEMENTS ON THE MACKENZIE RIVER, 1944

Place	Lat.	[1]Dis.	White Popul.	Streets	Stores	Houses	Ch.	Air	Wire.	Pol.
A. *Sub-Infantile*										
Tuktuk	69°N.	100	5	0	1	20	½R.	0	0	0
Brownings	61	65	2	0	0	2	0	0	0	0
Wrigley	63½	120	6	0	1	12	1R	0	1	0
Arctic Red	67	30	11	0	2	12	1R	0	0	2
Good Hope	66	110	14	1	3	30	1R	1	1	2
Hay River	61	70	16	1	2	30	2	1		
Macpherson	67	30	17	1	3	30	1½	0		
B. *Infantile*										
Fitzgerald	60	16	20	2	3	25	1R			2
Providence	61	70	39	2	4	25	2	1		3
Norman	65	30	63	4	3	30	2	1		4
Simpson	61½	120	76	2	3	30	2	1	2	4
Aklavik	68½	60	167	4	5	60	2	0	1	4
Waterways[2]	57	3	200	7	5	50	2	1		
Smith	60	150	240	8	6	60	2	1	1	12
Norman Wells	65	30	500	3	1	100	0	1	1	6
C. *Juvenile*										
McMurray	57°N.	140	500	10	10	150	3		1	4

[1]Distance from nearest similar settlement. [2]Twin with McMurray.

The sixteen settlements are classified in the above table primarily in regard to the total white population and the plan of the streets. (The last two columns refer to wireless and police stations.) Three groups are indicated, which are labelled A, B, and C. The writer some years ago put forward a scheme for the classification of towns according to stages of their evolution. He has used this with some success in a number of papers dealing with cities and towns in various parts of the world; and the series discussed in the above table is an interesting one, concerned almost wholly with the very beginnings of what must in some cases develop into important towns. At this stage in our investigation it will be enough to enumerate the three first classes; and indeed most of the settlements fall into the 'sub-infantile' or 'infantile' groups.

	Classes of Villages
A. Sub-Infantile	Few buildings, arranged haphazard along one indefinite lane, or street.
B. Infantile	Haphazard distribution of buildings, several streets, no factories.
C. Juvenile	Differentiation into zones of houses and shops starts.

C. Sub-infantile settlements on the Mackenzie

The evolution of these pioneer settlements can best be understood by studying some of the earliest stages. About half-way between Fort Providence and Fort Simpson, near where the Trout River enters in latitude 61° 15′, there is a very interesting primitive settlement called Browning's Farm. Here we had to stop—as was usual every two or three days—to take on board the cordwood logs which form the fuel of the steamers on the Mackenzie River.

Browning's Farm is a clearing about 200 yards square in the spruce forest, but of this area only a small field, about 100 x 240 yards, is devoted to grass and alfalfa. About half the clearing consists of the sandy beach and the steep grassy scarp (Fig. 11). The farmer had several acres of potatoes about four miles away in the forest, which I did not see. His alfalfa was a dense crop about two feet high, much of which had been already cut once. Several cows and horses were feeding in the grassy paddock near the barn. His house was a neat log cabin with an attic in the ridge roof. Several small sheds bordered the house, as well as a cache elevated on tall posts after the Indian fashion.

I made a full survey of this farm, because I was unable to see how the land differed from any other section of the same area for miles along the river. There is no market near at hand for his products, for his farm is about sixty miles from either Providence or Fort Simpson, which he supplies with potatoes and hay. In empty Australia the infrequent settlers live where there is a sufficient supply of water, for elsewhere no settlement is possible. In empty Canada the chances of widespread settlement —even if it be not much above subsistence level—seem much more promising. Soil surveys will no doubt enable those in authority to pick out the most worthwhile portions of these wide croplands of the future.

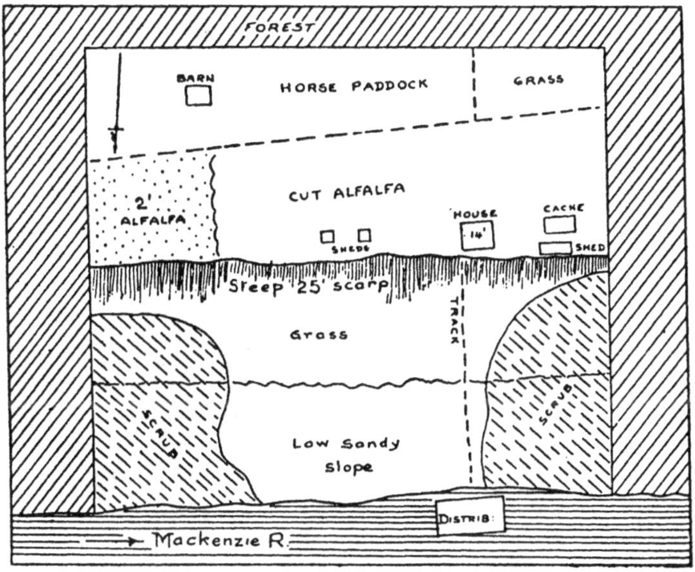

FIG. 11. A sketch survey of Browning's Farm (61° N.), looking south. The farm is about sixty miles west of Providence, and is 220 yards wide. (See *A* in Fig. 4.)

Our second example is Tuktoyaktuk which lies in the Arctic Ocean about twenty miles to the east of the mouth of the Mackenzie River. Tuktoyaktuk (Tuktuk or Tuk) has only been opened as a port for a few years, and this was the second voyage made there by the *Distributor*. It was started as a convenient meeting-place for the river and ocean steamers and schooners. The river boats are too flat-bottomed to venture far into the ocean, while the ocean boats have too deep a draught to tackle most of the channels of the delta.

The sea has drowned many of the small depressions, giving rise to the much embayed coasts near Tuktuk. Indeed the harbour at Tuktuk is but the protection afforded by a cluster of such hillocks, which are grouped into several islands and promontories as appear in the chart. The topography in the vicinity of the little port is shown in the map in Fig. 12 which represents a pace-and-compass survey made during our six days' stay. It was checked by the naval chart, which has only recently been completed for this area.

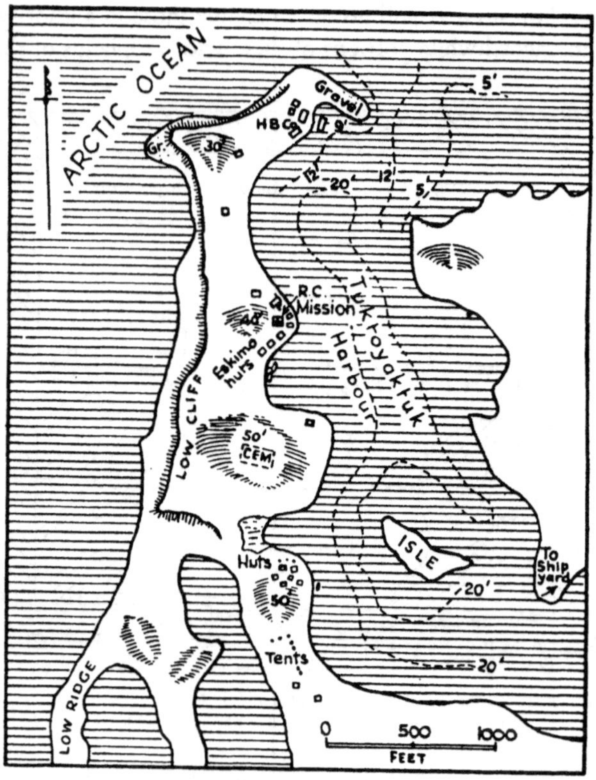

FIG. 12. A sketch survey in 1944 of Tuktoyaktuk (Tuktuk), the ocean port, at 69° 30′ N. (*Depths from the official chart.*)

A long narrow promontory runs almost due north for about a mile; and shuts off a similar-shaped elongated harbour to the east, between the promontory and an island also marked by the characteristic hillocks. There is about twelve feet of water at the narrow mouth of the harbour, which deepens to twenty feet to the south. There are really two settlements at Tuktuk, for a small ship-repair yard has been developed about two miles to the south-east of the main settlement, and here several small ocean steamers were drawn up on shore for needed repairs. At this shipyard there was a small stream, from which the settlers at the mouth of the harbour received their water supply by boat as needed. This is obviously a considerable disadvantage to the future of Tuktuk.

The new Hudson's Bay store was a neat building as usual, with a residence and several sheds surrounding it. The manager is accompanied by his wife and small daughter, and apart from two Catholic priests these are the sole white residents. There is no church building; but an interesting enclosed altar can be unveiled in the main room of the French mission house, and the room then serves as a small chapel.[1] There are about a dozen Eskimo families living at Tuktuk, some in huts near the mission, others about half a mile to the south. There were a number of small sailing boats with auxiliary engines drawn up on the beaches of the little bays. These belonged to the Eskimo, who went out catching the small white whales, seals, and fish in the ocean to the north. In summer, there is, of course, no vestige of snow or ice on the land, but it was interesting to see again the 'ice blink' far to the north, which was familiar to me in the Antarctic. This is a sort of silvery glare due to reflection from the pack ice, which indicated that the latter was present about thirty miles to the north of Tuktuk.

One of the most interesting features was the little garden maintained by the wife of the manager so far north as 69° 27′ N. In one small patch of cultivation she had radishes, but only one inch high, since it had been a very cold spring. She counts on getting lettuce and spinach, and occasionally a cabbage or two. But there is a big change in the possibilities of growing crops to the north of Aklavik, as indeed the natural vegetation would indicate. There is of course permanently frozen soil only a foot below the surface, even in summer, but the conditions vary a good deal with the nature of the soil. On the Yenesei the northern limit of open crop-lands is placed at Potavpo in latitude 68° 30′, about the same as Aklavik; so that there is not much hope of agriculture at Tuktuk. The settlement is clearly in the 'sub-infantile' stage.

Our third example is Fort Wrigley which is a very small settlement, containing only six whites. It was founded about 1880, and the houses straggle along the top of a twenty-five-foot scarp as shown in Fig. 13. It clearly represents the earliest stage of a river post, for the faint track through the long grass can hardly be termed a road. Some day, however, this may well become the 'front street' of a fair-sized town, for there is no other settlement in the three hundred miles between Fort Simpson and Fort Norman.

[1] ½R, in the preceding table, means that the Roman Catholic service is occasional.

There is only one imposing building in Fort Wrigley, the post office and radio station, a nice two-and-a-half-storey frame-house painted white. It is supplied with a prominent wind-motor propeller, used to charge batteries for transmitting radio messages. The Hudson's Bay store was naturally small, consisting of only one room surrounded by shelves. Some distance along the cut-bank to the north is a pretty little Catholic church, built of sawn planks with a shingle roof. It has a little belfry and a porch at the south end, but is only occasionally opened for service. Somewhat isolated at the north end are four log cabins, with a few elevated 'caches' and some tents. Here many of the Indians of the post spend the summer, and there are altogether about seventy-seven Indians in the immediate district. There is a similar cluster of Indian shacks at the south end of the settlement. The jungle of grass, raspberry canes, willow, and poplar, presses close on the settlement in midsummer, and almost impedes the path at the north end of the post.

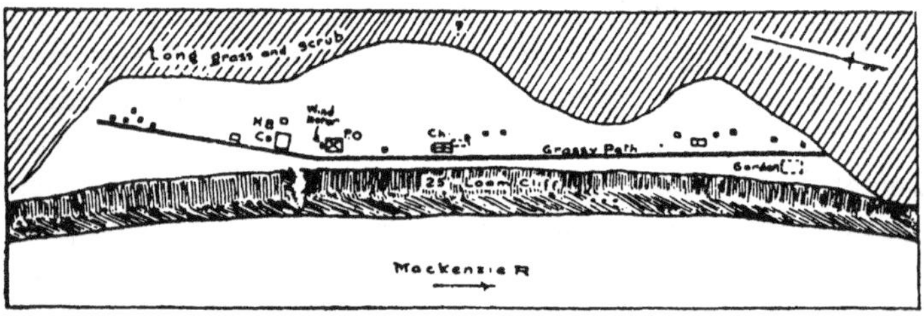

FIG. 13. A sketch survey of Fort Wrigley (62° N.), looking west. The settlement is half a mile long.

The chief interest in this sketch survey is that it charts a settlement about fifty years old, which has not yet advanced beyond the 'sub-infantile' stage. Indeed there is hardly a road, and the church is only intermittent in function. The juxtaposition of a modern wind-motor with the Indian shacks is worth emphasizing.

Arctic Red 'post' is situated at the mouth of the Red River on the east bank, just south of its junction with the Mackenzie. It has the most picturesque situation of all the posts, for the houses extend along a high curved ridge which encloses a pretty little pool, about 200 yards long, some twenty-five feet above the river level. Calla (arum) lilies, with large green leaves and white bract-shaped flowers, border this pool. There is a swampy area behind the pool, and the whole slope to the south-east is a blaze of purple fireweed in July.

The general arrangement of the post can be gathered from the sketch survey in Fig. 14. A novel feature is the distance away from the river of the Hudson's Bay buildings and of the R.C.M.P. post; both of which are on the hill, the latter half a mile from the water's edge. The most commanding position is that of the Catholic church on the end of the promontory nearly 100 feet above the river. The mission house and cemetery

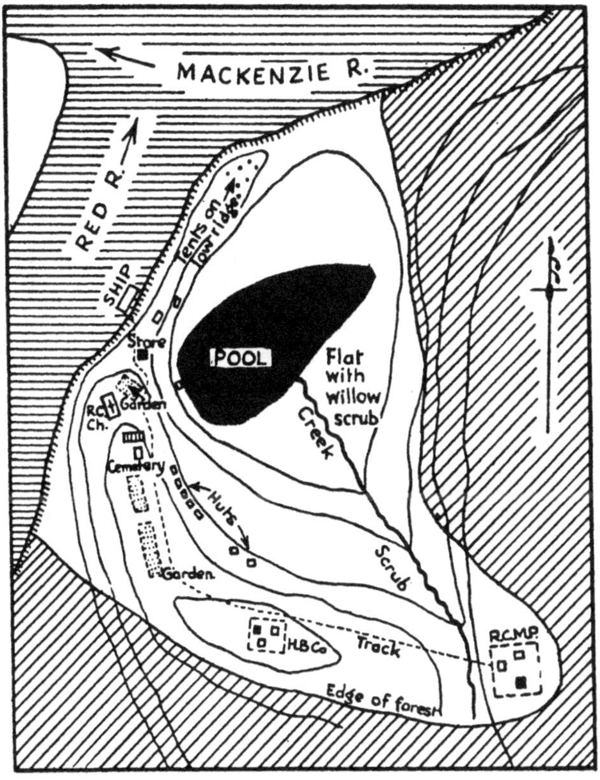

FIG. 14. A sketch survey of Arctic Red River, looking north. The pool is about 200 yards long. The form-lines are about 20 feet apart in elevation.

are close beside the church. There is a small store close to the landing place, with a few huts nearby. A row of small Indian huts south of the pool, completes the settlement.

There are only half a dozen white folk at the post, including two priests and two police. About three families of Indians live here, but there are about two hundred Indians in the vicinity who make this post their headquarters and visit it during the summer. Many of their tents were erected on the shingle bank north of the pool. The small gardens were interesting owing to the latitude of 67° 30′ N.

D. Infantile Settlements—Aklavik and Fort Smith

As a sample of the larger 'infantile' settlement let us consider Aklavik. This settlement (pronounced A-klàv-ik) is the metropolis of the north, for there are about 170 white folk, 213 Indians and 377 Eskimo in the district. It marks the southern limit of the Eskimo.

Aklavik became a small fur-trading post about 1917. In 1923 the police moved their station from Herschell Island, about 150 miles to the north-west, to this spot in the middle of the delta. The Hudson's Bay Company soon made it their centre, since it offered more contact with Canada, and there was a far better supply of fuel and timber.

Aklavik has been built where the Peel channel comes into easy communication with the west channel of the Mackenzie, i.e., about forty miles north of the other link, which we navigated much nearer Fort Macpherson (Fig. 15). Originally the post was erected in 1917 on the south bank of the distributary, here about 300 yards wide; but it was transferred to the north bank about 1920. It occupies a rounded promontory jutting to the south-east, and the main road borders the shore in the form of a crescent. Two other roads near the point help to form a square, and lead to the grassy clearing at the back of the settlement where a dozen or more Indian shacks have been built.

The little post office is at the corner, with the radio station on the west and the police station on the north. Then in the north-east quarter of the town is the large Anglican community, much the most impressive of any such group in the whole north-west. Along this part of the river front a reserve has been left about fifty yards wide. Much of this has been ploughed in 1944, and sown with oats. This rarely ripens (since the town is on the edge of the forest, 122 miles north of the Arctic Circle), but is cut for cattle food in late September. Behind the large radio station there is a community hall and a large private store. There is also a smaller store on the point near the post office, and yet another (with a café), between the Catholic school and the Livingstone barn. In the rear of the Catholic school is a pool or slough about 150 yards across.

Though Aklavik is not large compared with Igarka (with 15,000 inhabitants in 1937)—which the Russians have built in a similar latitude on the Yenesei River—yet it has some impressive buildings. For instance the Anglican cathedral consists of a nave with a well-proportioned square tower at the north-east corner. It is a frame building painted light yellow,

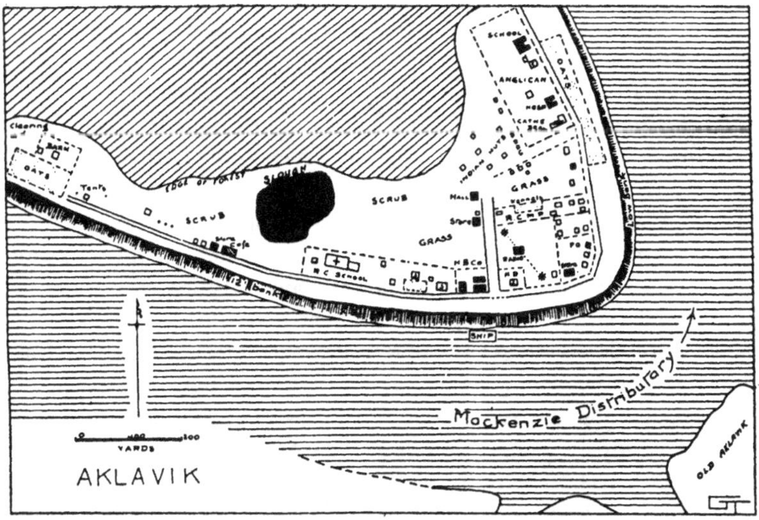

FIG. 15. A sketch survey of Aklavik (68° N.), looking to the north. The figures refer to storeys.

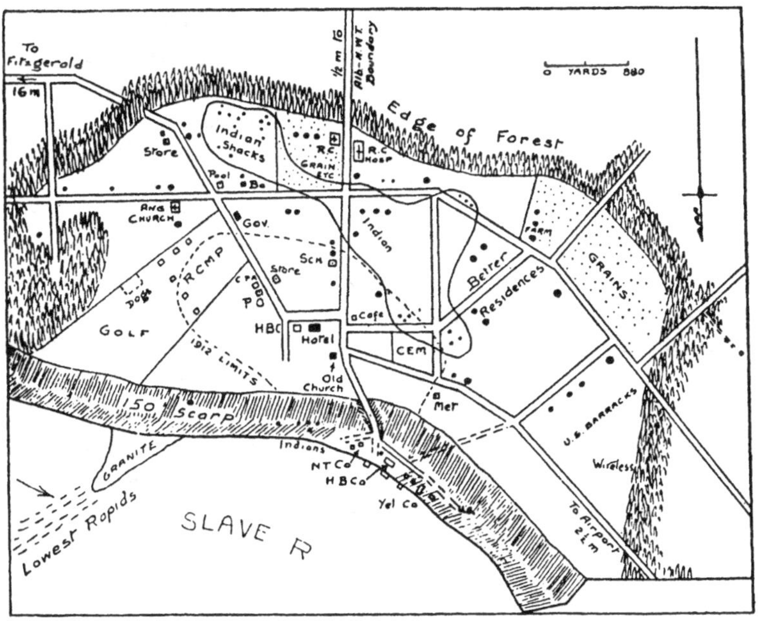

FIG. 16. A sketch survey of Fort Smith (see Fig. 4 at *A*), showing the limits of the town in 1944 and in 1912. Looking south. Black dots are houses.

as are all the buildings in the Anglican quarter. The remarkable altar picture—about eight feet wide—represents the Wise Men presenting gifts. The various figures in the painting, which was executed by an Australian artist, are clothed in the garb of polar Canada.

To the south of the cathedral is the two-storey house occupied by the clergyman; while to the north is the large two-storey hospital, built with two wings projecting in front of the central portion of the building. Further to the north is another neat two-storey dwelling where the senior teacher lives. Then comes the large boarding school of a somewhat similar style to the hospital. Here about fifty Indian children and Eskimo children live a large part of the year.

The Catholic buildings in the west of the settlement were of a similar character, but the school was larger, comprising three storeys. A portion of this building is used as a church. There is also a Catholic hospital and a large house for the priests, of whom there are half a dozen, as well as a number of sisters. The number of children in the school was about the same as in the Anglican community. Anchored in the stream was an auxiliary yacht which belonged to the Catholic mission.

Aklavik has been described as the metropolis of the far north. Its possession of a cathedral, the seat of the Bishop of the Arctic, would, according to some definitions, entitle it to be classed as a 'city'! But a reference to the classes described earlier in this memoir, places it among the 'infantile' type of settlements. It has the beginnings of a town plan, in the shape of the square of streets behind the promontory. However the shops are scattered, there are no factories, and there is no definite separation of better and poorer houses among the white settlers. The Indian area is, however, as usual somewhat segregated from that occupied by the Europeans. Most of the latter live in one-storey farm houses, but there are a few of two storeys, while other are mere log cabins. As noted, the religious houses are quite superior.

For our second example in the 'juvenile' class we may consider Fort Smith. It may perhaps be described as the capital of the North-West Territory, though it lies on the extreme border of the latter. The line of latitude 60° N. runs a few hundred yards south of the little town, and Fitzgerald is some fifteen miles within Alberta. The reason for founding Fort Smith is still quite apparent, for just above the wharves a broad belt of granite crosses the river, and the rapids can be seen from where the boats tie up (Fig. 16). Here in the early days the canoes were landed on the west bank, and here a small Catholic mission was established. The first church was built on the top of the upper cut-bank, which here forms a high sandy scarp reaching 150 feet above the river. The town is built on this level terrace or ancient flood-plain above the river. Between the scarp and the river is a lower terrace about 150 yards wide where the large sheds of the several trading companies have been erected (Fig. 16.)

The old portage road from Fitzgerald now winds rather irregularly through the little town from the south-east corner obliquely across to the scarp, and then down a deep cut in the sandy bluff to the former Indian halting place. Most of the chief buildings of the town are still on or close to this road. Thus the centre of the town is the rather large hotel, which is the property of the Hudson's Bay Company and is next door to their store. Two private stores are situated on this road as well as the government office, post office, pool room, Airlines office, and the very large Mounted Police reserve.

The Catholic church has been transferred about a mile inland to a large reserve, which also contains several mission buildings, barns, etc., as well as a large three-storey hospital. Most of the Indian huts and tents were at the time of my visit clustered in the vicinity of the Catholic mission, as is clear from the map. A few of their tents were canvas tepees, most however were ridge tents. The cemetery is near the old church just above the high sandy scarp. Until 1912 the settlement clustered round the hotel, as shown by the broken line in Fig. 16, but it has grown extensively in the last thirty years.

The second main road in Fort Smith runs parallel to the river front, but about one mile therefrom, following the curve of the river. Four or five cross streets link this second road to the river, including one which runs due south from the old church, and so reaches the Alberta boundary in about half a mile (Fig. 8). The largest building, the Catholic hospital, is at the intersection of the south road and the second main road. The Anglican church is a pretty painted wooden buillding where the second main road intersects the old portage road. The rectory is a pleasant wooden residence nearby. A half-dozen better residences (Hb class) are to be found mostly in the south-west quarter of the town in the region so labelled (Fig. 16).

E. Conclusions with regard to the Mackenzie Settlements

The reader may, with profit, also turn to the account of a somewhat larger settlement, that of McMurray, which has been described on page 22. It offers a picture of a village in the early juvenile stage, as indicated in the table on page 45. These seven descriptions all refer to settlements with much the same environment, except for Tuktoyaktuk (Tuktuk), which belongs to the Tundra. The same primitive necessities were obtainable at all the settlements in their very earliest years, but some have progressed some considerable distance on their evolution to larger and more complex settlements.

Brownings shows the earliest settlement of all, with no streets and no stores. A suitable place for the steamer to fasten her cables to a 'deadman' buried in the earth, seems to be the deciding factor in this selection of a site. It is approximately half way between older posts, and this is another factor. Tuktuk shows the accretion of a store and of a simple type of church before any street or semblance of a plan develops. Wrigley—though half a century old—has not yet developed a main road, though its position is indicated by the path along the top of the cut-bank. It has a church, which is only intermittently used. Arctic Red, with a white population exceeding ten, has a permanent mission house and a fairly large church Here also is a police post, but there is still no specific road, though the trail is clearly marked. All the above are definitely 'sub-infantile'. Good Hope is a transition to the next class, for it has a road well marked with fences in places, as well as three stores and a fairly large church. Its white population is only fourteen.

With the development of one or more cross streets, which seem to be correlated with a population of more than a score, we pass into the 'infantile' type of settlement. I chose two of the better known settlements of the north to illustrate this group. Aklavik with a white population of 167 has developed several cross streets. It has five stores and an assembly hall. The large size of the Anglican church depends on its being the religious centre for the Arctic areas. There is also a large Catholic community, and efficient wireless and police departments. Fort Smith, with about double the population, has twice as many streets and a number of administrative offices, etc., which places it at a higher level than Aklavik. However there is no differentiation of business and residential areas, so that it is still to be placed in the 'infantile' class.

McMurray has already been described in some detail. The chief difference is in the development of a 'shopping block', which is somewhat distinct from the residential quarters. This parallels a population of about 500, and the number of shops has also doubled as compared with Fort Smith. Churches are more numerous, and a large airport has recently been carved out of the Taiga in the vicinity. It is of course connected with the rest of Canada by rail. McMurray is therefore to be classed as an early stage of the 'juvenile' town.

CHAPTER IV

TOWNS, TOWNSHIPS, AND COUNTIES

The Origin of the County and the Township

The relation of the town to the township and county is not well understood by the layman, nor indeed by most geographers. Yet it is necessary for us to know something of the way in which the land upon which a town is built is subdivided and sold or leased to the pioneer, later settler, or citizen. A few words may first be given to the larger division of the land known as the 'county'.

The term 'county' was first used in England about the time of the Norman Conquest. Before that date the usual division (of about 1,300 square miles) was called a 'shire' (i.e., share). In England the counties belong to five distinct classes, as follows: A. *Early Folk-Shires* in the south-east, each of which was the home of a distinct culture-group; i.e. Norfolk (North Folk), Essex (East-Saxons); B. *Early Wessex Shires*, with place-names like Somerset (Settlers of the sea-meres), Cornwall (Horn of Wales); C. *Midland Shires* named after the chief towns; D. *Border Shires* (Cheshire, Lancashire and Yorkshire); E. *Latest Norman Counties*, the four in the extreme north.

Probably in the earlier divisions the boundaries enclosed the settlements of a kindred clan, such as the East Saxons. For instance, the County of Essex goes back to about A.D. 930, but the choice of the chief town, Chelmsford, occurred about 1225. In the Midland Counties the divisions were arranged about the main towns; such as Gloucester, Derby, Lincoln, etc. The boundaries were about twenty miles from the town—at such a distance that it could be easily reached in a day's journey over the poor roads of early England.[1]

The smaller division in Essex was called a 'hundred', there being about a score in that county. 'Hundred' corresponds somewhat to 'Township' in Canada or U.S.A.; and the term 'hundred' is still used in parts of Australia; whereas 'township' is there almost synonymous with 'village', and does not mean a definite area of land.

The township as a basis for land division—and also for most rural

[1]See the author's *Environment and Nation* (p. 224-229), Toronto, 1946, for a full discussion.

settlements—is seen in its simplest and most widespread form in the Ohio-Missouri region in the United States. In the rather hilly country to the east of this huge region, the land-divisions and the roads bear a close relation to the ridges and valleys of the Appalachians. But as soon as we cross westwards into Ohio we reach the lands where the 'checker-board' system dominates lands, roads, and towns, often to their considerable disadvantage.

About 1785 the 'North-west Territory' was being surveyed; and, in the words of A. K. Lobeck, 'the essential idea was to divide the western United States into a huge gridiron.'[1] This was done by establishing a number of principal meridians at various intervals, one being along the

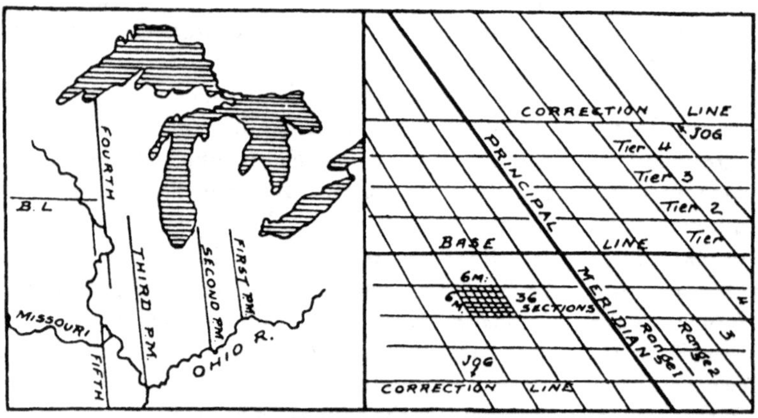

FIG. 17. The subdivision of the Prairie States into Townships along the Principal Meridians (*after A. K. Lobeck*).

western Ohio boundary (Fig. 17). East and west of the principal meridians, north and south rows of townships called 'ranges' were laid off. Each principal meridian, together with the system of townships based upon it, is independent of every other principal meridian, and where two systems come together irregularities are found.

A base line is run at right angles to each principal meridian. *Tiers* of townships, each six miles square, are laid off north and south of these base lines (Fig. 17). The eastern and western boundaries of townships are, as nearly as may be, true meridians; and when they have been extended northward through several tiers, their convergence becomes considerable. To prevent this diminution in the size of townships to the north of the base line, about every four tiers a new parallel is run, along

[1]See *Airways of America* p. 43, New York, 1933.

which six-mile measurements are made for a new set of townships. These are known as 'correction lines'.

Public roads are usually built on section and quarter section lines; and hence, wherever a north-south road crosses a correction line, there is a 'jog' in the road as the accompanying diagram (Fig. 17) shows. Each township under this plan consists of thirty-six square miles; and each square mile (or section) consists of 640 acres. A quarter-section of 160 acres is one-half mile on a side. This in turn may be subdivided into quarters of forty acres each. Many farms in the middle west of U.S.A. comprise forty acres or multiples thereof.

The townships into which the Prairie States were divided in the early days of settlement were each thirty-six square miles in area. Somewhat the same pattern of subdivision was used in the Dominion of Canada in the early days of settlement around 1800, but there were a number of variations. Moreover the French system—in operation before the conquest—was entirely different; and for a time the English settlers used the French plan of *Seigneuries*, divided into very narrow strips at right angles to the rivers or to the Great Lake shores.

In Canada the township may be defined as an area of land usually about 100 square miles, which has been plotted on a map with a view to the ready subdivision of the area into small blocks for settlement. The township is thus, in a sense, the *surveyor's* approach to settlement, rather than the pioneer's, though it is drawn expressly for the benefit of the latter. The townships over large areas were determined long before any settlers occupied them; and when the latter arrived the pattern of the subsequent village and town very closely followed the gridiron pattern of the blocks in the township.

Early Settlement in 'Upper Canada'

By the provisions of the Canada Act of 1791 the Governor or Lieutenant Governor was authorized to create districts, towns and townships. It is the character of the latter area with which this section is at present concerned. The first system of land settlement in Canada, as stated, was that of the French Seigneury. The explorer, Frontenac, built a fort at Cataraqui (Kingston) in 1673, and later La Salle added a stronger fort and also started stations at Niagara (1678) and Detroit. The French pattern near Detroit is illustrated in Fig. 18 at 1. But the conquest by the British in 1759–60 changed the pattern of settlement almost entirely.

The instructions to Governor Murray as regards Lower Canada stated that townships were to be laid out which 'were to contain about 20,000 acres each, having as far as possible natural boundaries extending up into the country, and comprehending a necessary part of the River St. Lawrence'. In the most convenient part of each township there were to be town plots (Fig. 18 at 5) as close as possible to navigable water. (A. Fraser, 1921). In 1779 settlement was begun along the Niagara River in the vicinity of the fort, as a means of supplying the needs of the garrison. In 1781 the first survey of the district above the Ottawa

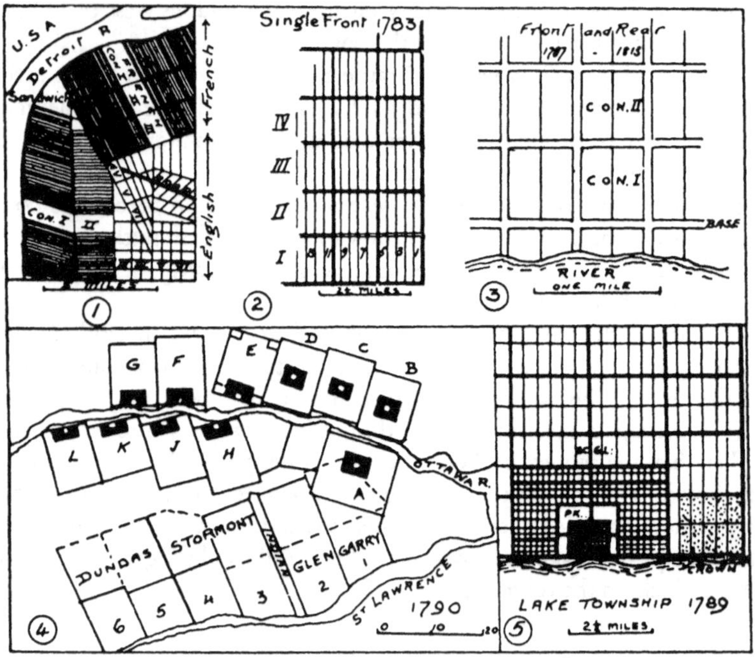

FIG. 18. Sketch maps illustrating early methods of sub-dividing new lands. At *1*, the French and English systems are contrasted near Windsor. At *2*, is the 'Single Front' and at *3*, the 'Double Front' method of subdivision. At *4*, the suggestions for town-plans along the Ottawa River are sketched. At *5*, such a town-plan is enlarged. School, Glebe and Park are indicated. (See *Economic Geography*, April, July, 1945, for further details.)

River was begun, and by 1784 some ten thousand of the United Empire Loyalists had reached Upper Canada. At first these blocks were called 'seigneuries' or 'fiefs', not 'townships'.

About 1788 several regiments had received grants; the non-commissioned officers getting 400 acres and the private 300 acres each. About this time Southern Ontario was divided into four districts; named, in consideration of the large German element among the United Empire

Loyalists, Lunenburg, Mecklenburg, Nassau and Hesse. The method of holding the land now gradually changed from the French to the British practice. A Land Board gave a certificate to a satisfactory applicant, and within two days after this was in the hands of the local surveyor he was to assign a lot of about 200 acres, giving also a signed description of it.

The 'town plots' in each township measured one mile square; and usually, if an inland township, were situated in the centre. If a water township, they were in the middle of the water front (Fig. 18 at 4). Each town plot was laid out on a prescribed plan, with town-lots of one acre, town-parks of twenty-four acres, and squares and streets of stated dimensions. The Crown reserves in the corners of the townships consisted of eight farm lots each. It is a pity that none of these projected towns ever materialized. (Charlottetown in Prince Edward Island is one of the few large towns, which still shows something of such an original plan.) When a body of immigrants decided to settle on a block of new land, surveyors were dispatched to establish the baseline of the township—usually parallel to the lake front. This done, a second line parallel to it was run at about one and a quarter miles to the rear, followed in due course by similar parallel lines at the same intervals. These *lines* were termed 'concessions',—a term also used for the *areas* between the parallel lines.

The allowance for roads was usually sixty feet along the first concession, and forty feet along the succeeding lines. The farm lots along these concessions approximated 200 acres in area, having a frontage of about eighty rods (440 yards) and a depth of 400 rods. At intervals of two or three miles, a space of forty feet between lots was left for a side road running at right angles to the concessions. Usually the baseline was more or less parallel to the waterfront. (Fig. 18 at 2). This explains the remarkable pattern of the blocks of lots in Ontario, and also explains the presence of triangular '*Gores*' between the main patterns. A good example of a Gore is to be noticed just west of Toronto in the vicinity of Etobicoke.

Prior to 1791 the settlements were confined to the upper St. Lawrence, Quinté, Niagara, and Detroit; for no grants had been made near Toronto, Long Point or to the south of the Thames Valley. At this time the total population was under 25,000 (Fraser). In 1793 Quebec was divided into Upper and Lower Canada, and the method of tenure in Upper Canada was made much the same as in England. In July, 1792, the province had been divided into nineteen Counties, which extend along the water front from Glengarry in the east to Kent in the extreme west. Usually the counties were two or three townships wide. At the same time the boundaries of the four districts were revised and they were renamed Midland, Home, London, and Western (Fig. 18 at 4).

By 1798 much more was known of the province, and a number of changes in the major boundaries followed. Russell and Prescott were marked out as new counties along the Ottawa River, while Niagara and the Western District were further subdivided. 'Thus it was hoped that no settler in the province would be at a greater distance than an easy day's journey from the local centre' (Fraser).

Whitby, Toronto, London and Guelph

The early history of the district near Whitby illustrates the procedure about this time. William Willcox from Cork secured a grant of 200 acres for himself, and then arranged for a settlement by himself and associates in a township of 1,000 acres. Large grants were to be allotted to the four leaders and 200 acres was to be given to each settler he could induce to settle at Whitby. In 1795 he collected thirty-three settlers in Ireland, but on his return journey most of them abandoned him and settled in U.S.A. Finally in 1795 he reached Ontario; and as compensation 1,200 acres each were allotted to his wife, three daughters, son and son's wife on condition of residence within the province. Such grants to 'associated companies' were given up about 1797.

In the vicinity of Toronto, Yonge Street and Dundas highway were built about 1795 by a group of Rangers under the command of Simcoe. These roads were primarily designed for military purposes, but naturally, greatly helped to open up the country. Indeed north of Eglinton Avenue (Toronto), Yonge Street became the base-line of the townships to the north. In July, 1794, the plan for the town of York (now Toronto) was approved. Each person occupying a town lot was required to build a house on it within three years, or forfeit the patent. By the end of the year there were about seventy families in the vicinity of the projected town. For a time it seemed likely that London on the Thames would replace Newark (Niagara) as the capital of Upper Canada, but in 1797 York was made the permanent capital.

Various attempts to found towns occurred in the last few years of the century. The most important of these early settlers was Colonel Talbot, and since his experience illustrates one of the early methods of opening up Ontario, a few words may be devoted to his history. In 1803 he received an officer's quota of 5,000 acres in the township of Dunwick on the shore of Lake Erie to the south of London. (An interesting map of Port Talbot is given in Guillet's book.) It was agreed that for every family he settled on fifty acres in the township, he was entitled to receive 150 acres for himself. For thirty-five years Talbot engaged in settlement, and he placed some 30,000 settlers permanently on the land. By 1824 he had control of twenty-eight townships, which contained some 50,000 people by 1837 (Guillet). He acted quite independently of the Government; but as a result of his energy, the London and Western Districts became flourishing settlements with good roads, long before most of the rest of south-west Ontario. He retired after a most successful pioneer career in 1837.

The most ambitious attempt to develop Ontario by private funds was the settlement of the Huron Tract (between London and Goderich) and of adjacent areas near Galt and Guelph (Fig. 23). The Canada Company purchased about two million acres, and John Galt, the Scottish novelist, played a major part in their development. In 1827 the town of Guelph was founded, and a brief description of its interesting plan follows (Fig. 21). The road to Goderich was cut through the forest from Guelph in 1828. By 1838 Guelph contained a population of nearly 2,000; while the Huron Tract to the north-west had received some 4,500 settlers.

The following table (based on Fraser's book) gives the amounts of land granted in the early years of settlement. The figures refer to thousands of acres.

EARLY SETTLEMENT (AFTER FRASER)

Year	1798	1799-1800	1801	1802	1803	1804	1805	1806	1807	1808
Area	1170	178	479	605	438	253	138	140	244	180

The Allotment of Land

The various systems of subdividing the lands of southern Ontario were as follows. The 'Single Front' system was used in the earliest days, in which the Concession lines were surveyed, and roads were planned along these (Fig. 18 at 2). About every fifth lot it was proposed to make cross roads, and the lots were about five times as long as they were broad. The 'Front and Rear' system followed from 1787 to 1813, and in this cross roads were planned around each two lots (Fig. 18 at 3). This method was used chiefly in the Niagara Peninsula, and led to a costly network of roads which was expensive to build and maintain. The 'Double Front' system (*ca.* 1818) followed, in which the concessions were much narrower, so that the east-west roads were closer together. Cross-roads were made every five lots, much as in the first method. The next system was the '2,400 acres section' type, in which alternate concession-lines became roads, and the side roads were placed about six lots apart. This gave blocks of twelve farms between the roads, each comprising about 200 acres.

The Canada Company around 1835 cut their land into somewhat similar arranged blocks, but there were ten farms of 100 acres each in a sections in a township (Fig. 17). Some five per cent of the area was reserved for roads which bordered each square section. The corners of the sections farms formed a section of a square mile, and there were thirty-six such sections in a township. Some five per cent of the area was reserved for roads which bordered each square section. The corners of the sections were carefully surveyed, and many of the errors of the earlier surveys were obviated. Fraser and Whitson (see references) give many other details of these subdivisions, but the main methods in use in the early days are summarized above.

Whitby Township and Town

The diagram of the Whitby district given in Fig. 20 will help to make clear the early procedure in planning a town in Ontario. The surveyor was the first official to appear on the scene, in this case before 1800. He marked a line parallel to the lake front, which became the southern boundary of Concession I. Thus he cut off from the main survey the irregular lake frontage, which was known as the 'Broken Front'. This is indicated in Fig. 20. A road was laid out along the south of the first Concession, but in the case of Whitby this did not become the main road of the little town, which ultimately developed between Concessions I and II. The concessions hereabouts were laid down 100 chains (2,200 yards) apart. Since the surveyor approached the area from the east, the south-east corner of each township was in general the datum for marking and numbering lots. In Fig. 20 the datum (or origin) of the next township to the west (that of Pickering) is indicated, but the datum for Whitby was about ten miles to the east, and does not appear in Fig. 20. The lots were about twenty-five chains wide, with a depth of 100 chains, and those in the 'Broken Front' (numbered 16 to 35) are labelled.

According to the Ontario County Atlas of 1877 (Beers, Toronto) the town site occupied a series of lots symmetrically arranged about the main road leading from the little port on the Harbour to the north. It comprised sixteen full lots and eight half-lots. No suggestion of parks is apparent, though such appeared in the older plans elsewhere, as shown in Fig. 18 at 5. No doubt to the early settlers, when the total population was only a hundred or so, such foresight seemed unnecessary. In the case of Whitby, perhaps it was; since only a very small portion of the twenty lots of the town-site has been actually used for close-set houses. This latter area is shown by the black patch in Fig. 20. The original farm lots, with a frontage to the main roads of twenty-five chains (1,650 feet), were cut up into shop blocks or town blocks, which as usual would have a frontage of 50 to 100 feet, or even less.

The relation of town to township and country can be well illustrated by the evolution of the little town of Whitby—the first town to the east of Toronto. York was the fourteenth county to be proclaimed in July,

FIG. 19. The limits of the original county of York (1798), and the limits of the modern county of Ontario (dotted). There were 8 townships in 'Ontario' in 1852, but Whitby was divided in 1857.

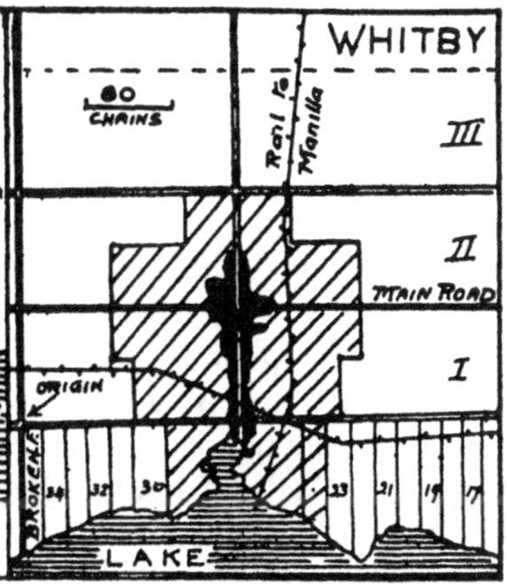

FIG. 20. The Town-site of Whitby (diagonal ruling) as given in a map of 1877. The concessions and lots of the Township are shown. The black patch shows the area occupied by houses in 1940.

1792, and was originally much larger than its present area, since it extended west to the headwaters of the Thames River. In the early days it was part of the 'Home' District, and had its centre at York (Toronto). In 1821 the county of Simcoe was cut off from the northern part of York County; and further agitation led to the creation of the county of Ontario in 1852 (Fig. 19). Since this county is bound up with the growth of Whitby it may be briefly discussed.

The township of Whitby was surveyed between 1791 and 1795, and originally also included East Whitby Township. Hence we see that townships as well as counties were subdivided as population increased. A Vermonter named Wilson settled in Whitby in 1794, and was the sole settler for several years. In 1804 Lynde's house was said to be the chief residence between Kingston and Toronto, while two other settlers had a house on the lake front nearby. A log school-house was in use by 1811. By 1821 there were settlers as far back as the 3rd Concession. A fulling mill was built a little to the east at Oshawa as early as 1822. The old township of Whitby was subdivided in 1857, and Oshawa (the chief town in the new East Whitby township) is now much larger than the older settlement of Whitby (Fig. 20).

An energetic individual—Peter Parry—moved into the district in 1836 from the eastern part of the Province. He succeeded in developing a main road running north, to link Whitby with the interior, after the manner of Yonge Street. He also advocated separation from York County, but died in 1851 just before the separation took place. In the 'Historical Sketch of the County' (1877) we find the following statement: 'Had he been spared ten years longer, the great railway from Whitby to Georgian Bay would have been built; and in all probability Whitby with its fine harbour and surrounding advantages would be to-day a city rivalling in wealth and importance the City of Toronto'. Its population around 1870 was about 3,000, and to-day is about 4,500. The main purpose of the split from York County was said to be due to their desire to 'work out successfully their own local concerns, untrammelled by the overwhelming centralizing influence such as they had to contend with in the City of Toronto.' (Beers).

Planned towns of Guelph and Goderich

A brief description of the efforts of the Canada Company to open up the region north of the Thames basin in south-west Ontario has already been given. The town of Guelph was surveyed about 1827, and is one of the very few examples of town-planning in Ontario. The Speed River makes a large bend where the tiny Eramosa River enters from the east. The latter occupies one of the numerous broad post-glacial channels, among the till and drumlins of this district. On the flat in the bend, at an elevation of 1,100 feet above sea level, the surveyors laid out the central streets of the future city (Fig. 21). The essential features are the central square and the radial streets leading to the river bank at this point. Thus they kept in mind two of the major principles of town planning, which have been ignored in almost all the other ('checkerboard') towns. The shopping centre and the chief public offices are still situated in or near this centre, which can be approached directly from the outlying areas by the numerous radial streets, from every angle of the compass. It is true that in the outer portions of Guelph the streets are based on the rectangular concession lines, determined by the base-line which forms the 'township limit' to the south-east of the town. Nevertheless Guelph has a plan better suited to town conditions than almost any other Ontario town.

Goderich is well known for its definite plan, which is illustrated in Fig. 22. It lies at the western end of the main road through the Huron Tract, just as Guelph lies at the eastern (Fig. 23), and was laid out in 1827. At the entrance of the Maitland River into Lake Huron the river has cut down some 200 feet into the Devonian formation, giving rise to interesting 'incised meanders' along the river. There is a steep cliffed

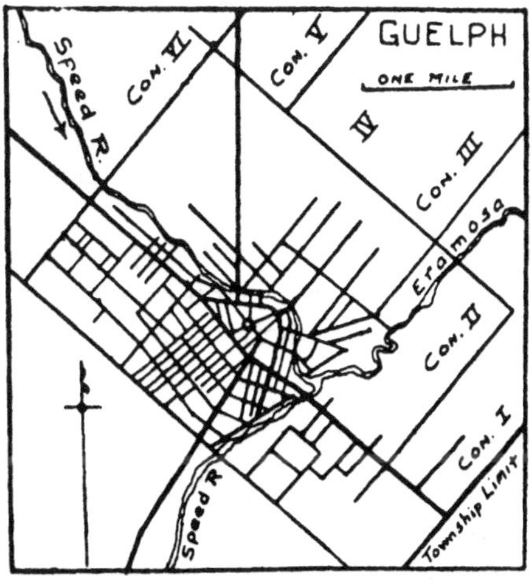

FIG. 21. A somewhat simplified road plan of Guelph, to show the radial streets in the centre of the town. Note the early concession areas.

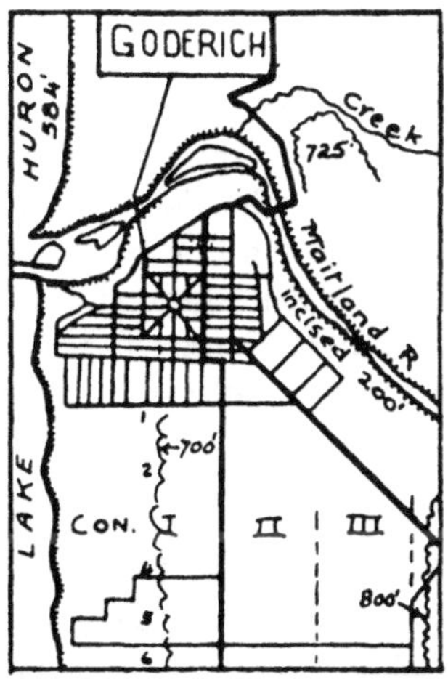

FIG. 22. A somewhat simplified road plan of Goderich, to show the 'octagon' with eight radial streets in the centre of the town. A few contours are indicated.

shore to the lake rising for 100 feet, and then flattening out to a gentle slope between the 700 and 800 feet contours. Upon this almost level ground in the bend at the mouth of the Maitland, John Galt located the new settlement. A central 'octagon' was laid out (later surrounding the Court House) with eight radiating streets; and shops soon occupied the outer edge of the octagon. The whole area was bounded by four streets forming a square with each side 600 yards long (Fig. 22). Four of these radiating streets are appropriately, if not poetically, named North, South, East and West streets. Lauriston tells us that in 1839 practically all the business of the town centred about the harbour below the hill, or on the lower part of West Street overlooking the harbour, while below the hill stood a string of residences. Already there was a foundry and a boat-building plant; and a tavern stood where the large flour mills are to-day. About 1866, men drilling for oil on the banks of the lake struck the saline layers of the Devonian formations, and thus inaugurated one of the earliest salt industries in Canada. In 1890 Goderich began to handle grain from the prairies, and now there are several large flour mills in operation. Many vessels use the port for trade or as a winter 'harbour of refuge', and to-day its population is about 5,000.

In Fig. 22 it is seen that the Concession areas start from the lake shore, which here runs north-south; i.e., there is no 'Broken Front'. The Concessions inland are a mile wide, with main roads between them. Two of these north-south roads are indicated in the south-east corner of Fig. 22. The east-west roads were surveyed at one-and-a-quarter mile intervals, and there were five lots in each rectangle so formed. The method of subdivision in the first block south of Goderich is indicated for Concession I in Fig. 22. It is to be noted that the plan for Goderich did not extend beyond the central block of streets already described. Beyond that the streets belong to the orthodox gridiron pattern, where the surveyor rather than the planner was the authority responsible.

Stratford and its radial streets

One of the most interesting little towns in southern Ontario from the point of view of town-planning is Stratford. Its plan in a somewhat simplified form appears in Fig. 24. Here we see the little artificial lake named after Queen Victoria, which has replaced the typical mill pond on the Avon River. Parklands fringe the lake on both sides, so that for a small town Stratford ranks high among its peers. But it is the radial plan of the streets which is of chief interest. No less than five main streets, from five different directions, meet near the former mill dam, giving a very convenient approach to the shopping centre and the public offices of Stratford. Who planned this little town? To a considerable extent it seems to have gained its radial pattern by accident.

The Huron Tract was surveyed as a vast triangle between Guelph and Goderich. The road from Guelph to the west crossed the Avon hereabouts (Fig. 23), so that it was a natural focus for a small settlement. (This road is now used by routes 7, 8, and 19.) It became the base for the township areas in the vicinity, so that the two Easthope townships were laid off to the east of this natural focus, and Ellice and Downie townships were surveyed to the west. Owing to the changing direction of the road, and to the triangular shape of the Huron Tract hereabouts

(Fig. 23) the roads between the various concessions meet at the oblique angles already mentioned.

In January, 1829, the first surveyor, J. MacDonald, reached the site of Stratford. The rough road had been cut through the forest before this date, as mentioned. In December of the same year the first settler —S. Fryfogle—located Lot 14 on Concession I, and built his hut there. Hence the road came first and determined the baseline; then followed the township survey, and the earliest settler was still later. This is rather typical of early Ontario. The population increased very slowly and had

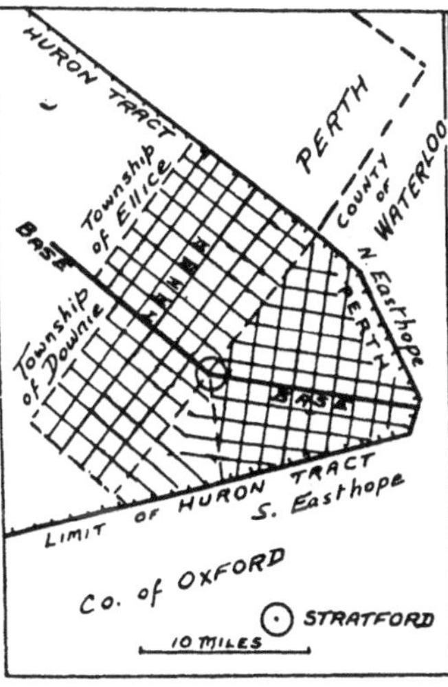

FIG. 23. The east end of the Huron Tract showing the boundaries of four townships. Stratford grew up at their common junction, shown by a circle.

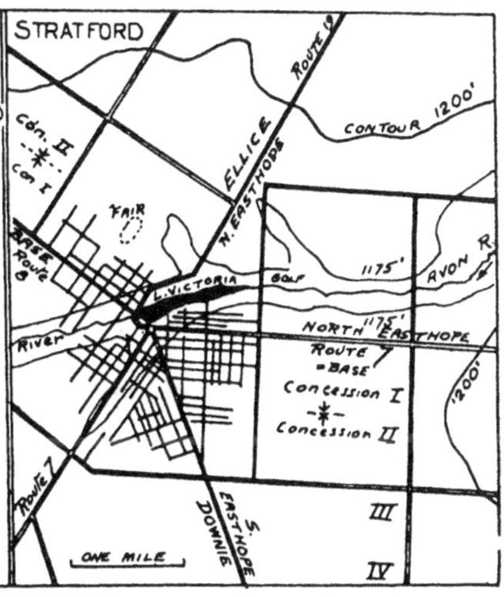

FIG. 24. A simplified road-plan of Stratford showing the main roads based on the irregularly arranged townships. Note that here the two main roads were the bases for subdivisions.

only amounted to fifty in 1841, but by 1850 it had grown to 900. Within two years the local residents were successful in their attempts to separate the country of Perth from Huron and Bruce, with the chief centre at Stratford. Two separate railways reached Stratford in 1856, and the public-spirited Park Board which gave to the town its lake and Park commenced work in 1904 (*vide Beacon-Herald*).

It may be noted in conclusion that the boundaries of the counties of Perth, Waterloo, and Oxford are rather intricate to the east of Stratford. This is due to their conforming to the wedge-shaped eastern end of the Huron Tract.

The Stabilized Plan of Charlottetown (P.E.I.)

It is only rarely in our very progressive and increasingly urban form of life, that a town of some importance reaches a satisfactory stage of growth, and then remains at that level for many decades. This, however, is what has happened in the case of the capital of the province of Prince Edward Island. Although it is a small town with a population of only 14,000 or so, yet it ranks with Montreal and Toronto as the chief city of a whole province.

Owing to the absence of coal, oil or water-power on the Island there

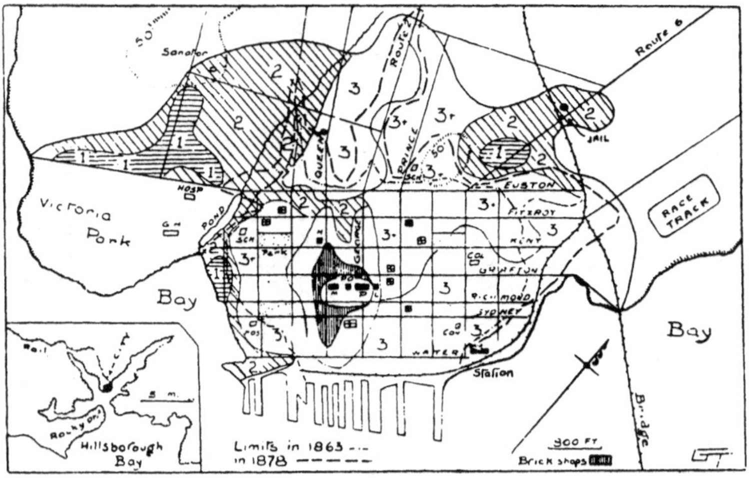

FIG. 25. A survey of Charlottetown (P. E. Is.) showing also the limits of close-set houses in 1863 and 1878. The figures refer to the class of residence. Areas with better-class houses are ruled. *Inset* is a map of the drowned valleys forming the harbour. (*Can. Geog. Jnl.* 1945.)

has been very little in the way of industrial development in the capital city. The whole island depends very largely on agriculture; and while Canada as a whole has increased her population 180 per cent in sixty years, in Prince Edward Island the population has decreased by six per cent in the same period. In 1881 Charlottetown had a population of 11,500; in 1911 it was only 11,200; while in 1941 it had grown slowly to 14,400. As a result of these somewhat stagnant conditions, we shall see that the pattern of the city has 'crystallized' in a very interesting fashion; and there is little of the proliferation and over-lapping that so complicates the zones in most growing American cities.

The first settlement was at Rocky Point in the French regime, and in 1728 there were about 115 folk living in the vicinity. The modern site was chosen about 1764, and the city is built on a promontory on the north side of the trident-shaped estuary opening into Hillsborough Bay (see inset in Fig. 25) This promontory nowhere rises to fifty feet, and not much of the hinterland is as high as this moderate elevation. A minor cape in the west was set apart for a park, and contains Government House.

The main streets formed a complete grid, but only three of these streets are indicated in Fig. 26. Four small parks were surveyed at the corners of the grid. These contain grass and large trees, and add considerably to the attractions of the little city. George Street was the axial street, but as is so often the case, the main shops have been built

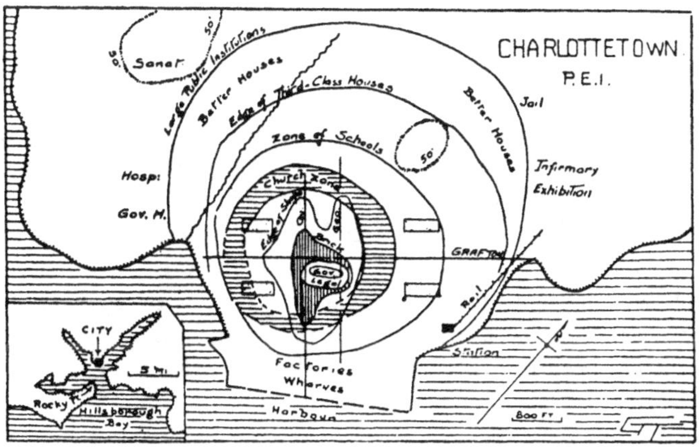

FIG. 26. A diagram of the functional zones in Charlottetown, somewhat generalized. They illustrate the regular arrangement in a town in equilibrium. (*Can. Geog. Jnl.*: 1945.)

in an adjacent street, Queen Street in this case. The large Provincial Building (where the Federation of Canada was decided upon) occupies the central position on the plan, with the Post Office and Markets to the west. By 1878 the houses extended to the north of what are now the third-class houses. In the last sixty years there has been very little change in the town pattern, save that a suburb of first and second-class houses has grown up along the northern edge of the town.

I have endeavoured to classify the zones in this old-established city, and my conclusions are given in Fig. 26 In the centre is the block of administrative buildings. Closely surrounding this—mainly on the south side—are the legal and professional offices. Clustered along the central parts of Queen and Grafton Streets are the more important stores, built of brick and usually three or four storeys in height. Outside of this zone is a belt of wooden shops of much less importance, which are found also in George Street. The many churches have almost all developed in a belt which is close to the four parks. Outside of this again, and about one-third of a mile from the centre is a zone which contains four or five large schools.

As usual the original houses within the city have declined in status, and are now mainly of third-class character. The better houses are naturally found towards the park, and in the higher land in the north-west and north-east. The wharves and the railway areas have attracted the factories and the poorer houses. This tends to give us a zone of better dwellings about one mile away from the centre of the city. For various reasons institutions such as jails, hospitals, race tracks, etc., are found in the outer zone beyond those so far enumerated.

Here then is a simple pattern which perhaps indicates the normal arrangement of the functional zones of an occidental city, when undisturbed by 'growing pains' (Fig. 26). The zones run as follows, commencing at the centre: 1, Administrative; 2, Professional; 3, Better shops; 4, Smaller Shops; 5, Church Zone; 6, School Zone; 7, Third-class Houses; 8, Second-class Houses; 9, First-class Houses; 10, Hospitals and Institutions. In terms of the classification used in the preceding part of this volume, we may conclude that Charlottetown, with no very distinct factory zones, etc., has a place among the *Early Mature* types of towns, though it has been in existence for nearly 200 years and is a Provincial Capital.

CHAPTER V

STAGES IN THE DEVELOPMENT OF A LARGE CITY

The Evolution of Toronto

LET us study the way in which a large city, such as Toronto, has evolved in about 150 years. Its topography is shown in Fig. 28; while the main stages of development are shown in Fig. 27, where only the salient characteristics are charted. In a large city we need to know the position of the commercial core, the industrial areas, and the general characters of the residential quarters. The latter may be fourth class (Hd), small inconvenient (or decayed) houses without gardens; third class (Hc), better houses but still without much space or convenience; second class (Hb), modern two- or three-storey brick houses with fair gardens; or lastly first class (Ha) with rather large gardens for a town, and of the type that at times might be called mansions. These types of residences are numbered 4, 3, 2, and 1 in the maps for 1885 and 1940 in Fig. 27.

If we ignore the other details we see that there is a constant change in the position of these zones. This change takes place in two ways. The best class houses of one decade become the second class houses of a later decade. But this change in zonal distribution is complicated by the building of new suburbs of various classes in the marginal hitherto empty regions. There is therefore no very close resemblance between the development of a city and that of an animal. But it is something like the way a young tree grows. Its trunk and branches increase by a sort of expansion, while quite new characters such as flowers and fruit appear as it reaches maturity.

The lower diagram in Fig. 27 shows Toronto in 1818. The original grid of nine streets was laid out about 1790, on well-drained ground at the head of the enclosed bay, near the Don outlet; where also the best protection from naval attack was presumably to be gained. The junction of King and George streets was the early centre of the city. By 1818 small industries such as tanning and brewing were scattered near the original settlement. A few larger houses were built on the margins of the town, but all but a few of the dwellings were still built of wood. The population was about 1,200 at this time, and the built-in area covered about half a square mile. As usual in such a young organism it is not possible to distinguish the zones, which will only differentiate at a later date.

By 1842 the town had about 20,000 inhabitants. The centre had

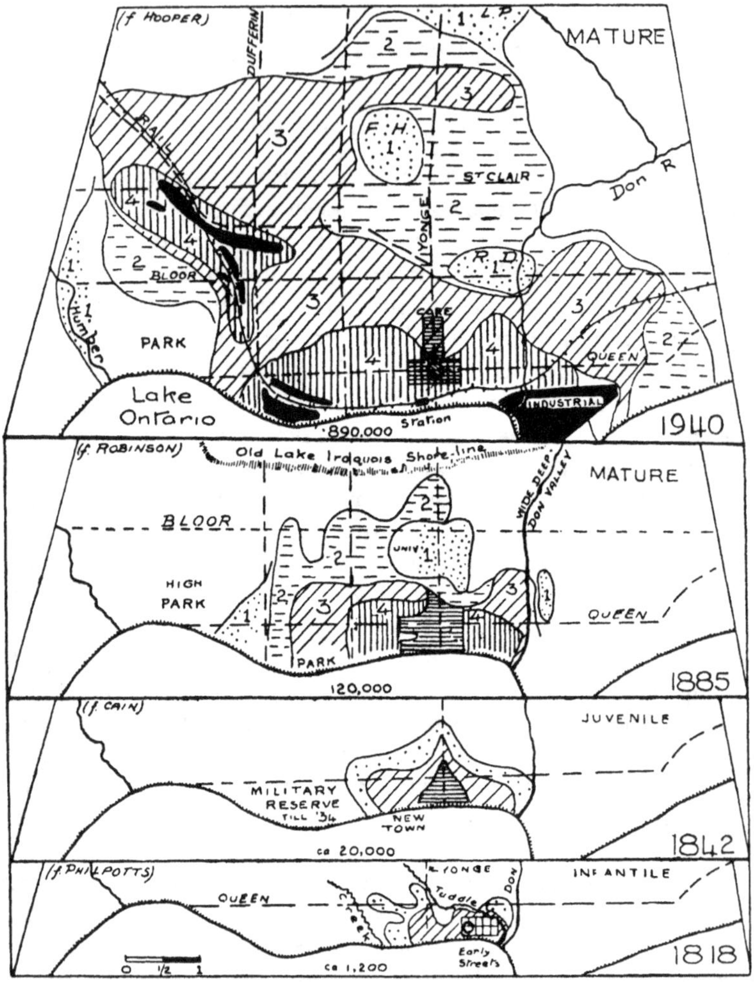

FIG. 27. Stage-diagram to show the evolution of the city of Toronto. The four maps are to the same scale, the earliest at the bottom. The rulings indicate the various elements of the city pattern. *1*, *2*, *3*, *4*, are the four types of city residences, varying from mansions (*1*) to the poorest types (*4*). The zones are, of course, somewhat generalized. The 'commercial core' includes the large Department Stores. (*Ann. Amer. Geog.*, 1942.)

shifted to the lower end of Yonge Street, the main corridor to the interior. Swamps near the foot of Yonge Street tended at first to isolate the old from the new portion of the town, but these were filled in by degrees; so that by 1842 this swampland was becoming the commercial core of the city. The houses were still small for the most part, but finer residences were being built on the northern margins. One of these was the Grange, which is now part of the City Art Gallery. The built-in area (diagonal ruling) now covered about two square miles. [Railways reached Toronto in 1855, running along the former 'Lakefront'. Much 'made land' extended the city to the south along the Lake by 1863. For growth of port see pages 273-7.]

In 1885 the town contained about 120,000 citizens, and extended over most of the City Plain, between the lake and the forty-foot cliff of the old Lake Iroquois shore-line. The commercial core has moved a trifle north, but still keeps to the Yonge Street axis. The houses are now numerous enough to show a fairly clear zoning, ranging from the poorest houses (4 or Hd) up to near-mansions (1 or Ha). The underprivileged live near the heart of the city, in old houses for the most part. The wealthy are building large residences on the margins, where parks (as in the west) or deep ravines (as in the east) promise security from further encroachment. The built-in area covers approximately eight square miles. The city is definitely growing to the north-west, mainly because the deep wide valley of the Don bars ready access to the east. The little stream-valleys also guide the expansion in the same direction. It is still not easy to differentiate between the commercial and industrial areas of the city. In a later section, we shall label such a stage of development as *Early Mature*.

In the modern city of 1940—which contains nearly 900,000 inhabitants, if adjacent satellites are included—there has been an enormous expansion of houses in every direction, but especially to the north and north-west. The old shore-line ceased to be the northern boundary about 1890. A great development of industrial plants has taken place, especially along the main railways west of the city, and on the filled-in lake areas to the south-east of the city. The commercial core has broadened somewhat, but is still clustered about Yonge, Queen, King and Bay Streets. Here also are the large department stores, which have not been differentiated in my discussion.

In the last half century the third-class houses have become the poor-class houses of 1940. The earlier first and second-class houses have been hemmed in by the growing city, riddled with small shops and with small industrial establishments. They have become the third-class houses (3) of to-day. To keep pace with the immense growth of the city, now seven times as populous as in 1885, vast areas of third-class houses cover most of the City Plain; but naturally many of the better houses (2) hold their own in the marginal portions of zone three. The modern zone 2 has developed mainly along Yonge Street, the main corridor to the north, especially near the fine west-east avenue called St. Clair. In the extreme west near the Humber, and the extreme east towards Scarborough Cliffs, there are other clusters of second-class houses. The positions of the four first-class suburbs (1, or Ha) are interesting. Partly they occupy the margins of ravines, as near Lawrence Park (L.P.) in the north, Rosedale (R.D.) in the east, and Humberside in the west. But the development of Forest Hill (F. H.) is perhaps due to the times of prosperity about twenty years ago. The high rolling hills of till, here about 300 feet above the lake, were beyond the city at that date, and were rapidly covered by large and expensive mansions. Possibly no similar development will occur again, since both the children in families and domestic servants are now rather rapidly declining in numbers.

Tentative 'ages' of a city

I have labelled the four stages shown in Fig. 27 with the names so familiar to geographers in regard to an evolving landscape. Most progressive cities of the occidental type go through similar stages. In the earliest stage there is no clear differentiation between industrial, commercial, or residential areas. The first tendency is for the bigger houses to develop near the margins, where gardens and privacy are available. We see Toronto in this *infantile* stage up to 1818.

In the next stage there is a fairly clear segregation of an extensive commercial quarter towards the centre of the town, including a number of streets without residences. But shops, offices and small industries are still rather mixed. A definite zone of better-class houses fringes the town. The poorer folk live near the centre; in many cases taking over the better houses of the former stage, but also building new small houses in any empty spaces. This may be called the *juvenile* stage, and is illustrated around 1842 in Toronto.

The beginning of maturity is shown by a definite differentiation of the residences. The various types are displaced outward as the years move on, though naturally examples of the early houses survive in the expanding zones. In 1865 the fourth classes of residences mentioned earlier were beginning to become evident. Poor quarters (4) clustered around the centre sufficient to form a connected zone. High Park in the west helped to determine the zone. The ravines at Rosedale attracted another area of good houses. This stage is *early mature*. [The early 1855 railway is charted in the top map.]

In the mature stage of Toronto about 1885 we see that the concentration of industrial areas along the railways and on the made ground is the chief change in the conditions.[1] Toronto is still growing, but the main axis of growth (i.e., the central line of the new industrial zone) is now along the railway to the north-west. There is a separation of the industrial city from the residential city, and the city is *mature*.

Perhaps when a definite zoning is carried out according to a laid-down plan (such as is indicated in Canberra) a city may be said to reach its zenith or climax; in geographical language it is *late mature*. The *senile* condition of a city is not indicated in Toronto. Such towns as Pekin (Fig. 5) or Nanking, with vast areas originally covered with houses but now heaps of ruins, may perhaps be described as partly senile. But the complexity of a large and ancient city is like the complexity of a long river like the Nile. Parts of the Nile are juvenile, others mature, others senile, depending on external factors developing during the life of the river.

Environment and the city's growth

A city is obviously an artificial growth. Nature produces nothing resembling a city. I have discussed this aspect of the problem as regards Toronto in several publications. Here therefore I will only introduce one diagram (Fig. 28), and point out a few examples where nature's lay-out might well have been considered in deciding man's design.

The Toronto environment consists of a topography built up wholly of glacial debris, either till or sands. These are 300 feet thick in the upper part of the town, and about sixty feet thick in the lower part. Rock outcrops are absent everywhere, except in the bed of the Humber. The City Plain extends from Lake Ontario (246 feet) up to the Iroquois shoreline (425 feet). Here a sharp slope, almost a cliff forty feet or more in height (which was cut out by the waves of former Lake Iroquois), separates the lower from the upper town. The Canadian Pacific Railway runs along the foot of the Iroquois shoreline and definitely blocks road communication with the upper suburbs.

The chief natural features of the town, however, are the ravines, which have been eroded in the till by the drainage from the former ice-sheet. These are fifty to 100 feet deep, and in general run from north-west to south-east (Fig. 28). The civic fathers followed the usual chess-board lay-out of the streets, with their axes near the cardinal directions.

[1]Obviously the mature stage of a slow-growing town is reached with a smaller total population than in the case of a quick-growing town like Toronto. In many towns the 'mature' stage is reached when the population is less than 50,000.

Hence the roads have been made without the slightest reference to the ravines, and without much regard for the main objective of the citizens' journeys (which was to reach the centre of the city). Some of the most

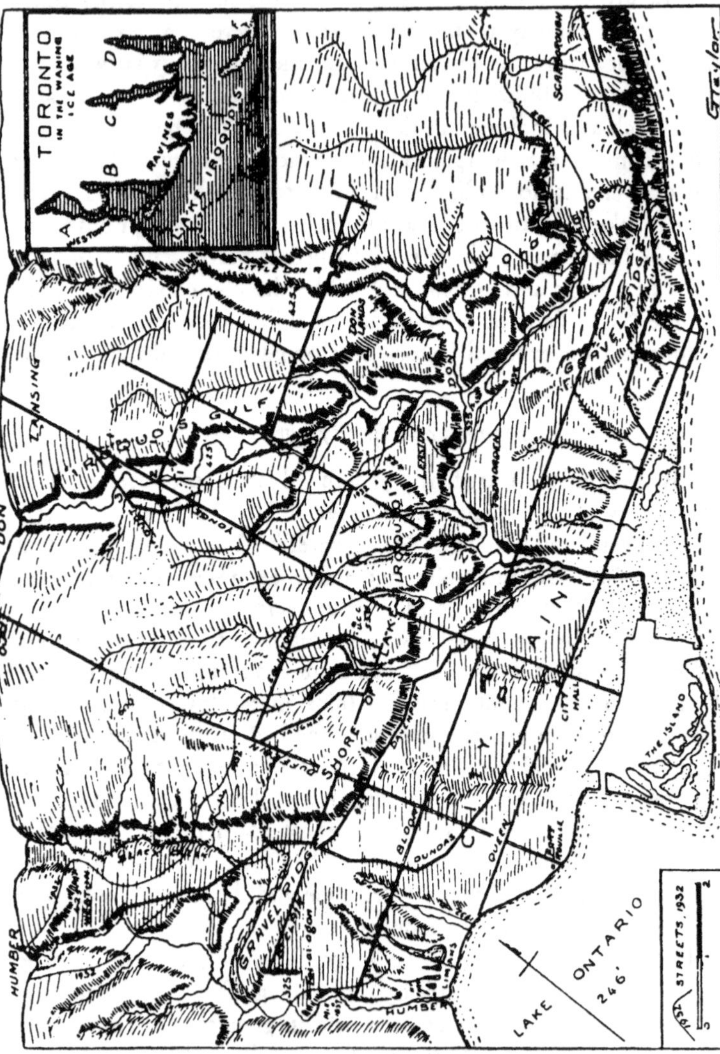

FIG. 28. A block-diagram of the city of Toronto, showing the topographic features which have affected the evolution of the city. *Inset* is a diagram showing the locality at the Lake Iroquois period. The fine line labelled 1932 shows the edge of the streets with close-set houses in 1932. (*From Environment, Race and Migration.*)

expensive city-viaducts in the world have been built to cross the Don and Humber valleys. These were inevitable, but others over the smaller ravines need not have been built if the roads had been built along the divides between the ravines. In the west part of the town such roads would have the added advantage that they led directly to the centre of the city. For instance, Vaughan Road running diagonally between ravines is a much more sensibly designed road than Spadina Road with its huge viaduct.

In Toronto, as in many other fairly modern cities, the chessboard was laid out on the surveyor's chart, and often the roads were named before construction. Thus we find a continuous Walmer Road in the middle of the city map, but actually it is still in three isolated pieces. These sections are separated from each other by the Iroquois cliff, or by deep ravines, or by the railway.

The most amusing example is Woodbine Avenue which has a gap three miles long in its continuity. It seems hardly likely that the whole of the Little Don Valley will be filled up to enable the surveyor's design to be maintained! Yet this is merely a large-scale example (of which we see many smaller illustrations) of the absurdity of ignoring the environments in planning the development of a large city. Lack of space forbids my enumerating many other examples of man's unsatisfactory choice of plan in this connection, but others are referred to on page 349.

The writer would like to enter a plea for the use of the block diagram (as in Fig. 28) to illustrate the setting of a city. Its history may be so easily summarized therefrom, as follows:

The first houses in 1793 were built about a mile southeast of the City Hall. These formed the nucleus of the city. By 1818 the little town had spread along the shoreline. The great growth of the city within the next century is shown by the isopleth labelled 1932 (Fig. 28), which indicates the outer limit of built-in streets. The large valleys of the Humber and the Don, the small glacial-water ravines and the Iroquois shoreline are clearly represented. This illustration should be used in conjunction with the stage-diagram (Fig. 27).

The 'Zones and Strata' concept applied to cities

In the case of large human movements—such as racial migrations—the use of the zones and strata technique enables us to deduce the cradle of human evolution and the order of development of the races. But it is also of considerable value in connection with the much smaller agglomerations of man which we call villages and cities. Here also we are concerned with clusters of folk about the original nucleus, about which the town gradually expands.

What value has this technique in city evolution? Perhaps in connection with house-types we see the process most clearly. The first houses at Toronto were log huts, of which one is actually still preserved in the Exhibition Grounds. This was originally erected where Queen Street crosses the Don River. The nearest log hut, in use to-day much as it was originally, is (as far as I know) in Queensville, thirty-five miles north of Toronto. Thus the first stage of house has now been pushed out for thirty miles or so from its original position. The second type was the small frame house built of horizontal planks. I recently photographed one of these houses in the heart of the city in Queen Street opposite Osgoode Hall. It has now been pulled down, but plenty of specimens still remain in corners of the city. They are now found fairly numerously in the little towns near Toronto; and we may say that they

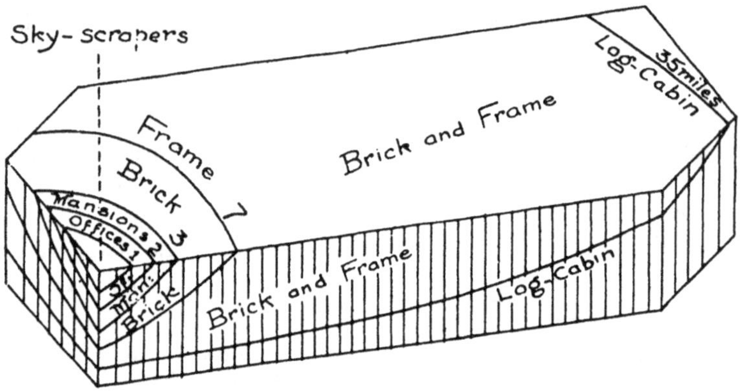

FIG. 29. A block-diagram illustrating the 'Zones and Strata' technique applied to the Evolution of Toronto. The first houses were log-cabins, now displaced 35 miles away from the city, but an early specimen still remains in the city. So also the later types of frame-houses, small brick houses, mansions, etc., have been displaced successively from the nucleus of the city.

have been dispersed to a distance of seven or eight miles from the nucleus (Fig. 29).

The farmhouse with a barn perhaps comes next. None is left in the city as far as I know; but they still occur on the margin, at a distance of about five miles, as for instance near Wilson Avenue. The fourth type is the two-storey brick house often with a snow-shedding gable in front. These are still common in the third- and fourth-class houses in the city. They are of course almost universal throughout Lower Ontario, being the usual type of farmhouse built during the last half century.

The remaining types are the mansions and sky-scrapers. The Grange was built in 1817 only a mile from the Lake. It is a large mansion still in use as part of the Art Gallery. To-day houses of this type and culture are not found nearer than Rosedale or Forest Hill, i.e., about four miles north. Finally in the busiest part of the city we find the latest type of building, the sky-scraper, which is almost wholly found in the vicinity of Queen, Bay, and Yonge Streets. They of course constitute the most striking buildings in the commercial core.

The zones are therefore fairly definite to-day. If we assume that the few relics of older types still remaining are fragments of ancient 'strata', most of which have been 'buried' (i.e., removed), then the evidence given above shows how the technique can be applied to city evolution. In other words all the types originated near the heart of the modern city (near the 'nucleus'); but the older the type the further it has been displaced to the margins as the city evolves. Conversely from the distribution of these zones we can deduce the way in which the houses have evolved.

The data can be summarized in a table as follows, with the oldest types of building at the foot of the table.

ZONES OF HOUSES

Order	Type	First site	Present Position	Displacement from centre
6.	Sky-scrapers	Centre of city	Centre of city	(Area 1 mile wide)
5.	Mansions	do.	Margin of city	3 or 4 miles
4.	Early brick houses	do.	Throughout province	?
3.	Farmhouses	do. ?	do.	5 miles
2.	Frame houses	do.	Especially in pioneer areas	7 do.
1.	Log cabins	do.	Obsolete	35 do.

The Circumspice Concept illustrated by Port Credit and Whitby

Darwin, while studying coral reefs, realized that amid the many examples in the Pacific there were likely to be reefs in all stages of evolution. In effect he said: 'Look around you, and you will be able to reconstruct the evolution of the coral atoll.' The epitaph of Wren the architect in Saint Paul's in a sense contains the same idea, 'Si requiris monumentam, circumspice.' We can usefully apply the same concept of 'looking around you' to city evolution. The village is father to the city; hence if we study a village in the same environment as the city, we shall learn something as to an early stage in the evolution of the city.

I have made a number of surveys of towns and villages in southern Ontario, and I think it will be of interest to discuss two of these. Port Credit is a little town on the lake thirteen miles west of Toronto. It has about 2,000 inhabitants, and in some ways resembles an early stage of Toronto, when the latter contained about the same number of folk. (It is not of much significance that modern features such as electricity in Port Credit had no counterpart in the early stages of Toronto.) My second illustration is the town of Whitby (5,000 inhabitants), which has grown up about twenty-seven miles east of Toronto with much the same topography and hinterland as the other two lake settlements.

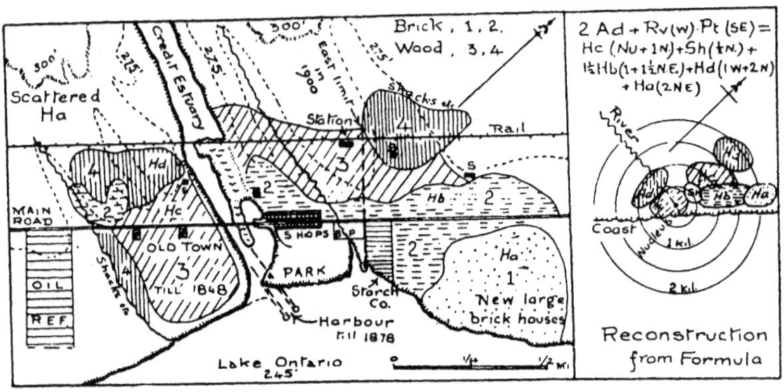

FIG. 30. (*Left*) A functional plan of the little town of Port Credit 12 m. west of Toronto. Two contours are charted. The houses are in four classes; Ha, Hb, Hc, and Hd; the latter are mostly shacks. (*Right*) A plan reconstructed from the formula shown at the top of the figure. It is to be compared with the actual plan on the left.

Port Credit was first settled in 1804 at the mouth of the Credit River (Fig. 30), and until 1845 the village was confined to the western bank. Almost all the houses were wooden, and were mostly two-storey gabled residences, of which several still survive. The next few decades were quite prosperous, and the little town had a considerable trade with U.S.A. in grain and timber. There was a busy port with quays, harbour works and a lighthouse. But tariff laws ruined the American grain trade, while all the timber was exhausted in the hinterland during the seventies. The storms around 1878 demolished the harbour works, and they were never rebuilt.

The railway station was placed on the east side of the estuary in 1855 and this led to the development of a new town nearer to the station, in which houses were built much more generally of brick. The population was about 450 in 1877, when the town contained three churches and three hotels; which is the same as to-day with a population almost 2,000.

In 1900 the eastern portion equalled the western in area, but since that date there has been a great expansion to the east, as Fig. 30 shows. Thus Port Credit developed in three stages; the old wooden town in the west till 1855 or so; the modern brick and wood town between the station and the little harbour till 1900; and after that date the newer portion right in the east.

There are two large industrial plants, each giving work to about 100 employees. The Starch Company commenced about 1889, and the large Oil Refinery in 1933. To-day the town contains two schools and a High School, two banks and a large brick post office. The two rows of shops shown on the main street comprise three stores, four cafés, three realtors, three barbers, two butchers, a drug store, and a picture house.

The town pattern is *early mature*, since the older wooden houses were long ago replaced by brick near the shopping centre. The cheap wooden houses (Hd) have grown up on the margins, partly in response to the two large industrial plants. The very slow growth of the town has prevented the change of early homes (Hc) into the near-slums (Hd in part) as in Toronto. The group of large brick houses (Ha) in the far east is not a natural out-growth of Port Credit, but a sort of forest-suburb of Toronto.

It seems likely that the oldest western part of the town, consisting largely of wooden houses, gives us some idea of Toronto about 1840, though the latter was of course far larger. The addition of the newer eastern portion of the town with its numerous two-storey brick houses affords a picture which to some extent resembles Toronto in the seventies, before the development of special industrial sections in the city. The large houses (Ha) in the far east, each surrounded with a large garden and often by small forest trees, are really outlying parts of Toronto; for their owners have in general not much cultural connection with Port Credit. However there are still many acres of farmland separating the little town from the metropolis. Not yet has it joined Long Branch and Mimico and become a mere satellite of Toronto.

A second example of a small town on the lake-shore near Toronto will show us the normal development of a town considerably larger than Port Credit. I have chosen Whitby, which is twenty-two miles east of the metropolis. Its site is much the same as at Port Credit, save that there is no river there, but only a small shallow bay which has been converted into a harbour by a breakwater across the mouth. The main

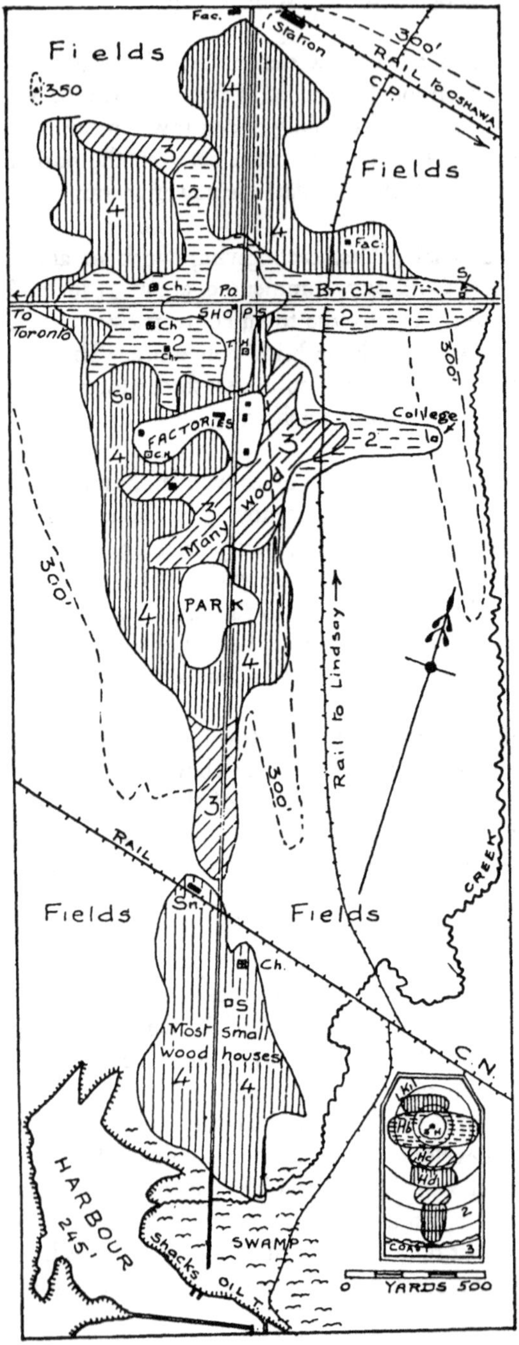

FIG. 31. A Functional Plan of the little town of Whitby, 23 miles to the east of Toronto, and in a similar environment. The house-zones 1, 2, 3, 4, correspond to the zones Ha, Hb, Hc, Hd, in Fig. 27. *Inset* is a reconstruction of the plan, derived from the formula given in the text. *S.* is school; *Ch.* is church; *Sn.* is station.

cross-road follows along one of the usual broad north-south ridges of this coast. The 300-foot contour is shown on Fig. 31. The swamps near the harbour have prevented house development near the lake. As at Port Credit, the days of shipping grain and timber have passed. Now oil is the chief import, and is stored in four tanks which take the place of bygone elevators. A suburb south of the Canadian National railway survives from these earlier days. The total population is about 4,500.

Whitby to-day has developed at the cross-roads, where the Toronto-Kingston corridor crosses the north-south road to the little port. The importance of Whitby as a port has almost vanished, but a number of small factories, dealing with brass-work, tanning, blankets, and canning, are adding to the vitality of the little agricultural town. The factories are not yet segregated, indeed several of them are not even on the various railways which meet at Whitby (Fig. 20).

The pattern of the town shows the usual inner zone of shops at the cross-roads together with the post office and Town Hall. Outside this is the zone of larger houses, nearly all of brick (2 in Fig. 31), and here are found the chief churches. Beyond these zones the houses are on the whole built of wood, the nearer being usually two-storey and the distant houses one-storey. However a few larger houses are of course to be found in these zones also. A large Ladies' College is isolated on a ridge to the east of the town and a hospital west of the harbour.

Although the town is twice as large as Port Credit it is not in quite such an advanced stage of urban development, mainly because the factories are smaller. The town may be described as in the *adolescent* stage, for there is no industrial zone and no zone of larger residences. Owing to the vicinity of Toronto, which tends to attract so much of the local trade, it does not seem likely that Whitby will soon advance much beyond this stage of development.

Formulae for towns

In a genetic approach to urban geography we are chiefly interested in the town as an evolving organism. We want to know its present pattern, how that developed, and where and why the town started. For the present we can ignore the street plan, the source of its trade, the extent of its umland, and a vast number of other interesting features of a city. I see the city as a series of zones, expanding from the original nucleus, and usually conditioned in its growth by the surrounding environment, unless we are dealing with a city of the plains. Several features therefore seem to be essential in our proposed formulae. We should know something as to the original site or nucleus; as to the reason for its founding; as to its growth with respect to the nucleus. As regards the zones, we should know their width and position. Further we might well include some clue to its stage of development, and of course a reference to the total population.

In 1938 I read a paper before the Association of American Geographers dealing with a number of towns in the Italian region of the Trentino. There I gave a preliminary description of formulae, including many of these terms. Since that time I have been trying to apply the same technique to American and other towns. I feel sure that such research is worth while, and equally sure that the technique which is suggested herewith will be greatly improved upon as time goes on.

ABBREVIATIONS USED IN THE FORMULAE

Directions. North (N.), North-east (N.E.), etc.; all from the nucleus (Nu).
Stage of Development. Infantile (In), Juvenile (Ju), Adolescent (Ad), Mature (Ma), Senile (Se), Nucleus (Nu).
Site of town, Capital (Cp), Fan (Fn), Fort (Ft), Mountain (Mt), Mining (Mn), Pass (Ps), Oasis (Oa), Railway (Ry), Rail-crossing (Rc), Road (Rd), River (Rv).
Population. Figures represent thousands.
Distances of zones, etc. Figures represent kilometres.
Zones. Apartments (Ap), Factories (Fc).
Houses—Largest (mansions) (Ha); Large (Hb); Small (Hc); Shacks or decayed houses (Hd).
Offices (Of); Shops (Sh); Stores, large (Sa).
Equation order. Population, Status, Site—on the left.
Zones, proceeding outwards—on the right.

Let us run over the description of Port Credit and see how our formula would work out (Fig. 30). The population is about 2,000. This would be represented '2' (the number of thousands) in the formula. The town was originally a lake-port on a little river. This is indicated by Rv Pt. With regard to the nucleus, the river is to the west, and the lake (or sea) to the south-east. So our formula becomes 'Rv(W.)Pt(S.E.).' The stage of development is between adolescent and early mature. Since it is so small a town, and the mansions are linked to Toronto rather than to Port Credit, it would seem to be adolescent (Ad). This general portion of the formula forms the left side of our equation.

Now we may turn to the particular zones, which appear in the right-hand side of the equation. The present shopping centre has been shifted from the original nucleus, about ½ kilometre to the north. Thus the shop-zone appears as 'Sh(½N.).' The house-zones should be given in order from the centre of the town outwards. Thus the second-class houses (Hb) form a zone with a centre about one kilometre north-east of the nucleus, i.e., Hb(1N.E.). Then comes Hc, the original large wooden houses which still surround the nucleus, but also cover a zone one kilometre to the north, i.e., Hc(Nu+1N.). The shacks Hd are found one kilometre west of the nucleus, and also two kilometres north and so we get Hd(1W.+2N.). Finally the large brick houses (not quite mansions) represent class Ha, and are being built two kilometres north-east of the nucleus, i.e., Ha(2N.E.).

The total formula comprises two sides of an equation as follows:

2Ad + Rv(W.)Pt(S.E.) = Sh(½N.) + 1½Hb(1N.E.) + Hc (Nu + 1N.) + Hd (1W. + 2N.) + Ha(2N.E.).

The factories are not arranged in a zone, hence are not mentioned in the formula; but the presence of a few factories may be gathered from the fact that the town is adolescent.

To reconstruct a rough plan from the formula, we proceed as follows (see the inset in Fig. 30). A large dot is placed for the nucleus, and concentric circles are drawn with radii increasing by half kilometres. We can add the river coming in from the west, and the lake shore along the south-east. The distance indices (in the brackets) show the position of the centre of a zone. Thus the shops can be placed half a kilometre to the north of the nucleus. If reasonable patches or zones be drawn contiguous to each other, in accord with the indices, we obtain a plan which is quite close to the actual pattern of Port Credit, and gives us almost all the characteristic features of the settlement.

The formula for Whitby can be obtained in the same fashion. It comes out somewhat as follows:

4.5Ad + CrPt(3S.) = Sh(Nu) + Hb(½E. + ½W.) + Hc(½S. + 1½S.) + Hd(½N. + 1S. + 2½S.)

We may interpret this as follows. A small town in the adolescent stage with a population of 4,500 has developed at a cross-roads with a little port three kilometres to the south. On the right of the equation we find the zones; i.e., the shops at the nucleus, with three house-zones at various distances as shown in the brackets. For instance there are three separate areas of Hd (smallest houses); at one-half kilometre north, one kilometre south, and again at two and a half kilometres south. A reconstruction from the formula is given as an inset in Fig. 31. It clearly gives us a fair picture of the actual plan of Whitby.

PART II

HISTORY OF URBAN EVOLUTION

After the Dark Ages we find a cowering countryside only gradually (in the 11th century) building walls about its settlements. In the 16th century the Baroque way of life develops, characterised at first by towns with stellate walls, and later by palaces of geometrical design. The Industrial Revolution (after 1800) gave us the Paleotechnic town, slowly changing into the overgrown Megalopolis.

LEWIS MUMFORD (*passim*)

CHAPTER VI

PREHISTORIC AND ORIENTAL TOWNS

Part A. Prehistoric Settlements in Britain, and Marginal Survivals

THE Zones and Strata concept may well be applied to the study of the earliest settlements in such a country as Britain. The archæologist studies innumerable excavations, and, as time goes on, obtains a fairly clear picture of how a technique—such as for instance the development of settlements—has evolved in a given environment. But a very useful *parallel* investigation consists in studying the 'zones of distribution'

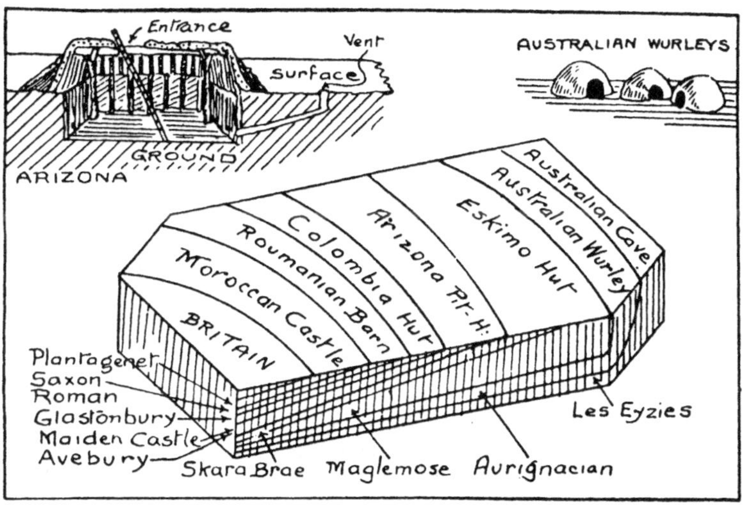

FIG. 32. The 'Zones and Strata' concept applied to settlement, showing bygone cultures in Britain paralleled by existing distant cultures. *Inset* are sketches of Australian and Arizona huts. The latter is seen in section.

about such a centre of evolution as Western Europe. This may be made clear by the following simple diagram.

Fig. 32 shows a simplified diagram of 'zones and strata' applied to settlement. On the front edge we see a series of archæological strata such as have been widely investigated in Britain. On the upper surface of the block diagram is a series of zones, admittedly much generalized, which suggests that by moving across the world to regions of primitive culture we pass through a series of zones which indicate the same line of development as do the strata.

Only a few of the stages passed through in Western Europe since Ice Age times, when man lived in caves, can be discussed in this brief account. The following table shows some of the main phases in this evolution of hamlets and villages.

ANCIENT SITES AND MARGINAL SURVIVALS

Approximate dates		Strata in Britain	Existing examples in marginal lands
Paleolithic	B.C. 20000 15000 10000	Neandertal Man (Les Eyzies) Aurignacian Man Magdalenian Man	Australian aboriginal cave (Devon Downs) Australian wurleys
Mesolithic	6000	Maglemose Man	Eskimo Hut
Neolithic	3000	Windmill Hill Man (Skara Brae)	
Bronze	1700	Beaker Man (Avebury) (Goidelic speakers)	Recent Pit dwellings of Arizona
Iron	800 100	Halstatt Man (Maiden Castle; Brythonic) La Tène Man (Glastonbury)	 Colombian wattled huts
Roman period Saxon period Plantagenet	A.D. 43 500 1200	Brythonic speakers English speakers Medieval Castles	 Roumanian barns Moroccan Castles of to-day

One of the earliest books, which showed that primitive man in England lived much as do some far-distant tribes on earth to-day, was that by Professor Sollas, which he called *Ancient Hunters* (London, 1911). Here he stressed the resemblances between the way of life of the Australian aborigines and that of the Paleolithic folk of Britain. Among these Australian tribes—which are fairly well known to the writer—we find the least complex settlements of to-day. A short discussion will not be out of place.

The aborigines in Australia never showed any love for deep caves, nor was there the need for cave-dwelling which was felt in Ice Age times in Britain; for the primitive dwelling in Australia was mainly necessary to keep off the sun's rays rather than rain or cold. However a few cave shelters in Australia were inhabited for long periods, and the best known example is at Devon Downs about sixty miles to the east of Adelaide (Fig. 34). Here the Murray River flows below more or less vertical limestone cliffs, and in the latter, about thirty feet above the river, is a rock-shelter which has been carefully investigated by Hale and Tindale.

The main features of the Devon Downs 'settlement' can be learnt from the vertical section given in Fig. 34. The hollow, dissolved out of

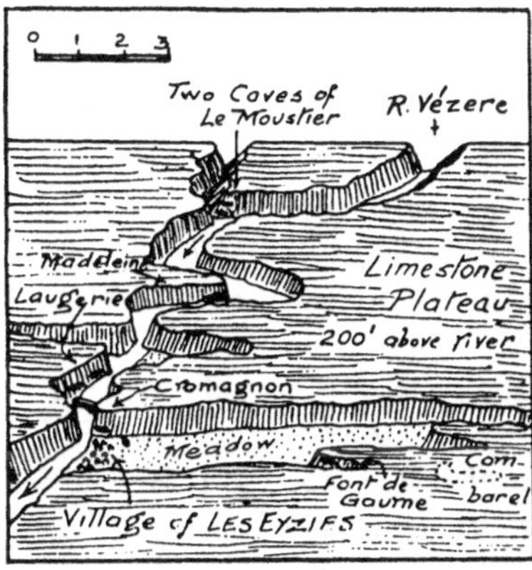

FIG. 33. Block Diagram of the Paleolithic caves at Les Eyzies in south-west France.

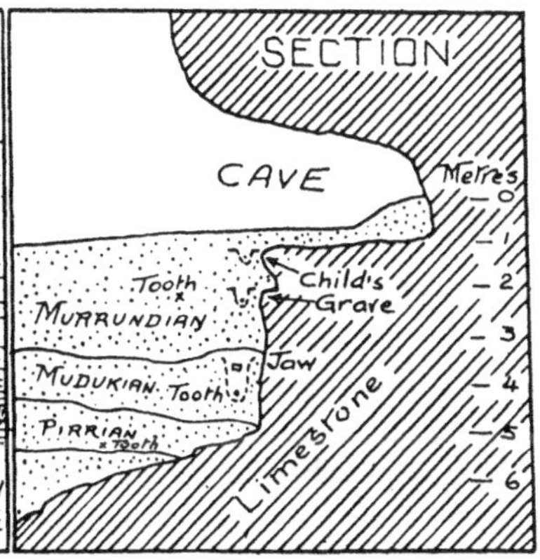

FIG. 34. Vertical section through a cave above the Murray River at Devon Downs, in South Australia (*after Hale & Tindale*).

the limestone at an early stage of the river's history, is about fifteen feet deep, and about eight feet high. It is indeed exactly like many of the caves along the Vezère in France, to which reference will shortly be made. On the floor of the Australian cave are several feet of debris, while on the slopes just below the mouth of the cave, some twenty feet of debris have accumulated. This deposit is deep enough to hold three graves at different levels, and it represents the accumulation of centuries; since the shells found therein have changed in character in the various layers. In the lower Pirrian deposits, these shells obtained from the adjacent river indicate that the climate was wetter then than now. Curious pointed bones called 'Muduk' marked the middle layer. They were perhaps a crude form of fish-hook. In the Murrundian layer many small flint adze-edges were found. The skeletons seem to be rather more primitive than the average aboriginal of to-day; but we have not yet enough data to decide how old the upper layers may be. Possibly the cave was occupied until quite recently, though there are few aborigines left now in that part of Australia. Clearly it represents an early phase of human life, before food-producing (by agriculture, etc.) was thought of. No pottery was known to the aborigines, and their stone weapons were about in the Mesolithic stage. The dog was domesticated, so that the culture was to some degree like that of the Tardenoisian folk, who entered Britain in the Boreal Phase of post-glacial time.

To gain some knowledge of settlements during Paleolithic times (previous to Tardenoisian) we cannot do better than take a journey to the south of France. Some seventy-five miles to the east of Bordeaux we find ourselves in the foothills of the Cevennes of central France. Here is a somewhat plateau-like landscape, where the surface rocks consist of limestone. The tributaries of the Garonne have cut rather narrow valleys to a depth of 200 feet in this plateau, and one of them, the Vezere, runs through the most interesting region in Europe from the point of view of the student of Early Man.

The surroundings of the village of Les Eyzies can be deciphered from the sketch given in Fig. 33. The railway from Paris to Agen runs down the valley of the Vezere hereabouts, and crosses the river at Les Eyzies. In the bluff at the east end of the bridge is the famous cave of Cromagnon, where the type skeleton of late Paleolithic (Magdalenian) times was discovered. About five miles up stream is another famous cave, where the flint implements of this period are unusually abundant in the cave debris. This is the cave of La Madeleine, which one reaches in a boat from the farm on the east bank.

There are many other caves in the limestone cliffs bordering the Vezère, and these have been inhabited throughout the whole period of Paleolithic Man. This period possibly extended over 100,000 years. The earliest phase of the Paleolithic is named from the cave which occurs on the west bank of the Vezère about ten miles north of Les Eyzies. Here are the two caves of Le Moustier, hidden behind a number of stone houses in the hamlet of that name. In these caves the crudest stone implements have been found, and they are given the name of Mousterian. We know that they were used by Neandertal Man—a much cruder and more animal-like specimen of humanity than Cromagnon Man.

Finally on the south wall of the broad meadow just to the east of Les Eyzies are several tunnel-like caves, such as Font de Gaume and Combarel, where one can penetrate hundreds of yards into the limestone along former solution channels of the Vezère. Here the lamp of the guide shows innumerable drawings (in charcoal and ochre) of bison, reindeer, cave-bears, etc., which bear witness to the artistic skill of late Paleolithic Man in this corner of France.

To the student of human settlement it is clear that these caves offered unique advantages to Paleolithic Man, especially as we know that during much of the long period of his life in the Vezère valley, the climate was far colder than it is now. It is not perhaps too far-fetched to think of the Vezère valley at Les Eyzies as a sort of metropolis of our ancestors in Paleolithic times. In no other site in western Europe, so far as I am aware, has any such series of Paleolithic sites been discovered; but there is nothing to show that the total population at any one time was particularly dense, even in this favoured site. In Magdalenian (Creswellian) times in Britain, Clark doubts if there were more than a few hundreds in the whole population.

During Paleolithic times there were long periods when the climate in western Europe was not unlike what it is now. During these milder centuries, there is little doubt that the settlements would be of a different character, and some crude sketches on the cave walls suggest that in Aurignacian times man lived in beehive huts, which were very much like those still used by the Australian aborigines to this day.

Throughout Australia the native 'wurley' is of much the same pattern. It has been aptly compared to an inverted teacup with a circular bite taken out of the rim to represent the entrance. A collection of half a dozen wurleys is as close as the aboriginal life gets to a 'village'. In the south there is a tendency for the entrances to be on the east side, away from the prevalent west winds. But there is no arrangement in rows, and the settlement is usually abandoned after the floors of the wurleys and adjacent ground become unpleasant from offal and debris.

The wurleys consist merely of a ring of boughs, whose pointed stems are driven into the ground about five feet from the centre of the hut (Inset in Fig. 32). The tops of the boughs are bent over to form the roof, and lashed in that position with fibres or roots. There is no attempt to form a door, though in some places a sort of crude porch—also built of boughs—shuts off some of the wind. The fireplace is usually out of doors, and there is no furniture in the hut, except perhaps some skins and a few wooden dishes (coolamons) shaped like curved shields. The spears of the men are stuck in the ground near the wurley, and numerous dogs lie around the cluster of huts. We may picture our Aurignacian ancestors in Britain living in settlements of much this type.

It is natural that elaborate grave-chambers and mounds shall persist in much better preservation than examples of houses and settlements. Gordon Childe's recent volume on *Prehistoric Communities* (London, 1940) gives a very clear picture of the development of cultures in Britain. The region around Avebury in Wiltshire, some twenty miles north of the more famous Stonehenge, is particularly famous for early Neolithic monuments. Windmill Hill, near the impressive circles of Avebury, is accepted as the classic site in early Neolithic times. Here is a 'causeway camp' one of many which extend from Devon to Brighton, which has yielded a great variety of Neolithic artefacts. The folk lived here within a camp enclosed in three concentric circular ditches, each with its stockade. The outer ditch had a diameter of 1,200 feet. Their chief occupation

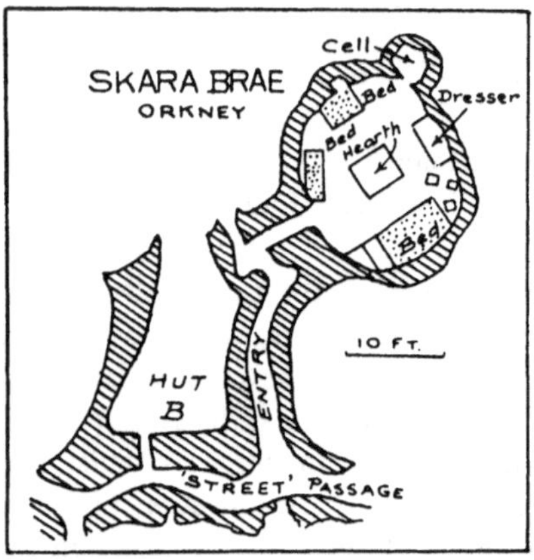

FIG. 35. Skara Brae, a late Neolithic village in the Orkneys, Scotland. (*After V. G. Childe.*)

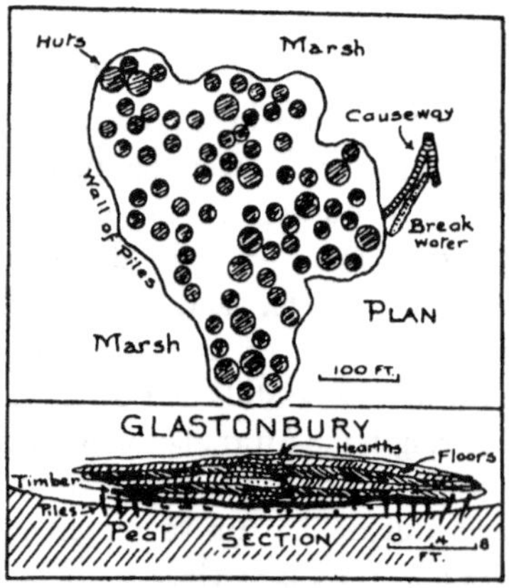

FIG. 36. Glastonbury, a late Celtic village in Somerset.

was herding cattle, but they also grew grain in small plots, which they cultivated with digging sticks or deer antlers. Traces of small huts have been found, but the details are not clear. Pit houses, like those formerly in use in Arizona (inset, Fig. 32) were in use in Britain about this time.

Succeeding these peoples were invasions of tribes who built imposing 'Megalithic' tombs. Here chambers made of stone slabs were covered with enormous masses of earth, and they are particularly numerous on the open cuestas of limestone in Wiltshire and Yorkshire, as well as in north-east Ireland and north and west Scotland. In some cases, where there was a scarcity of timber, their houses were built of flagstones, and some of these have been preserved. Perhaps the best known are in the Orkney Islands off the north coast of Scotland, and a sketch of such a settlement appears in Fig. 35.

The Orkney Islands are almost treeless, and there are large areas of pasture. Far back in Neolithic times herds of cattle were grazed, and in those early days there was little or no hunting of deer or fishing in adjacent waters. At Skara Brae there was a permanent settlement for many generations. The stone huts were built in the shelter of sand dunes, and in some cases these overwhelmed the huts and so preserved them to this day. Fig. 35 shows portion of this settlement, in which the seven huts were in a cluster connected by paved alleys, which might almost be termed the earliest streets in Britain! The huts were drained by sewers built of flagstones.

The houses at Skara Brae were square or irregular in outline with a width of about fifteen feet. Each was entered from a passage only four feet high, and this could be closed by a barred stone door. In the centre was a square hearth on which a peat fire burned. The roof was made of stone slabs, overlapping in corbel fashion. Bedplaces of stone were present, in which heather and skins made comfortable beds. Small luted stone cists were used to contain liquids. Stone shelves and cupboards were numerous. Stones in the walls are in places carved with lozenges, triangles, etc. Ribbed pottery occurs, which is very like that found in 'Copper Age' caves in Catalonia; but Childe believes that this Skara Brae site antedates the 'Beaker' period. Perhaps it may be dated about 2000 B.C. Quennell and others have pointed out that these stone huts with their covered entrance passages, show considerable resemblance to the architecture which has persisted in certain Eskimo houses to this day. This would seem to be another example of marginal survival.

Somewhere about 1700 B.C. Britain was invaded by a group of people who occupied most of western Europe from Gibraltar to the Vistula. These were brakeph (broad-headed) tribes, quite distinct racially and culturally from the dokeph (narrow-headed) peoples who built most of the *long* barrows of the previous migration. The Beaker folk carried bronze daggers; and vases of special 'Beaker' shape are found in their graves. They were usually buried in a contracted position under *round* barrows.

It is almost certain that the most remarkable stone monuments in Western Europe—those at Avebury and Stonehenge—were erected by the Beaker Folk. They were a pastoral people, and not much is known of their villages, although their 'temples' were on such an elaborate scale. A brief description of the complex monument at Avebury is of interest, since to the writer the plan resembles in some respects that of a 'bora' ground, as used by the aborigines of Australia (Fig. 38).

The outer circular ridge of Avebury has a diameter of 1,400 feet, and encloses an area of twenty-eight and a half acres. Within this ridge or 'vallum' was a ditch whose floor was fifty-five feet below the top of the vallum. A ring of megaliths was erected within this ditch. Within this ring are two separate circles of large megaliths about 300 feet in diameter, and one of these megaliths is seventeen feet high. Less than a mile to the south, near the Kennet brook is Overton Hill, and an avenue of paired stones led south to this spot from the main 'temple'. On the hill was a 'Sanctuary' formed of two concentric rings of stones, the outer having a diameter of 130 feet. Wood posts alternated with the stones of the inner circle.[1]

The aboriginal tribes to the south of Sydney, Australia, held their main ceremonies in Bora Grounds of which the circular ridges and

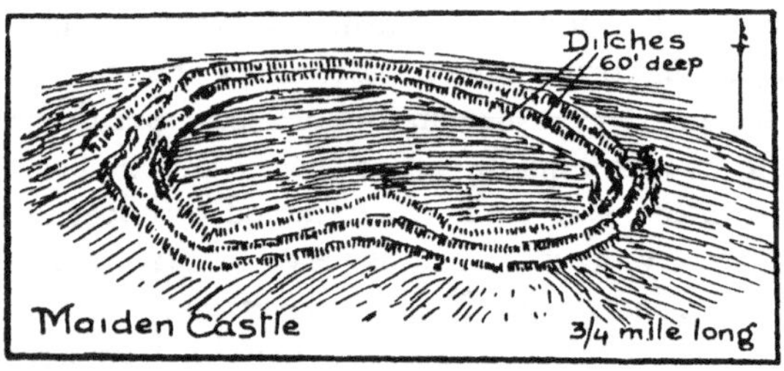

FIG. 37. Maiden Castle in Dorsetshire, a late Bronze Age fort built about 1000 B.C.

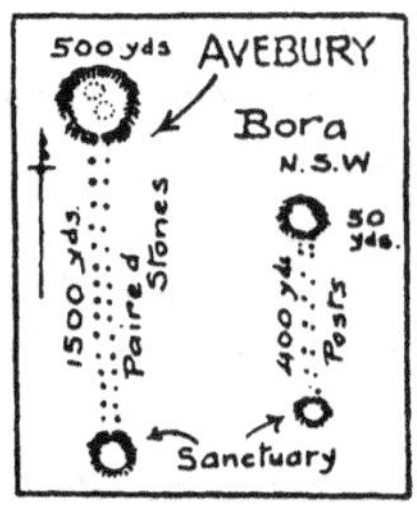

FIG. 38. Avebury, a temple erected by the Beaker Folk (*ca.* 1700 B.C.) near Oxford. Also a sketch of a similar ceremonial ground of Australian aborigines.

the carved wooden posts are familiar to the writer. To-day nothing but the low circular ridge, about fifty yards in diameter is to be seen, with perhaps a few of the carved posts (like those at Woodhenge) which surrounded the 'bora' ground and bordered the track leading to the 'sanctuary' some 400 yards away. Here was a smaller circle, where the more secret portion of the initiation of the youths was carried out. It seems probable to the writer that this form of 'temple' was carried out of Asia to the far east by Australoid folk of Mesolithic culture, much as the similar pattern reached that outlying portion of the Asiatic cradle-land to the far west which we call Britain (Fig 38).

With the dawn of the Bronze Age in Britain somewhere about 1500 B.C. we find not much change in the domestic architecture; but there seems to be a general tendency to form small settlements within lines of

[1]*Prehistoric England*, Grahame Clark, London 1940 is the best popular book.

defence such as earthworks and palisades. The best existing examples naturally survive in the sparsely settled moors of Britain; for it is the later inhabitants themselves who are responsible for the ruin of the ancient monuments, as witness the wrecking of Avebury by the modern villager.

One of the best illustrations of such a primitive hamlet is to be seen at Grim's Pound about fifteen miles south-west of Exeter. Its main features are illustrated in Fig. 39. The outer wall encloses an area of four-and-a-half acres. This wall is built of dry stones and is ten feet wide. It encloses a narrow passage along the axis, but in many cases the inner

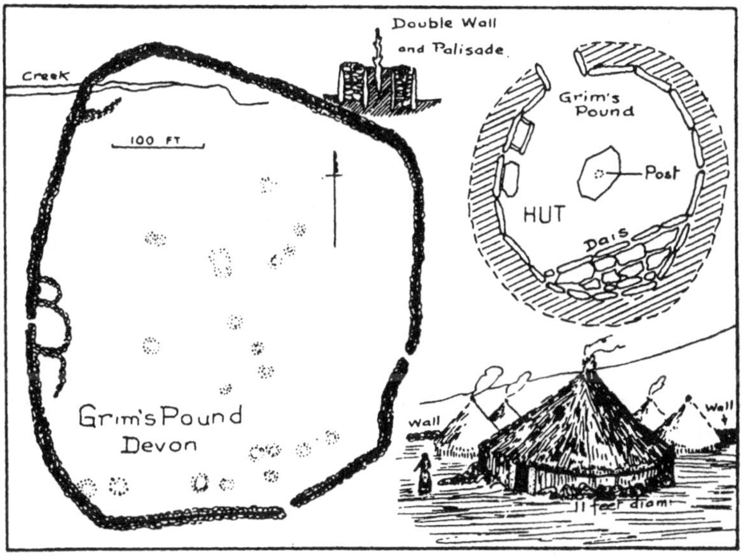

FIG. 39. Plan of the Bronze Age settlement at Grim's Pound, Devonshire. Details of the double wall, and a plan and view of a hut are also given. (*Based on Allcroft and Quennell.*)

portion was filled with earth, and carried a palisade as suggested in the inset in Fig. 39. Within the wall are the remains of about a score of circular huts, each about twelve feet in diameter. Some of these, however, may have been cattle-pounds. The huts were built on somewhat the same plan as the defence wall, i.e., two concentric layers of stone with an infilling of smaller stones and earth or turf. The doorway was only about three feet high, and sometimes had a sort of porch, while the floor was of clay or flagstones. A raised dais carried the bed, and a central pole supported rafters, on which were laid turf or rushes as indicated in the sketch in Fig. 39.

The later phases of the Bronze Age saw Britain rise to considerable importance in European trade, owing to the rich stores of tin in Cornwall, copper in Wales, and gold in Wicklow, Ireland. Probably there was little change in the domestic and town arrangements, but about the end of this period we find colossal earthworks being built in the south of England, such as that at Maiden Castle near Dorchester.[1] This takes the form of an oval camp on the summit of a hill, and is 1,000 yards long and 500 yards wide (Fig. 37). The main axis runs east-west, with very elaborate zigzag entrances through the concentric walls at each end. At the west end there are eight defensive walls separated by deep ditches. These were reinforced near the entrances with masonry, while timber revetments were also employed. Some hut sites have been discovered, but probably the camp was for refuge in times of stress, the cattle grazing on the forty-five acres included within the colossal ditches. Gordon Childe, however, thinks that Maiden Castle and Cissbury (in Sussex) belong to the early years of Hallstatt (Iron Age) Man in Britain.[2]

In the third century B.C. the second wave of Iron Age Man invaded Britain. These brought in a new type of culture, named from the site near Neufchatel the La Tène Culture. They built powerful fortresses which are best preserved in Scotland, especially near the Tay and along the Caledonian Canal. These forts had walls from ten to twenty feet thick, which were formed of squared stones tied together by stout horizontal timbers. The interior of the wall was filled with rubble and heaps of timber. Childe believes that these timber-filled walls were readily set alight, and such great heat was generated that the rubble infilling was vitrified.[3] He thinks this vitrification was, however, never designed, but arose through accident or by enemy action. Toward the end of this century Belgic invaders from northern France brought in the 'Arras' culture. They were buried in their chariots, and a dozen such graves are known in Yorkshire.

For our last example of a village in Britain before Roman times we choose Glastonbury in the marshy lands about thitty miles south of Bristol. Here arose a typical settlement amid the 'sea-meres', which is the meaning of the name 'Somerset'. Obviously other villages in England of this time—about 100 B.C.—with quite different environments, would not have the same pattern. But it is precisely owing to the marshy site that we owe the preservation of this late Celtic village in Britain. Nine miles away were ancient lead mines in the Mendips. Iron was obtained at Hunsbury, and the chalk downs of Wiltshire were not far to the east.

[1] *Earthwork of England*, A. H. Allcroft, London, 1908 with many maps.

[2] and [3] *Prehistoric Communities of the British Isles*, London, 1940 (fully illustrated).

A navigable stream led to the Bristol Channel some twelve miles away, so that at Glastonbury a notable trade-centre developed in the period just before Cæsar invaded Britain.

In these swamps an artificial island (or *crannog*) was laboriously built up including an area of about three acres (Fig. 36). This space was occupied with a platform of logs and brushwood, and on this about sixty huts were erected. The floors were of clay and the walls of wattle and clay. The roof was supported by a central post, and was probably thatched. At times the Glastonbury huts sank somewhat in the peaty ground, and so as many as a dozen hearths, one above the other, have been found by the excavators (see section in Fig. 36). The settlers were farmers and fishers in addition to being traders. They were expert weavers and iron-workers, and used a lathe for turning wooden articles. Among the most important finds were iron currency bars. This interesting settlement was wrecked by Belgic enemies about the time of the Roman Conquest.

Survivals of primitive settlements in Morocco

In many respects Morocco is worthy of study by those interested in the development of our town life. In this part of North-west Africa we can still see in being the sort of life which was common in Britain in Neolithic and early Middle ages. It is worth remembering also that in Morocco we find a group of tribes of 'Mediterranean' race (dark, narrow-headed folk of slender build) who are blood brothers of the folk of western Britain, especially of the so-called 'dark men of Devon'. Moreover it seems likely that this corner of Africa was never overrun by the migrations of Aryan-speakers, i.e., the Goidels and Brythons of central and western Europe in the first millenia B.C. Hence it may well be that the Hamitic languages of Morocco represent fairly closely what was spoken in Britain and Scotland before the arrival of the Gaelic and Welsh speakers.

Jules Blache has given us a most interesting study of the *Modes of Life in the Moroccan Countryside* (Geog. Rev., Oct., 1921) which should be consulted by all interested in this phase of our civilization. The '*nuala*' or earth huts possibly resemble those used by early shepherds in Britain. The *dar* (or house) of rooms built round a court is not very like the early houses of Britain, since the semi-arid environment of Morocco is quite unlike the cool, snowy climate of Britain. But the *kasba* or fortified town still preserves a way of life which was common in Norman and Plantagenet days in Britain. These are found in the Atlas Mountains remote from the Mediterranean area. Here the mountains have never been subjugated, and the tribes never paid tribute to the early sultans, and there is still bitter opposition to French and Spanish penetration. Readers are referred to the air photographs (Fig. 12, *op. cit.*) of these 'medieval castles' still in active life in the vicinity of Bu Denib on the Saharan slope of the High Atlas, about 250 miles to the east of the city of Morocco.

A Primitive Village in the Northern Andes

In 1930 I made an interesting journey in Colombia into the Andes to the south of Santa Marta, and spent several days in the vicinity of an

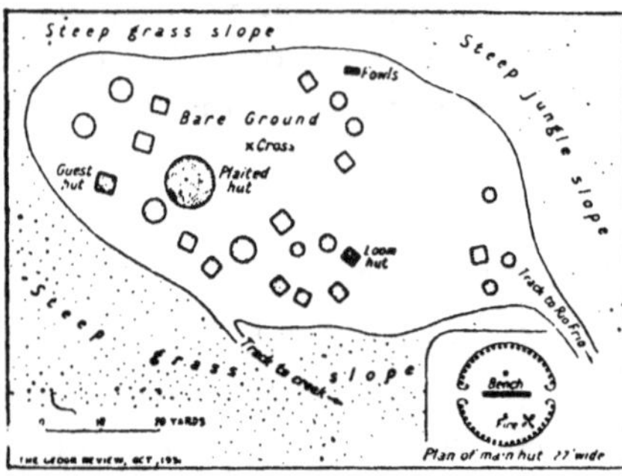

FIG. 40. Plan of the Indian village of San Andres on the north-western slopes of the Sierra Nevada de Santa Marta, in Colombia (S. America).

FIG. 40A. Sketch of one of the circular huts in the village of San Andres. It is about 15 feet high and 15 feet wide.

Indian village now known as San Andres. This was perched on a ridge about 4,500 feet above the coastal plain. The natives cultivated small gardens, where they grew beans and bananas, far below the village near the stream in a deep valley. The huts were arranged in no particular order so far as I could learn (Fig. 40). Of the two dozen that compose the village, half are circular in plan, and half are square.[1] They vary from fifteen to twenty-seven feet in diameter and are about twenty feet high. The walls are made of lumps of mud, some eight inches across, held together by a framework of laths (Fig. 40). Clearly this is close to the 'wattle and daub' of our ancestors. The roofs are made of palm thatch, and in some cases a buttress of mud surrounds the hut like a parapet. The doors are solid slabs of wood with projecting tenons which fit into holes in the sills, much like the doors of the Glastonbury village.

[1] "Settlement Zones of the Sierra Nevada, Colombia," G. Taylor, *Geog. Rev.*, Oct., 1931, New York.

There is little in the hut except a bench and a few stools, each carved out of a solid log, and each with four square legs. I saw only one pot in the village, though there were plenty of calabashes and many dishes and spoons made therefrom. A native loom, of the usual primitive American type, is the most elaborate article in the village. Its frame is square, about five feet each way. A bag of crude cotton hung nearby, and the fibre is spun by rolling a spindle on a man's bare thigh. There were about sixty adults and twenty children in the village. The sole objects of foreign introduction which I saw were several muskets and an axe. But the Indians herd oxen on the high grasslands (Paramo) far above the village, and these were of course acquired from the Spanish conquerors.

The huts, the loom, the absence of all but a little metal in this community, remind one quite closely of the primitive settlements in Britain during the early iron age.

Part B. Four Ancient Oriental Cities

There is little doubt that the first cities were built in Asia, somewhere in the region between Turkestan and Egypt. A generation or so ago it was usually supposed that the earliest civilization of note was that discovered in Egypt. This region was naturally examined by archæologists long before much attention was given to Mesopotamia, Persia or the Caspian region. To-day the investigations of experts seem to indicate that the oldest examples of city life are to be found far up the Tigris valley in the vicinity of ancient Nineveh. In another book (*Our Evolving Civilization*) the author gives his reasons for looking still further north, say to Askabad in Russian Turkestan, for the site of man's earliest city life. However, in this volume lack of space prevents our delving further into this problem; and it will be sufficient to consider the plan of one of the great Mesopotamian cities of the past—that of Babylon—in order to gain some idea of complexity of a city several thousand years before the birth of Christ.

Babylon is almost certainly older than 2250 B.C., when it became the capital of Khammurabi's empire. About 689 B.C. the city was destroyed by Sennacherib, but it was rebuilt almost immediately. Around 600 B.C. it became the capital of Nebuchadnezzar's Empire, and it is the plan of his city which has been excavated, chiefly by German archeologists, with the results charted in Fig. 41. To-day there are three vast mounds of clay, which represent the crumbled brick walls of the city and its buildings, and further afield various lines of ramparts can be made out.

The main buildings were on the east bank and were surrounded by double walls, of which the inner formed a square with each side measuring about 2,400 metres. These walls were said to be 300 feet high and eighty-five feet wide, but this is probably an exaggeration. The clay dug from the surrounding moat was used in the construction of the walls. The great Temple of Bel (E-Sagila) is placed by the Germans in the centre of the map (Fig. 41). Within this was a Tower ninety metres high. The Hanging Gardens have not been identified with any accuracy. They were built on the top of arches about seventy-five feet high, and water was raised to them by means of a screw. In the diagram (Fig. 41) they are placed between the Temple and several forts in the northern part of the town. A 'new town' was built on the western bank of the

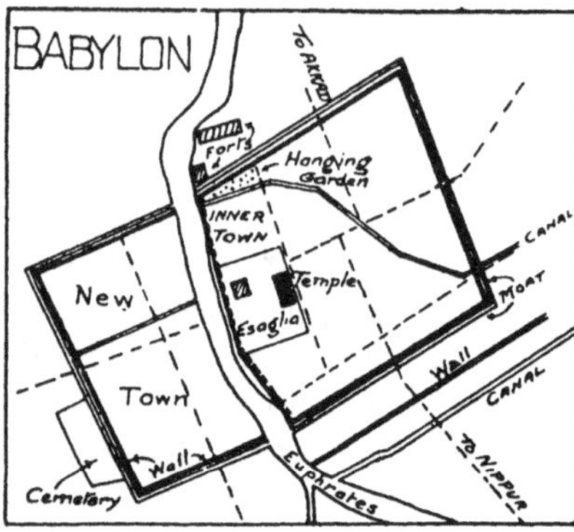

FIG. 41. A reconstruction of the vanished city of Babylon. (*After E. Unger.*)

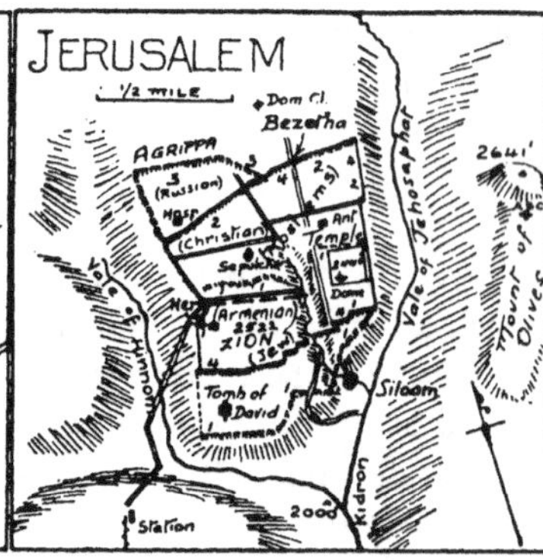

FIG. 42. A sketch map of Jerusalem, showing the four main series of walls (1–4).

Euphrates, with walls and moats like those of the earlier part of the city. There were said to be 100 bronze gates to the city; and it is interesting to compare this elaborate metropolis with the humble hamlets of Britain —such as Glastonbury (Fig. 36)—which were more or less contemporaneous with the last days of Babylon. It survived till about 275 B.C., when the inhabitants largely abandoned it and migrated to the city of Seleucia, about fifty miles to the north.

Our second example of an oriental city is Jerusalem, which also dates back far before the time of Christ, or even before the invasion of the region by the Israelites. Its authentic history perhaps begins about the time of David around 1000 B.C. This famous king occupied both the prominent hills (Fig. 42) which form a promontory between the Vale of Hinnom and that of the brook Kidron. Between the hills is the Tyropoeon Valley, which soon became enclosed within the city walls.

The soft limestone plateau (formed of Cretaceous rocks) on which Jerusalem has been built, is about 2,500 feet above sea level; while the valley just south of the pool of Siloam is five hundred feet lower. On the west side of the narrow Tyropoeon valley rises the dominant hill of Zion, where David built his city. Across the valley was Mount Moriah, and this site was occupied by the famous Temple constructed by Solomon. The city soon covered the Tyropoeon and the lower land just to the north of Zion, and this area is the centre of the city of to-day.

In the time of the Kings the city walls enclosed the area denoted by the numbers 1, 1, 1, on the map. In 168 B.C. the city was largely destroyed by Antiochus, who built a fort called Akra to dominate the people. The site of this fort is still uncertain. The Maccabean insurgents demolished the Akra, and held the city for some time. Herod the Great made Jerusalem his capital and added much to its strength. He built the Antonia Fort (Ant.) just north of the Temple, and constructed a fortified palace for himself on the north-west side of Zion. About this time a second wall (2, 2, 2) was built somewhat north of the old wall, and this extended from Antonia to the Vale of Hinnom.

Herod Agrippa built a third wall still farther to the north, to enclose the new suburbs of Bezetha (3, 3, in Fig 42). This was the condition of Jerusalem when the terrible siege ordered by Titus resulted in the capture of the city in A.D. 70, and in the migration of many of the Jews from Judea. Hadrian rebuilt the city in 130, and Constantine erected a church over the probable site of the Holy Sepulchre (Fig. 42). In 614 the Persians conquered the city, and the Moslems captured it in 637. A great mosque was built on the conjectured site of the Temple, known as the Dome of the Rock. Saladin repaired the walls, as did Suleiman around 1550. These are the walls (4, 4) which surround the city to-day, giving it a lozenge-shape of somewhat smaller area than it possessed in the earlier days of Rome.

Until the middle of the 19th century Jerusalem was a tiny city only half a mile across, and confined within Suleiman's walls, but since that time it has spread considerably. This expansion has naturally taken place mainly in a north-west direction along the promontory between the two valleys. Large colonies of Jews have settled in this area, so that modern Jerusalem extends for half a mile to the north-west of the Russian quarter marked in Fig. 42. In 1939 the population was about 130,000, of whom 79,000 were Jews. The city is divided into quarters, where a large majority of the various national groups tend to live. Thus the Jewish section is in the south-east near the Tyropoeon valley. The Arabs live mostly in the north-east and centre of the town. There is a large Armenian minority in the west, while the Russians hold a large area with churches and schools in the north-west. Here also are some French institutions, such as a large hospital. Scattered all round the adjacent hills are other institutions such as that of the Dominicans to the north, and of the Syrian and English to the north-west. On the Mount of Olives is a German hospital, and there are Carmelite and Benedictine Cloisters in the south-east.

The life of the city obviously depends on its character as a holy place, not only to Jews and Christians, but also to the Moslems. The surrounding region is rather poor in resources, though the rainfall is twenty-six inches a year, which is high for the coasts of the Mediterranean. The city has long been known unfavourably for its unsanitary condition; and the chief industries are the olivewood mementoes, etc., sold to the thousands of tourists and pilgrims who throng the city. To the student of urban geography it offers one of the best examples in the world of the influence of religion upon city sites and city life.

Probably the most interesting city in India is Delhi. It has grown up on the river Jumna in the relatively narrow neck of fertile land between the Thar Desert to the south-west and the great Himalayan barrier to the north-east. Here also a corridor diverged to the south-west to Baroda and the Gulf of Cambay. The numerous conquerors who poured into India by way of the Khyber Pass inevitably took the route through Delhi, so that this district has some claims to be called the 'Cockpit of India'.

Only a brief mention can be made of the early history of Delhi, which dates back to the 11th century A.D. A number of the conquerors destroyed their predecessors' cities and built capitals of their own. Some of these ancient sites are indicated in the map (Fig. 43) by the figures 1 to 7. The site labelled '7' was chosen by Shah Jahan about 1650. Most of the earlier cities were to the south of modern (Indian) Delhi, some of them as much as four miles to the south. The country is relatively flat, and the river itself is perhaps the main controlling factor in siting these towns since it has varied its course considerably. To the north is the 'Ridge' which played so striking a part in the Indian Mutiny (1857). It is an outlying spur of the Aravelli Hills, rising about sixty feet above the plains. Here the British troops under Nicholson were encamped during the siege, which lasted for three months. The blowing up of the Kashmir Gate enabled the British to capture the mutineers in the Indian City. The Hall of Audience is in the former palace, which is now called the Fort. Here was the famous Peacock Throne, removed to Persia in 1739.

Delhi to-day has a population of about half a million, of which some 50,000 live in New Delhi. The native city has been described as a huddle of mean houses in mean streets diversified with splendid mosques. The main street (Chandni Chauk), running west-east from the Fort, is however a broad avenue with a double row of trees down the centre. From Delhi radiate railways to all parts of India.

In 1911 it was decided to construct a new city as the centre of Indian administration. This interesting planned city is called New Delhi, and its general features can be seen in Fig. 43. It was largely designed by Sir Edward Lutyens, who has followed in some respect the pattern of Washington. A striking rock was chosen as the site for the main buildings, and a 'Mall' leads from this to the Jumna. Other features of this design will be described in a later chapter of this volume.

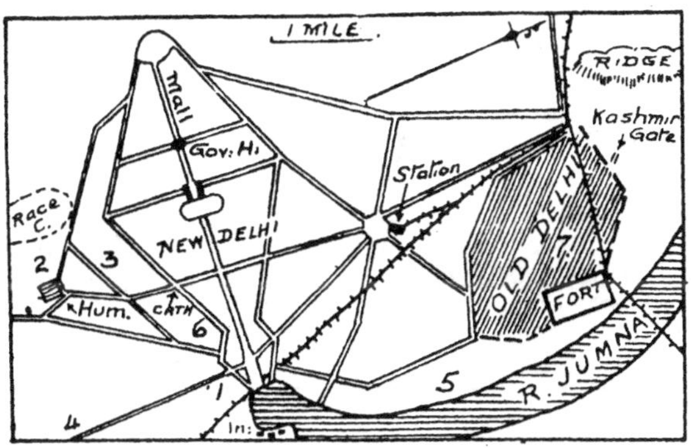

FIG. 43. The cities of Old and New Delhi in north India. (*After H. V. Lanchester.*)

Three oriental cities have been briefly described in this chapter, of which one, Babylon, has entirely disappeared. Our map is a mere reconstruction based on the piles of debris, as worked out by the German archæologist, E. Unger (*Babylon*, Berlin, 1931). It is futile to speculate as to why the great city was sited here, since there are numberless sites of equal value in this vast area of plains. Perhaps, as in the case of Pekin, the site was chosen by a necromancer—surely the acme of 'possibilist' planning! To-day only a few small villages occupy the vicinity of mighty Babylon.

Jerusalem has had a long and eventful history, but the religious factor has been constant throughout the centuries, so that the centre of the city has not changed its position much in 3,000 years. Here topographic control has always been of considerable importance; and since, until lately, no very progressive nation has been in charge we do not see the development of a modern city in or near Jerusalem. However the growth of modern suburbs, mainly to the north-west, is a natural and common experience in such an environment.

Delhi carries us one stage further, in that here we find a quite new city, designed on modern principles of town-planning, while the old native city exists a mile or so away. Such examples will be discussed further in later pages. Our last example offers us (in Kano, Nigeria)

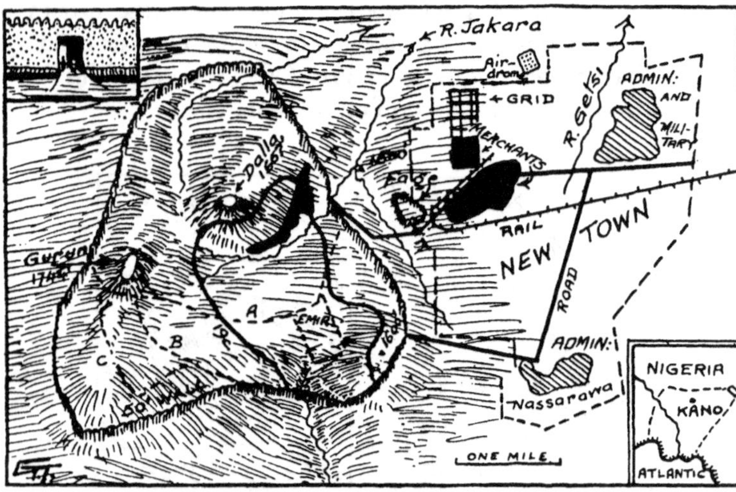

FIG. 44. The Old and New City of Kano in British Nigeria. The black patches are the industrial areas. The heavy black line shows the built-up area in the 19th century. (*After D. Whittlesey*, 1937.) Inset are a gateway and a map of Nigeria.

a city somewhat like that of Delhi, but where the native section is perhaps larger and more important than the modern 'new town', where the British administrators have their quarters. (Fig. 44.)

Kano has a permanent population of about 89,000, and is sited about 600 miles inland in latitude 12° N. It has a rainfall of thirty-six inches, so that it is a flourishing agricultural region, growing sorghum, millets, and groundnuts. It lies at the junction of a number of caravan routes connecting the Sudan with Tripoli, and seems to have been founded about A.D. 1000. Kano's site was undoubtedly determined by the presence of an 'acropolis' or rocky knob, called Dalla Hill, rising about 150 feet above the banks of the little Jakara stream.[1] This rock was first surrounded

[1]Based on the remarkably full description by Derwent Whittlesey in the *Geographical Review*, April, 1937, New York.

by a circular wall labelled 'A' in Fig. 44. Apparently in the early days there was a good deal of wood for fuel in the vicinity, so that Kano was noted for primitive iron manufacture. But after the timber was all burnt, the city turned to weaving, which is an important occupation to-day.

During modern times the city walls have been considerably enlarged by the building of new lines as shown by 'B' and 'C'. The Emir's Palace is in the south of this walled town. The walls are built of mud, and rise about forty feet above the plain. In the inset is shown one of the gates, with a bridge crossing the moat at this portion of the wall. The interior of the walls encloses about eight square miles, but most of this area was pasture land for the herds of the citizens during times of warfare. The headwaters of the Jakara were enclosed in the northern part of this enormous wall. Only about one-third of the area within the walls, mainly around the Palace, was occupied by mud dwellings. The houses were made of mud, largely derived from ancient walls, but also in part dug up from 'borrow pits' near the market, which is shown as a black patch in Fig. 44, near the Dalla Hill. The roofs are made of mud reinforced by the fronds of fan-palms.

Three culture groups are found in the walled city. The Fulani dominate the southern portion near the palace, the Haussa the north, and the Yoruba the north-west of the occupied sections. When the British took charge of Nigeria they placed their barracks in Nassarawa. To-day, however, a new town, almost as extensive as the ancient walled city, has grown up to the north of the latter. There is no defensive wall, but a space 1,200 feet wide has been cleared all round the 'New Town'. The administrative buildings are in the north-east, while the native merchants have congregated near the station, where the railway to the coast at Lagos serves the two towns.

Kano illustrates the type of town life which is common throughout much of the primitive part of the world to this day. There are many parallels with European cities; for instance the open spaces inside the walls were common in medieval walled cities in Europe, though they were filled with slums at a later date. So also the growth around a hill fort was very common in early towns, as we shall see later in this volume. The crude sewage arrangements, and the moat around the wall are also reminiscent of early Europe, though at Kano the moat was filled with thorny bushes rather than water. The wide use of mud is of course due to the environment, since the long dry period enables such houses to last much longer than they would in rainier northern climates.

CHAPTER VII

EVOLUTION OF CLASSICAL GREEK TOWNS

In the preceding chapter the evolution of some early types of settlement in north-western Europe has been considered. But it must be remembered that civilization began in Europe, in all probability, in the island of Crete, where the archæology of Cnossus has been investigated by Evans and others. Settlements by traders of this Cretan or Mykenean culture

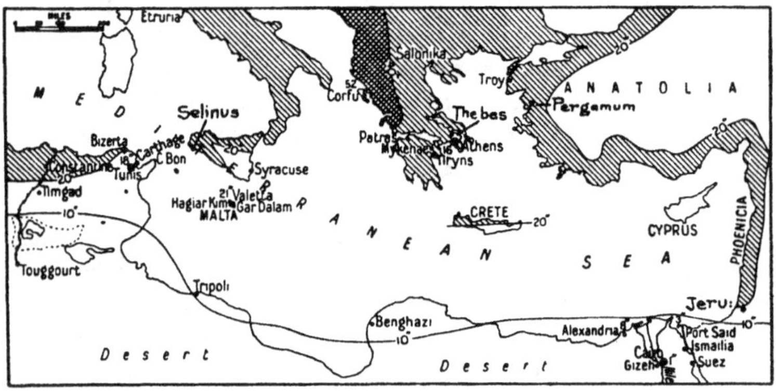

FIG. 45. Map of the eastern Mediterranean Sea, showing places mentioned in chapters VII and VIII. Annual rainfall is indicated.

are to be found all round the eastern and central Mediterranean; as at Troy (near the Dardanelles), and Tiryns and Mykenae near Argos, and perhaps among the Etruscans in western Italy (Fig. 45). Most of these sites have been visited by the writer, but perhaps before discussing Mykenean settlements some mention may be made of the relics which have been discovered recently in the island of Malta, which recall settlements already described in the British Isles.

Malta and Mykenae

Malta is an oval island about eighteen miles long, which lies almost at the centre of the Mediterranean Sea at its narrowest point. Here during Pleistocene times a land-bridge connected the island with Africa, and perhaps at a later date with Europe. The famous cave of Gar Dalam, about eight miles south of Valetta, contains relics of African hippopotami, of European deer, as well as of Neandertal Man; and should certainly be visited by every traveller to Malta. The most interesting relics, however, are the buildings at Tarshien in the vicinity of Valetta itself, which give us a fair picture of life at the end of Mesolithic times in the island of Malta.

Tarshien is only a mile or two to the south of Valetta, and here, amid the characteristic small walled fields, with their occasional platforms of grey limestone, a series of subterranean temples was discovered as late as 1914. A number of these somewhat complex temple-sites has been opened up, and three of them are shown in Fig. 46. To-day they appear as open courts with roofless corridors. When in use many of the smaller rooms and the corridors were probably roofed with slabs. The floors of corridors and rooms alike are about eight feet below the general level of the ground. The general plan of these Bronze Age buildings consists of two large outer courts of oval outline, with a smaller court at the end furthest from the entrance. The walls are usually made of vertical stone slabs some six or eight feet high. These are often carved with spiral decorations, or with animals such as bulls. Along the sides of the corridors are cupboards made of stone; and huge bowls, some three feet high, once contained olive oil or other liquids. Certain hollowed stones are supposed to have collected blood from the sacrifices. In Fig. 46 it is clear that the larger temple is the latest of the three, and to some extent occupies the area formerly used in an earlier temple. At Hagiar Kim in the south-west of the island are many other oval courts connected by walled corridors. In general however, these are above ground, and are probably dwelling places. These buildings in Malta certainly go back several millennia before the time of Christ, and their

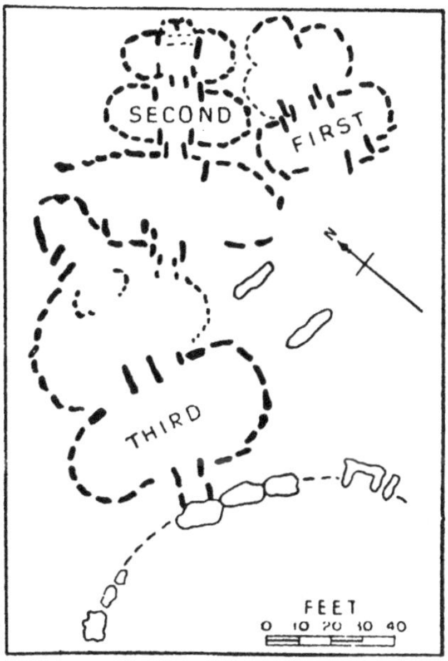

FIG. 46. Plan of the three temples at Tarshien, Malta (*after Zammit*).

plan and construction remind us of the less elaborate structures already described as existing in Britain in the later years of Neolithic times.

In recent years classical authorities seem disposed to give more credit than of old to the Mykenean strains in early Greek culture. Under these circumstances it seems useful to pay some attention to the type of settlement characteristic of the Mykenean civilization, which flourished perhaps from 2500 B.C. to the period when it merged with the developing Hellenic (Greek) culture, say about 1000 B.C. The two towns which are briefly described in this section are Mykenae and Tiryns, which are situated about sixty miles to the south-west of Athens, near the present town of Argos.

Late one evening in June, 1938, I descended from the train at Phictia, which is a tiny station about two miles west of Mykenae. It was an eerie walk in the dark up the hills to the village near the ruins; the more so since my knowledge of Greek was confined to about a dozen words. However, I made out some folk sitting round a fire, and a small boy guided me to a clean if simple inn. Next morning I set out for the famous hill-fortress, which lies on a knob in the pass between two mountains. These are about 2,400 feet high, while the fortress is about 900 feet above the sea. In summer Greece is very hot, dry and bare. Limestone crags and heathy slopes are characteristic of Argolis, and boulders cover the ground, with much blue thyme growing between them. There were a few shepherds about, and patches of stubble showed where grain had been harvested.

The road runs along a narrow ridge as it approaches the famous Lion Gate. This is mainly formed of three giant slabs of limestone. The door-posts are ten feet high, while the lintel is sixteen feet long and three feet thick (Inset Fig. 47). A triangular slab of brown limestone above the door has been carved to represent two lions with a pillar between. Probably the heads were made of bronze, but these have vanished. This is undoubtedly one of the earliest large sculptures in Europe. The walls are suggested in Fig. 47, and though no cement was used, they have withstood man and weather for over 3,000 years. Passing into the Acropolis by a short corridor (Fig. 47) I followed the north wall and emerged by the North Gate. Near here is a subterreanean stairway leading to a well. The walls of the passage were made of the same uncemented cyclopean blocks. At the east end of the fort is a fairly sharp ridge of limestone, and nearby was a delightful spring hidden under thick bushes. This was one of the few springs I found in my Mediterranean wanderings, and was correspondingly welcome. Its waters supplied the reservoir in the Mykenean City.

Turning back over the grassy plateau inside the fortress, one passes many remains of prehistoric rooms. Some very rough stone jars are still extant, but little is left of the Palace but the outline of the rooms. The most interesting features are the Treasure Tombs just south of the Lion Gate. There is a double enclosure wall of thin slabs about six feet high.[1] This enclosure is about thirty yards across and surrounds five circular pits about fifteen feet deep. Within these pits, Schliemann, in 1877, found the skeletons of seventeen persons. Buried with them was an extensive collection of gold and silver ornaments which is now in the Museum of Athens. Gold masks are common, while brooches, model animals, cups and weapons are also included in this marvellous find.

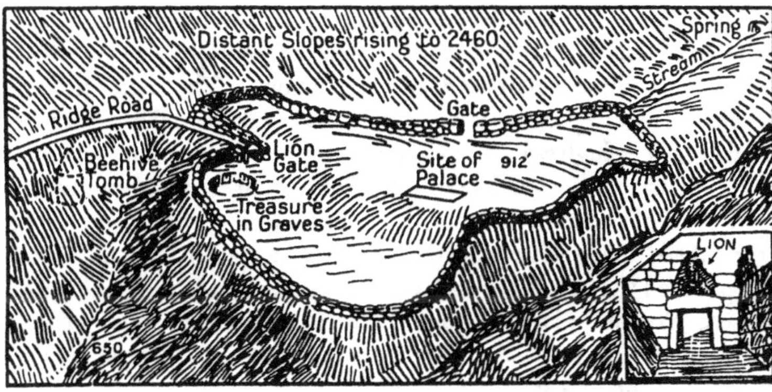

FIG. 47. Sketch of the upper city (Acropolis) of Mykenae, looking north. It is 1,000 feet from west to east, and the ravines are 200 feet deep. *Inset* is a sketch of the Lion's Gate.

The lower town of Mykenae extended about half a mile to the south-west of the Acropolis, along the broad ridge traversed by the present road. Only fragments of the walls remain above ground, but the 'Bee-hive Tombs' are the most perfect prehistoric monuments which the writer has seen. There are half a dozen near the Acropolis, and the size and shape of one of them is suggested in the hill slope just below the road in Fig. 47. The Tomb of Atreus is 1,500 feet south-west of the Lion Gate. Passing inside one finds that the tomb is about fifty feet high, and is circular in plan with a diameter of about forty feet. It is built up of about thirty horizontal courses of curved stones. The size and accuracy with which they fit together, and their state of preservation, are amazing.

[1]This type of double wall is present in some British Neolithic structures, e.g., Grim's Pound, (p. 99).

9

For instance, the door is seventeen feet high, and its inner lintel is a single stone said to weigh 113 tons. Yet this huge mass has been carved to the exact curve to fit the circular wall. Bronze nails have been used in ornamenting the tomb, and no finer monument of the Bronze Age is known in Europe. It may be dated about 1300 B.C.

From Mykenae I travelled south about twelve miles over the fertile plain of Argos to a second Mykenean site (Fig. 45) which is on the coast. Tiryns is not so impressive, for it is a brick-shaped ridge of limestone about fifty feet high and 1,000 feet long. It was perhaps in Mykenean times an island, but to-day it rises from the coastal plain about three miles from the sea. The relics of the walls are, however, more imposing than at Mykenae, for originally they were perhaps sixty feet high and twenty-six feet thick at the base. The entrance is by a ramp on the east side, along an imposing corridor of gigantic blocks of limestone. The top of the Acropolis is nearly level and quite bare, though one can still trace the outlines of the Palace and other buildings. The most interesting corner is the gallery at the south end. This is a passage thirty yards long and fifteen feet high which is included in the massive walls. There are six rough triangular windows at the side (built as usual without any knowledge of the arch) from which one can view the plain of Argos. For centuries sheep have used the gallery as a shelter, and their bodies have polished the lower portions of the limestone slabs in a very striking manner. Probably Tiryns is older than Mykenae, but both were destroyed about 460 B.C. Both of them are referred to in the songs of Homer, which date around 1000 B.C.

Shrunken Greek Cities: Thebes, Syracuse and Selinus

It is interesting to speculate as to the reasons why certain cities, once of paramount importance in the region, have in modern times become of little significance. Such cities are relatively common in the Old World, but have hardly had time to develop in our North American civilization. However Jamestown, Annapolis, and Tadoussac show that even in America conditions have changed drastically in the last 350 years, and so have brought about the waning of towns which started with much promise.

In Greece there are few areas of level agricultural land, owing to its structure, which consists of a series of parallel folds with a general N.W.-S.E. direction. These folds end in the long fingers on the south-east coast (Fig. 45). The largest plain is that of Boetia to the north of Athens, from which it is separated by mountains of some elevation. Lake

Copais occupies the centre of this plain, which is about sixty miles long and twenty miles wide. When the northern invaders entered Greece somewhere about 1200 B.C. they realized that the site of Thebes on the southern edge of this plain offered considerable advantages from the point of view of defence; and accordingly a settlement of varying importance has been here throughout historic times.

The main road north from Athens to Livadia reaches the Boetian plain over a gap in the hills at Thebes (Fig. 48). Here several parallel ravines lead down to the plain from the south, and a rocky promontory with a somewhat level surface (at 200 metres) stood out between two

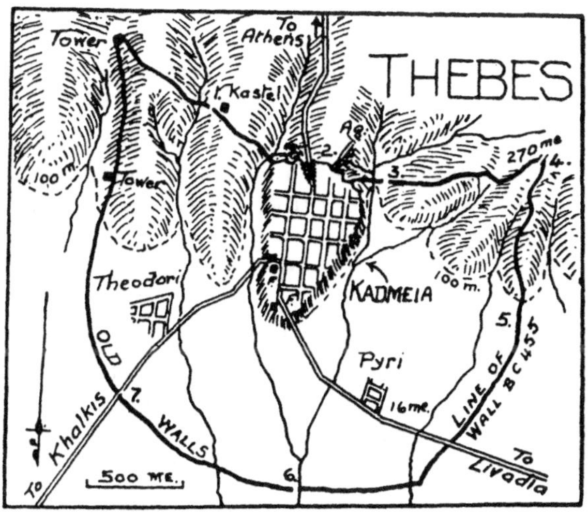

FIG. 48. The shrunken city of Thebes now occupying only the ancient acropolis. Heights are only approximate.

of the ravines. This tiny plateau is called the Kadmeia, and is traditionally linked with Cadmus the inventor of the alphabet. It was soon crowned with a primitive settlement, and became the capital of the city-state of Boetia. As such Thebes waged war with Athens and Sparta after the manner of those early days. About 455 B.C. the city had grown to great size, and is said to have had a population of 40,000. The city was enclosed in a wall with a circumference of about five miles.

The remains of this wall can in places be traced to-day, though the city now holds only about 4,000 people, and has shrunk once more to the small area of the Kadmeia. Thebes was famed for its seven gates and these have been traced with some accuracy, and are indicated in Fig.

48. Several of the original towers along the wall can be identified in the hills to the south-west of the present city. Many springs gushed from the hills behind the town, and some of these are still flowing.

In 336 B.C. Alexander punished the Thebans for their rebellion by destroying almost all the buildings, and Thebes never regained its former eminence. It was held by Crusaders from France and Catalonia for a time around 1300, and the two towers on the north-west edge of Kadmeia date from this period. They also built an aqueduct from the south to the town which is still in use. The map shows clearly how the city has shrunk from an area of about two square miles to a tiny group of streets occupying perhaps one-eighth of a square mile. Two outlying villages, Theodori and Pyri, are within the ancient walls, and border the main roads to Khalkis and Livadia respectively.

An age of widespread colonization preceded the Golden Age of Greece. Merchants set up trading stations all round the Mediterranean and Black Seas. Soon daughter colonies of Greece sprang up, from Massilia (Marseilles) in the west to Crimea in the north-east and Cyrene in the south-east. We have only space to refer to a few examples in Sicily such as Syracuse and Selinus.

Syracuse was founded as a colony from Corinth in 734 B.C., and soon became one of the chief of their colonies in the island of Sicily. At its zenith it contained half a million inhabitants, and its walls were fourteen miles long. The ancient city covered all the low plateau to the north of the present town of Syracuse, which is only one-tenth the size of its fore-runner (Fig. 49). Indeed it is interesting to note that here, as in Thebes, the folk to-day dwell mainly on or near the site of the earliest part of the town, in this case the rocky promontory seen in Fig. 49. At the maximum there were four or five suburbs enclosed in the great Walls of Dionysius. He flourished from 406 to 367 B.C., and extended the sway of Syracuse over Sicily and much of Greek Italy. Syracuse took the Carthaginian side in the Punic wars, and in 212 B.C. was defended against the Romans by the famous mathematician Archimedes. It was conquered by Rome, and never became of much importance in later days. It has a population to-day of about 53,000.

Some mention may be made of the once flourishing town of Selinus, which was a Greek colony founded 628 B.C. on the south-west coast of Sicily, and constituted the western outpost of their culture in the island. It had a position much like that of Thebes, i.e., the first town was placed on a rocky promontory between two ravines (Fig. 50). Later the town spread backward along the ridge to form the 'Citta Antica' of to-day. To the east the citizens constructed one of the largest temples known from classical times, which may have been dedicated to Apollo. The length of the temple was 371 feet and its breadth 177 feet. Near at hand were two other large temples, while in the acropolis itself there were three others of smaller area. In the early city (the Acropolis) there was one main street along the axis of the hill which was thirty feet wide, and seven cross-streets, each twelve feet wide. This indicates that even in the early days of Greek planning, a definite rectangular pattern was adopted; though of course the same design had appeared long before in Eastern cities. This economy of design is ascribed to Hippodamus, who was the Greek 'Haussmann' of the fifth century B.C. (H. I. Triggs).

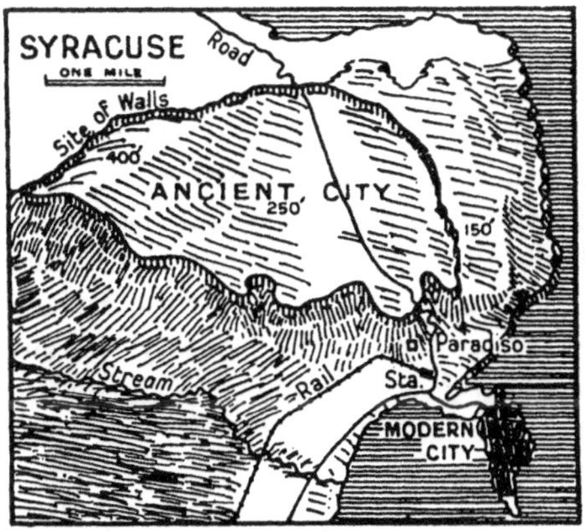

FIG. 49. View of Syracuse (in Sicily), showing the contrast between the small modern city and the large ancient city which also included the port.

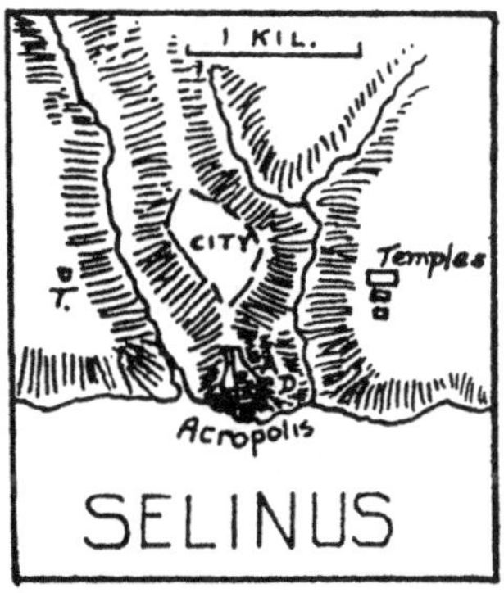

FIG. 50. Selinus an ancient Greek city, which flourished in S.W. Sicily about 600 B.C.

The three cities just described offer good examples of settlements whose importance has greatly diminished through the ages. Selinus has vanished, and the nearest town to-day is Castelvetrano some six miles to the north. We see here perhaps evidence of a general tendency for centralization of a farming population, coupled with the decline of the advantages of promontory-fort sites. The latter had great merits in classical and medieval times, but they have not survived in the present industrial age.

Syracuse has a good harbour, which seems to have been wanting at Selinus, and so it still persists as an important port, though Catania forty miles to the north is a very much larger port and city. The fertile soils of Etna seem to have been a factor in the growth of Catania which is lacking at Syracuse. Thebes is an illustration of the changes in a country's economy, for the Copais Lake basin seems to be less flourishing to-day than in classical times. The centre of power in this part of Greece has moved very definitely to Athens; though the latter city was of much less significance in the days of Turkish domination, and by no means so far ahead of Thebes as it is to-day. Naturally in the days of Epaminondas (around 365 B.C.) when Thebes was the leading city state in Greece, this settlement was far more flourishing than at any later period, for it never regained the political importance of those days.

Ancient and Modern Athens

No city is of more importance in a study of the development of European civilization than Athens (Fig. 45). It lies about six miles from the sea on a plain about four miles wide, watered by the Cephissus. This stream receives the Ilissus from the east, and the two are separated by the most striking of the Athenian hills, that of Lycabettus (900 feet). To-day, in summer, the Ilissus is sometimes a dismal stream about three inches deep and jet black in colour, but no doubt after the winter rains, it is more attractive. In the south-west is the hill named after the last king of Athens, Philopappos, and nearby are the low Observatory Hill and the Areopagus, where Saint Paul preached. All round this corner of Athens are innumerable ruins of houses, but the modern city has spread to the north-west away from the Ilissus.

The most interesting hill is the Acropolis which rises abruptly about 330 feet above the plain. It was originally a hill-fort very like those which we have studied at Mykenae and Tiryns. It is about one-quarter of a mile long and is surrounded by low walls. There are still several wonderful temples on the Acropolis, which have been restored in large part. The map (Fig. 51) shows that the earliest walled city, going back to the times of Peisistratus, was of oval shape and surrounded the Acropolis on all sides.[1]

After the great Persian invasions around 500 B.C. the city was nearly destroyed. It was rebuilt with great splendour in the time of Themistocles (about 450 B.C.), and now included a much larger area of the large plain to the north of the Acropolis (Fig. 51). The market-place was in the centre of this city. Enormous walls were built to connect Athens to its port of Piraeus (pronounced Piréfs), and these also are indicated in

[1]My colleague Professor Homer Thompson is working on the archæology of Athens. He tells me that the popular names and dates assigned to the relics of the various walls (which I have used) will probably be altered in the near future.

the sketch. Greece was conquered by Rome in 146 B.C., but many of the Roman emperors visited Athens, and Hadrian greatly beautified it. He added some of the walls in the south-east of the old city.

Until about 1830 Athens was only a small town like many others in the former Turkish territory. But after the War of Independence a city-planning programme was undertaken on a larger scale. The

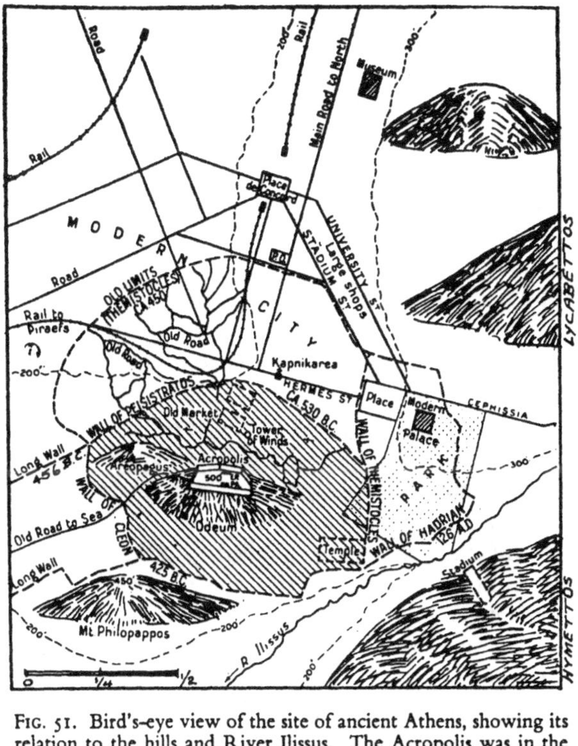

FIG. 51. Bird's-eye view of the site of ancient Athens, showing its relation to the hills and River Ilissus. The Acropolis was in the centre of the oldest town. The new main streets of the modern city lie to the north-east of the Acropolis. It is just half a mile from the Tower of the Winds to the Palace.

German Schaubert laid out much of the modern city, which reminds one somewhat of the plan of the north-west of Paris. There are two main squares, or 'Places', one near the Palace, and the other in the north near the main railway stations. Streets radiate from the latter in four directions. The quite fine department stores, etc., are mostly to be found along the two wide streets leading from one 'Place' to the other. Here also is the University, while the Museum is rather away from the centre of the town in the north. In summer the most pleasant place to visit is the large park surrounding the Palace. Here only in my limited survey of Greece did I see anything approaching dense tree growth, and this is due to the constant irrigation of the gardens. It is very interesting to note the contrast between the modern shops on Stadium Street and the primitive little booths along the street of Hephaestus in the 'Old Town' which extends just to the north-west of the Acropolis (Fig. 51).

To the scientist probably the Tower of the Winds will be as interesting as anything in Athens. It is surrounded by a small garden, and is forty-two feet high. In outline it is octagonal, and on each face is a fresco illustrating one of the main winds. Boreus is portrayed on one side, and Zephyr, the gentle west wind on another; and so for the other directions. Inside are the remains of a water-clock, while various sun-dials once adorned the various walls. It was built by a Syrian philosopher in the last century before Christ, and is better preserved than almost any monument.

The Acropolis is entered from the west by way of a winding road which brings us to the majestic steps and portals of the Propylaea. This monument was built about 435 B.C. On the south side of the main portion of the Acropolis stands the Parthenon, which was erected about the same time. It is usually described as the most perfect monument of ancient art. The writer is no judge of classic architecture, and preferred the smaller Erectheion, built in 407 B.C. It is an Ionic temple which originally contained a famous statue of Athene. One could write pages describing the unique relics to be studied in Athens, but space is not available. It must be understood that the modern houses in the city of Athens extend far to the north of the area shown in the diagram.

Hadrian added many temples and public buildings as well as an aqueduct which has been in use for centuries. The Goths conquered the city in A.D. 267, and it finally lost most of its importance in the time of Justinian (580). In Byzantine times Athens enjoyed some fame, and the tiny church of Kapnikarea (Fig. 51) dates from this period. During the Crusades there were various Dukes of Athens, and it fell to the Turks in 1456. The Venetians captured it in 1687, and during the ensuing Turkish attack a bomb fell into the Parthenon and ruined the magnificent monument. After many vicissitudes Athens in 1835 became the capital of the newly independent kingdom of Greece. At the close of the Greco-Turkish war in 1922 thousands of Greeks left Smyrna in Anatolia, and founded colonies in the suburbs of Athens. Here new industries have been created, and before the recent World War these industrial colonies were fairly successful.

Summary of the evolution of Classical Greek Towns.

The reader cannot do better than refer to the small book *Ancient Town Planning* by F. Haverfield (Oxford, 1913) for a discussion of the general development of Greek towns. The following summary is based on his study. Town planning seems not to have been much employed in the early days of Greece; but towns laid out before 400 B.C., as we have seen in the case of Selinus, had a regular rectangular plan. Even earlier was the use of a 'processional way', but the main cities like Athens and Sparta seem to have been slow to adopt new methods. Early Greek literature speaks of the streets of Athens as narrow and tortuous, unpaved, unlighted, and more like a chaos of mud and sewage than even the usual Greek road (Haverfield).

Hippodamus, whose name appears earlier in this volume, laid out Piraeus and Rhodes, and both had rectangular streets, though we do not know that he laid much stress on the plan of the city walls. About 350 B.C. inscriptions show that city officials were interested in clean markets and streets, and also in the 'overhang' of upper storeys over the streets. After Alexander's conquests (*ca.* 333 B.C.) a large number of cities were founded, and in most of these the 'Chessboard' pattern was adopted. This was in part derived from the layout of military camps, as we shall see in the next chapter. The last stage of Thebes, already described, belongs to this type of town. Pergamum in Anatolia was one of the finest of these later Greek towns (Fig. 45). Its plan consisted of a semi-circle of great edifices crowning the crest and inner slopes of a high crescent-shaped ridge. Below the palaces and shrines a theatre was hewn out of the slopes; and lower yet, beneath the market-place, the town of common men covered the lower hillside in such order as its steepness allowed. Here at Pergamum it was not till Roman days that the lower city was laid out in chessboard fashion.

CHAPTER VIII

ROMAN TOWNS AND THEIR BEARING ON LATER TOWN PATTERNS

Terramara, Pompeii, Aosta, Timgad

IT is important to remember that Roman civilization and culture followed a long period of remarkable achievement by the Greeks in much the same Mediterranean area. Selinus and Syracuse in Sicily, and many famous towns like Naples and Sybaris in Italy long antedated the rise of Rome as a city-building state. But there were special factors involved in the planning of the early Roman towns, of which one must be briefly described.

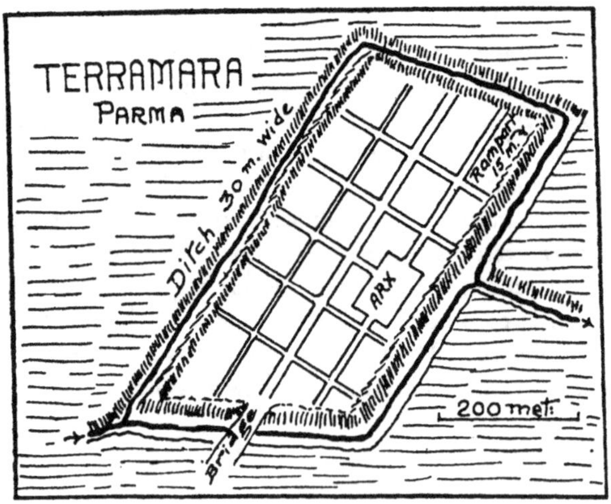

FIG. 52. The Terramara of Castellazo di Fontanellato near Parma built about 1000 B.C. It shows a plan on which those of certain Roman towns may have been based. (*After T. E. Peet.*)

One of the most fertile areas in Europe is the Lombard Plain in the north of Italy. Here the finer silts carried down from the Alps have been deposited in somewhat swampy flats near the meanders of the Po river. Here to-day are the cities of Parma, Mantua and Modena. At the dawn of the Bronze Age a remarkable group of settlements known as *Terramara* was established in this rather swampy land. They seem to have been derived from the Lake Dwellings of the Alps, but their peculiarities were developed for reasons of defence, rather than to find dry land amid swamps.

A typical Terramara is sketched in Fig. 52, and shows an area of about thirty acres enclosed within a rectangular ditch which is thirty metres wide. A flowing stream enters this ditch on the west, and leaves it at the eastern side. The ditch is crossed by a rough wooden bridge, which leads to a gap in a definite rampart built up on the inner side of the ditch. This rampart is about fifteen metres wide. The chief feature in the enclosed settlement is a small citadel or Arx, which had its own smaller rampart. Three longitudinal streets and five or six cross-streets traversed

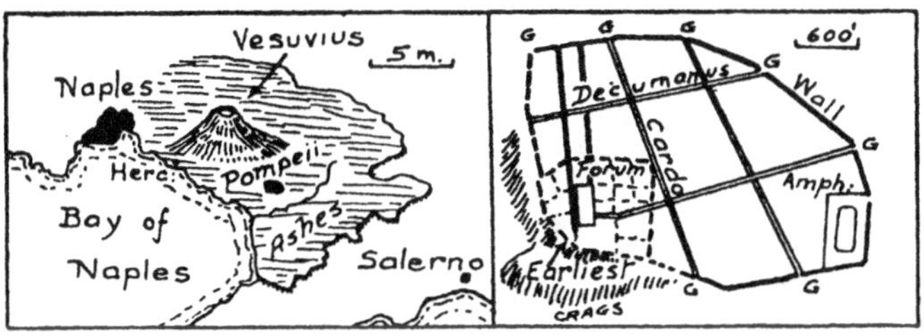

FIG. 53. (*Left*) The vicinity of Pompeii showing the ashy area due to Vesuvius. (*Right*) A simplified plan of Pompeii.

the settlement. Numerous huts, of the usual late Neolithic type, were arranged along the 'streets', and it is supposed that several thousand folk inhabited each Terramara. Some authorities think that the builders of these Terramara had migrated from the Alps towards the south, and carried out on relatively dry land the lake type of settlement they had built in the Alps. They were a hunting people, who had domestic animals and could cast bronze in moulds. They flourished about 1000 B.C.

One of the best known of Roman towns is Pompeii, which lies under the cone of Mount Vesuvius as illustrated in Fig. 53. It is about two miles to the east of the Bay of Naples; and well within the danger zone from the ash and lavas of Mount Vesuvius, as is suggested in the illustration. Some eight miles to the west is the site of Herculaneum, another ancient Roman town of great interest, while Naples is about twelve miles to the west of Pompeii.

Pompeii was perhaps founded by the Oscans, a rival tribe to the

Romans, and probably it was for a time under Etruscan rule like Rome itself. In the struggles against Rome during the time of Sulla it was conquered by this Roman soldier (in 89 B.C.), and was soon made into a military colony by the Romans. During the early years of our era it was a prosperous seaport town, though it suffered from earthquakes. However, in A.D. 79 occurred the cataclysm which preserved the evidence of its culture in such remarkable fashion for nearly 2,000 years. The very site of Pompeii was unknown in the middle ages, and systematic excavation began about 1800. Under the deep cover of ash and lapilli almost every detail of a Roman city has been discovered, and where necessary, reconstructed.

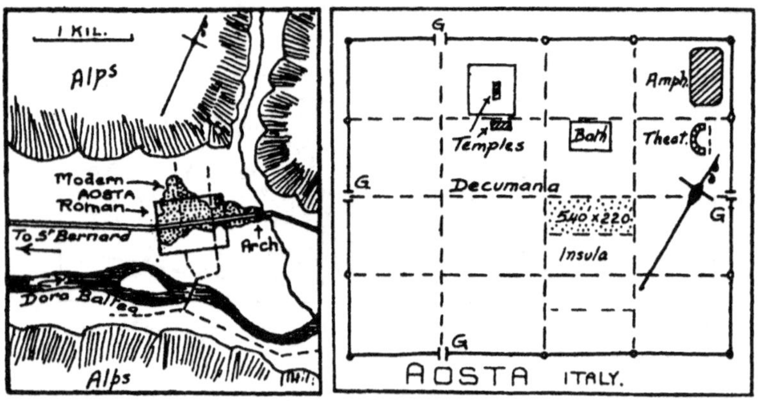

FIG. 54. (*Left*) The vicinity of Aosta in the Italian Alps, showing the change from the Roman town to the modern town. (*Right*) Plan of the Roman town.

The plan of Pompeii was somewhat lozenge-shaped, recalling that of the Terramara. A wall protected three sides, but seems to have been levelled on the seaward side, to permit expansion of the city in that direction. Towers and gates occurred at intervals along the walls. The street plan seems to show a fairly definite main west-east thoroughfare (Decumanus), and a similar north-south Cardo. (Much of the city, however, still awaits excavation.) There is an amphitheatre in the south-east corner. The oldest portion of the town seems to be the south-west portion where the Forum was situated. Haverfield points out that there is not much approach to symmetry in any portion of the plan.

The evidence seems to suggest that the irregular streets of the south-west corner formed a sort of 'Old Town' surrounding the Forum; and that this first settlement was in large part associated with the defensive character of the cliffs on the seaward face of the settlement. Perhaps the 'quarter' to the north of the 'Old Town' was next laid out, and finally the eastern half of the town. The street-plan of the latter portion agrees neither with that of the Old Town nor with the diagonal streets characterizing the north-west quarter.

For our second example we may turn to the little town of Aosta, which lies under the shadow of Mont Blanc in the far north-west of the Lombard Plain (Fig. 54). This district was inhabited by the Salassi, who were subjugated by Rome about 25 B.C. Some 3,000 soldiers were soon settled as a garrison on this route, leading to the two passes of Saint Bernard across the Alps. The new town was named 'Augusta' in honour of the Emperor of that date.

The ancient town-walls are fairly well preserved, and I noted on my visit in July, 1908, that the wall seems to be built of huge boulders laid in cement. Part of the three-tiered amphitheatre is extant, though modern houses as usual are 'plastered' on its face. The walls enclose a space of 793 by 624 yards, and are twenty-one feet high. The gates and the towers are well preserved, and their arrangement *across* the main road, shows that the chief purpose of the garrison town was to protect this route across the Alps. A triumphal arch in honour of the Emperor Augustus is one of the interesting sights near Aosta. It is not quite certain how many street 'blocks' (or *insulae*, as the Romans called them) were contained in this area of about 100 acres. The block shown with dots in Fig. 54 would not be an unusual size for such a Roman *insula*; and we shall see other samples later.

On the plateau of eastern Algeria are several Roman cities which have been less despoiled than most of the contemporary places in Europe. Their position as garrison towns on the desert border of the Roman 'Granary' has left them somewhat free from many of the disturbances of more attractive regions (Fig. 45).

Timgad was built about the year A.D. 100 at the order of the Emperor Trajan, and was occupied by a legion of Parthian soldiers. It lies twenty-four miles to the east of the modern town of Batna (Fig. 55). In the summer of 1938 we drove there along a well-made road, which passed through undulating country just north of the Aures Mountains. All along the route were wheat crops in a very poor way. Often the plants were only about eight inches high, but were bearing grain. Our driver said they had suffered bad seasons for the last ten years. We saw Arabs reaping with their tiny sickles, and in places the plants were so far apart they were gathering the grain by hand!

About eight miles from Batna we passed through the first Roman town, called Lambesi. Here a huge modern prison cuts right across the ruins of the old Roman settlement. There are still preserved several triumphal arches and a large building called the Pretorium. We did not linger here, however, but hurried on to Timgad. On the way we passed a motor lorry laden with large logs cut in the adjacent mountains. Nowhere in Algeria except near Constantine did I see any other large trees.

Timgad is situated in rather flat country just south of the hills. There is little settlement near nowadays, since the land seems to be farmed for wheat by various Arabs in the employ of a large French development company.

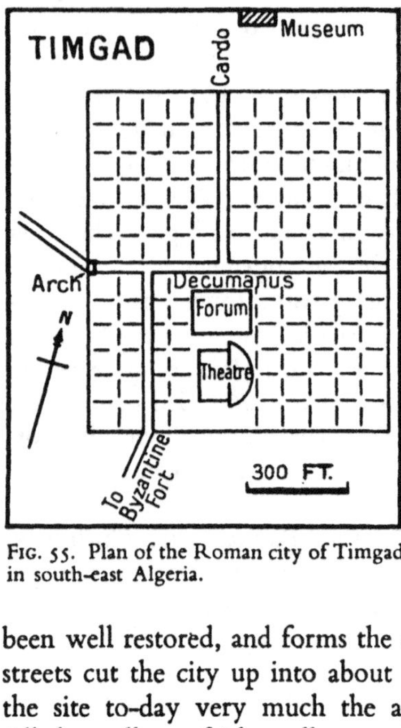

FIG. 55. Plan of the Roman city of Timgad in south-east Algeria.

The plan of Timgad is that typical of a Roman garrison town, and remains unaltered since it was first built, since no later town has trespassed on its site (Fig. 55). It is in the form of a square of about 350 yards on each side. The two chief streets are the Cardo running north-south and the Decumanus Maximus running west-east. The Forum is in the centre of the town with the large open theatre just to the south. Six large Thermes (Baths) are scattered close around the city, and there are relics of several of them in the city itself. The Arch of Trajan has been well restored, and forms the entrance on the west side. The cross streets cut the city up into about 100 equal squares (*insulae*), and give the site to-day very much the appearance of a gigantic chessboard. All the walls are fairly well-preserved to a height of about three or four feet; but there are no roofs or any part of the buildings in most cases above this level.

About a quarter of a mile to the south is a Byzantine fort built by Justinian in the sixth century A.D. It has thick walls, and relics of the eight towers survive. In the north of Timgad is a small museum, where are collected innumerable relics found on the site. These include statues, ornaments, lamps, bronze utensils, and money. Just outside the museum is a simple fountain which flows strongly even in summer, and is probably the same source which supplied the Roman soldiers 1,800 years ago.

Haverfield makes some interesting comments on the density of population in such garrison towns as Aosta and Timgad. In the latter town there were 132 insulae, separated by the various cross streets shown on the plan. Of these insulae eight were taken up by the Forum, four by the Theatre, three by various baths, and one each by a market, a library and a Christian church. The insulae here were square and about seventy feet on a side, and may have contained three or four houses each. The streets were fifteen feet wide, while the main streets were bordered by colonnades. There seems to have been no marked distinction between richer and poorer houses or streets.

Our plan does not show the later growth of the town, for it later spread to a width of over 800 yards, and 'as it expanded it broke loose from the chessboard pattern. Even the main streets wandered away in uncertain curves, so that we may say that a town-plan was binding on the builders of the "Colonia", but lost its power within a very few years'.

The Evolution of the City of Rome

The Tiber has never been an important stream for navigation, though in the early days boats brought goods to the market at the Forum. The word 'Rome' is said to refer to the fertile red soil deposited around the town by the Tiber (Fig. 56). The river has built out three miles of coastline since Ostia was a port of the Caesars. The buildings preserved from early times in Rome are, on the whole, disappointing. They have been destroyed not only by barbarians and by the tooth of Time, but also by folk tearing down the magnificent memorials to build insignificant later buildings. Thus Timgad (on the northern edge of the Sahara) seems to the writer to give a much better idea of a Roman city than does Rome!

The first 'city' of Rome was the small square area surrounding the Palatine Hill. This constituted the so-called *Quadrata*, which was soon enlarged by the inclusions of the spurs of the Esquiline and Caelian Hills. Later a Sabine village on the Quirinal Hill became a suburb of the early settlement. Servius Tullius, the second of the Etruscan kings of Rome, not only built the walls charted on the map (Fig. 56), but also divided the city into four wards, each associated with one of the hills. About this time the Capitol Hill formed the citadel of the young city.

The Servian Wall was built of blocks of tufa about two feet thick, and rose to an average height of fifty feet. A ditch 100 feet wide and thirty feet deep added greatly to the defensive value of the wall. The original Forum was built on what was formerly marshy ground. But the remarkable sewers (Cloaca) built about this time drained the city, leading the water into the adjoining Tiber. These giant sewers, ten feet across, are still in use for their original purpose. There were of course several other Forums built by later rulers, in the lowlands between the western hills of the city. Each was surrounded by magnificent temples, and embellished by triumphal arches in honour of various rulers.

The Forum Romanum is the most conspicuous of these monuments, and with the Coliseum occupies much of the central valley between the Caelian and Capitol Hills. For many years the ruins of the Forum

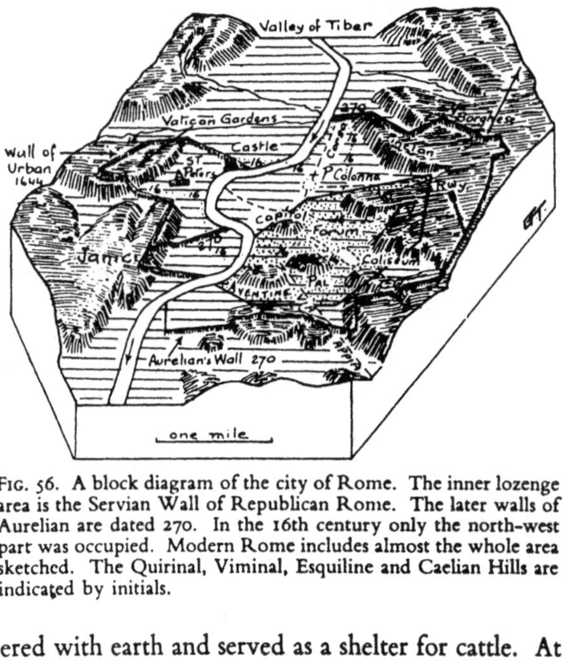

FIG. 56. A block diagram of the city of Rome. The inner lozenge area is the Servian Wall of Republican Rome. The later walls of Aurelian are dated 270. In the 16th century only the north-west part was occupied. Modern Rome includes almost the whole area sketched. The Quirinal, Viminal, Esquiline and Caelian Hills are indicated by initials.

were covered with earth and served as a shelter for cattle. At the south-east is the Arch of Titus erected by the Emperor in honour of the capture of Jerusalem in A.D. 79. The seven-branched candlestick of the Temple is visible among the carvings. The Forum proper is now a bare rectangular space littered with broken columns, and is about eighty yards long. A little farther east is the gigantic Coliseum, an amphitheatre, approximately 600 feet long, erected about A.D. 80. It could seat 80,000 spectators, and here took place the combats of the gladiators during the closing years of Imperial Rome.

Republican Rome grew up around the Palatine Hill, and legend states that as early as 550 B.C. the Servian Wall was erected to enclose a large part of the 'Seven Hills of Rome'. (Fig. 56). In later Imperial times, about A.D. 270, another great wall was built by the Emperor Aurelian, and this also is indicated in the block diagram. To-day the city extends over most of the area shown in the whole diagram.

In the early days of Imperial Rome under Augustus, the city had grown far beyond the four wards and beyond the Servian Walls. Augustus divided it into fourteen regions or wards, each containing from seven to seventy parishes. These covered almost the same area as was enclosed in the Aurelian Wall about 150 years later. This second wall is still well preserved on the northern side, but has vanished in places, especially near the Tiber. It was built of concrete to a height of fifty feet, and was twelve feet thick. Towers were erected along the wall at intervals of forty-five feet, and there were many imposing gates, that leading to the Appian Way in the south-east being perhaps the most striking.

During the Middle Ages Rome greatly decreased in importance so that it is described as a huddle of humble dwellings clustered about the strongholds of the barons. At this time the total population was perhaps less than 30,000. The better houses were built of tufa or brick, and were often of two storeys with an arcade in the upper portion. During the 16th century only about one-third of the area within the Aurelian Walls was occupied by close-set houses. The limits of the city at this time are indicated by the line marked 16 in the map. Across the river near the ancient Castello was erected the Church of Saint Peter, the largest in the world. The modern building was finished about 1626. Behind the Cathedral on the hill are the gardens of the Vatican Palace in which the Pope has lived since the close of the 14th century. (Fig. 158.) In recent times the chief changes have been the curbing of the floods on the Tiber by embankments, etc., and the cutting of fine avenues, such as the Via Cavour or Via Nationale, through the congested slums of the modern city. In the last seventy years the population of Rome has increased from 226,000 to 1,327,000.

Roman Towns in Britain; Colchester, Silchester, Chester

By the time the Roman conquests had spread to Britain the Pax Romana was well established. The great city of classical times was carrying its urban way of life to all parts of western Europe. The general interest in trade also greatly increased the development of towns as opposed to the subsistence way of living, which had obtained hitherto in many regions. It would be wrong however to think that there were no towns in Britain before the conquest by Claudius in A.D. 43. As we have seen earlier the Britons were largely a pastoral people, and their largest settlements perhaps were hill-forts of the type described (Fig. 37). But the Belgae had migrated to Britain long before the Roman Conquest, and there were at least three large towns in south-east England in their times. Of these Verulamum (now St. Albans) was at first the chief, and it was classed as a *municipium* by the Romans for that reason. This town was at the head of the Lea valley and was the capital of the Catevellauni. By degrees the port of Colchester (Camolodunum) to the east, gained a greater share in the trade with the continent, and at the time of the Roman Conquest was the chief town. Hence it was soon attacked by the Romans. The conquerors established a military 'colonia' here, where many discharged soldiers settled.

The River Colne runs along the north and east of the Roman site, which occupied a sort of promontory, open to attack mainly on the west. The plan shows a rectangular set of walls, about 1,000 yards from west to east and half that distance from north to south. The main road seems to have entered at a very strong western gate, which is said to be the largest of any Roman city in Britain. The main cross street apparently ran rather close to the western wall, and not across the centre of the town as usual. The reconstruction of the Roman streets is suggested in Fig. 58, and the relics of houses and pavements are shown by the dots. Of course the modern town of Colchester has covered much of the evidence, and the modern streets are shown by heavy lines.

Few parts of the world show a better example of the control of settlement by structure than Roman Britain. In the west is the 'Oldermass' a rim of ancient rocks which underlies the younger formation of the Paris-London Basin. These same younger rocks (the Youngermass) form in eastern England a series of cuestas (ridges) and vales running from north-east to south-west across the country. The map in Fig. 57 shows the Oldermass by the diagonal lines, and it is seen that this area was never closely settled by the Romans, but was held by military forces centred on the great garrison towns of York, Chester, Caerleon and Exeter.

Across the Youngermass were driven the famous Roman roads, of which that from London to Chester, later known as Watling Street, was perhaps the chief. Roughly along the main cuesta of England, which takes the form of a ridge of Jurassic limestone, runs the Fosse Way. Ermine Street and the Akeman Way are other ancient routes which are indicated. It is believed that the Romans held their military zone by some 150 forts, but these did not for the most part develop into important settlements. However many of the new Roman towns did become quite important centres, as we shall see; while they established many other towns in the vicinity of the pre-Roman villages.

Of second rank, after Colchester and Verulamum, was a group of towns built to command these cuestas of which the routes and agriculture have already been discussed. Nearly all show by their names (-caster, -chester, etc.) that they were for a time military camps. Among these are Silchester, Cirencester, Winchester, Chichester, Dorchester, Wroxeter, Exeter and Leicester. Most of these are shown on the map of Roman occupation which appears as Fig. 57. Of these Silchester may be especially mentioned, since it has never developed much since Roman times, and the relics of the early Roman town have been carefully investigated.

FIG. 57. Structure and settlement in Roman Britain. Note Avebury (A.), Stonehenge (St.) and Salisbury (S.) on the Chalk. (*Partly from E. W. Gilbert.*)

For much the same reason that Timgad gives us a good picture of a Roman garrison town, so Silchester preserves a rural town without the usual destruction consequent on later growth. Silchester to-day is of very little importance, whereas in the first century of our era it was one of the chief centres of southern Britain. Here the road from the east to Bath ran along the 'Kennett Syncline' between the outcrops of open chalk a few miles to the north and south. The Roman camp was built on the harder Bagshot beds above the clay. At Silchester the important road leading south to Winchester and Southampton branched off. Another led to the north-west to the 'Camp of the Corini' or Cirencester. However Silchester was not sited on any large stream, and its dominant position soon gave way to the better site at Reading, a river-junction town, where the Kennett joins the Thames (Fig. 57).

Roman Calleva (Silchester) was founded at the former capital of the

Atrebates, a Belgian tribe. There was no local stone except flint, and a good deal of oolite was imported from the west. Wells gave a water supply, for there was no useful stream at hand. Although the street-plan is rectangular, the walls were built diagonally to the roads for the most part, as is apparent in Fig. 59. Probably the stone walls were added long after the street plan was in use. The area included in the walls was about 100 acres, of which the central part is probably the oldest, forming a square about the Forum, as indicated in the diagram. This square includes sixteen insulae, each occupying about two acres.

The most interesting feature about Silchester, as pointed out by Haverfield is that the houses were not built regularly along the streets but often seem to have been arranged in a haphazard fashion, and were

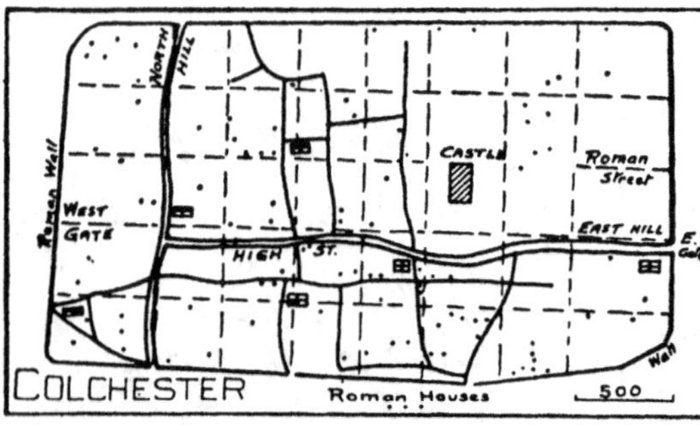

FIG. 58. A reconstruction of Roman Colchester, in which the ancient streets appear as broken lines. The modern roads are curved heavy lines. Modern churches are inserted. (*After T. Adams.*)

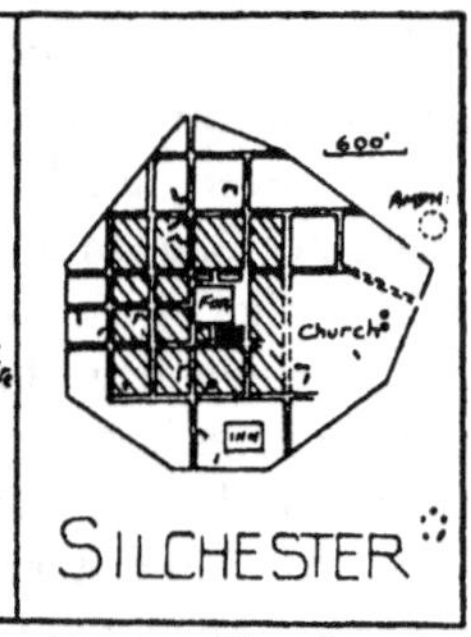

FIG. 59. The plan of the rural town of Silchester as occupied by the Romans.

undoubtedly surrounded by fairly large gardens. 'When his town had been Haussmannized' and equipped with Roman Forum, Basilica and the rest, the Callevan continued to live in his irregularly grouped houses and cottages. Possibly there were only about 100 houses in this early 'Garden City'.

Only in one region along the western coast of England does the Youngermass reach the sea in any wide area, and this is at the estuary of the Dee. It was natural therefore for the Romans to push to the coast along the Deva (Dee), and to build a military camp at what has since been known as Chester (i.e., the Camp). In Roman times it seems probable that the estuary of the Dee was a more important opening than the Mersey or the Ribble.[1] A vast amount of silting has taken place in the vicinity of Chester since Roman times, for in those days fair sized

[1]See the valuable account of Roman England by E. W. Gilbert in *Historical Geography of England*, Cambridge, 1936.

vessels were able to anchor under the walls near the ancient Water Gate (W.G. in Fig. 60).

The site of the Roman camp is on the north bank of the Dee at the lowest crossing of the river. It was about 100 feet above sea level on an outcrop of Triassic sandstone. It commanded important routes north to Lancashire and west to the copper and lead mines of North Wales. Here then for several centuries was the headquarters of a Roman Legion. The camp was rectangular in plan and enclosed an area of about fifty acres, as indicated by the thick black line in Fig. 60.

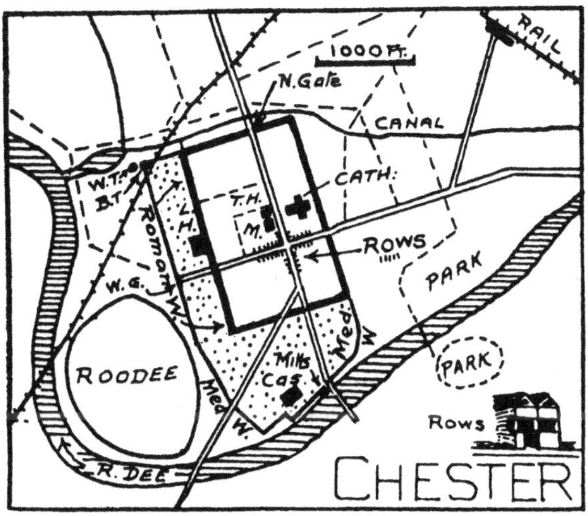

FIG. 60. The city of Chester in western England, showing the site of the Roman Camp (black) enclosed in the Medieval walls—which still survive. *Inset* a sketch of the arcaded 'Rows'. *M*=Market; *L.H.*=Linen Hall; *B.T.*=Bonwaldesthorne's Tower.

Chester has had a very eventful history since Roman times, which may briefly be described. It was one of the last Saxon positions to yield to William the Conqueror, and in the 13th century it became a special possession of the King. To-day the great attraction of Chester depends on a wall which was built in the 14th century, and is still almost perfect in its construction. On the north and east the medieval wall followed the plan of the earlier Roman wall, but it took in much more territory on the west and south, as the plan (Fig. 60) makes clear. However the original centre of the Roman Camp is still the Market Cross, where the two main streets intersected. Here are the famous 'Rows', which are arcaded houses built of timber, and perhaps the finest of their kind in England. (See the inset in Fig. 60.)

Lanchester in his book on Town Planning[1] gives a very interesting account of the origin of these elevated arcades. He believes that the road level of to-day in the centre of Chester is much as it was in Roman times, but that an immense amount of debris gradually accumulated (after the various sieges and fires, etc.) at the sides of these roads. Rebuilding took place on the top of this debris of Roman times, so that the first houses stood on the tops of banks on either side of the road. Subsequently little booths lined the lower frontages, and eventually the lower and upper structures were combined, hence we get the Chester 'Rows'. We may compare conditions in London, where the Roman roads are in places eighteen feet below the level of modern traffic.

To-day Chester has a very different appearance owing to the variations in the course of the winding Dee. A large level expanse, the Roodee ('Cross Isle'), has developed, which is used as a racecourse. The drop in the river to the south of the town, still works the famous Mills on the Dee near the main bridge, and these have been in operation for 800 years. The famous Cathedral occupies the site of earlier churches, but the present building does not contain much erected before 1200. A rocky knob close to the bridge was early occupied by a Norman Castle, but most of this has been removed in modern building operations. To-day the city of Chester spreads half a mile or so to the north-east of the old medieval city, and the present population is about 45,000.

Roman Towns in France and Germany, Autun and Cologne

In a later chapter dealing with Medieval towns further reference will be made to the influence exerted upon such settlements by their original Roman pattern. We may well, however, consider the development of two especially interesting cities which were founded in the days of the Roman Empire. The first of these is Autun in central France, and the second Cologne in western Germany.

In 58 B.C. the Helvetii decided to migrate from Switzerland into western France. They crossed the Saône and had reached Bibracte, the capital of the Aedui, on the northern slopes of the Cevennes, when Caesar came up with them and completely defeated them. Augustus believed in transplanting rebel tribes, and so he moved all the Gauls from Bibracte (now Mont Beuvray), and brought them down into the valley of the Arroux about ten miles to the east. Here the river received two streams from the south-east, and on the flattish lozenge between them Augustus in 12 B.C. marked out a very extensive settlement, whose plan

[1]H. V. Lanchester, London, 1925.

can be observed in Fig. 61. This new town was named Augustodunum, and this time has softened to Autun.

The new town covered an area of about 500 acres, so that it was much larger than some of the Roman settlements which we have investigated. The general pattern consisted of rectangular *insulae*, as usual with the streets parallel to the walls. The chief interest to students of urban geography is that the modern city of Autun is only about half the size of the Roman town, and by no means fills the space within the Roman

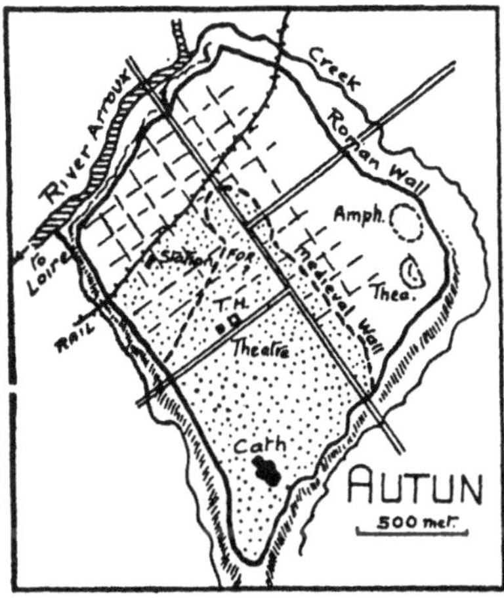

FIG. 61. The Roman town of Autun (southwest of Dijon) was much larger than the medieval walled town, as shown by the walls. Roman roads are indicated by solid lines and by broken lines. The modern town is the dotted area.

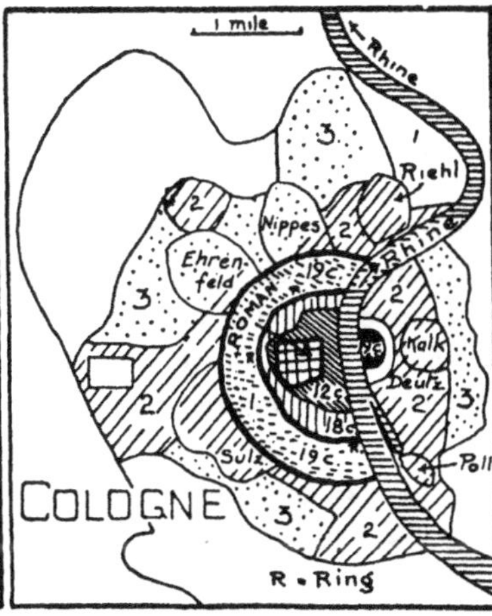

FIG. 62. The evolution of the Roman colony of Cologne.

walls. Many portions of the latter, including some of the chief gates, have survived, and can be compared with the later medieval walls, which are indicated in the map. The cathedral and most of the modern buildings occupy the southern angle of the Roman 'lozenge'. Since much of the old Roman settlement is no longer covered with buildings, it has been relatively easy to excavate the northern portion of the area. The Roman streets are indicated in the plan by broken lines. It is supposed that the Forum was in the centre of the city (Fig. 61), but the relics of the theatre and amphitheatre in the east are quite well preserved. The insulae seems to have been about two or three acres in area. Autun was for several centuries the capital of the Dukes of Burgundy. It has sustained several sieges, notably in 1591 when it was attacked by Henry of Navarre.

To-day Autun has a population of about 16,000, and it is much less important than the coal and iron town of Le Creusot, about ten miles to the south. Chalons to the east soon replaced Autun as the chief Roman town, while to-day Mâcon—also in the fertile Rhone Valley—is the capital of the Department of Saône et Loire.

Our second example is one of the most famous of the Roman 'colonia' planted in the north of their empire. It is Cologne, whose name means 'the colony', and it is situated on the Rhine just where it emerges on the northern plains after cutting its way through the ancient plateaux of the Ardennes and Sauerland. Claudius laid out the city on the site of the earlier settlement of the Ubii in A.D. 50, and it was the chief town of this part of the Empire. It was very early visited by Christian teachers, and has had a very important place in German religion for that reason. By the Golden Bull of 1356 the Archbishop of Cologne became one of the seven Electors of the Holy Roman Emperor.

The diagram given in Fig. 62 shows the main features in the growth of Cologne. The Roman site was naturally on the western shore, since this was in contact with the other regions of the Empire in Gaul. The original plan seems to have been almost a square, and this is still the chief section of the great city of to-day. A suburb Deutz had grown up on the east bank, and this was protected in the 19th century by semicircular defences. The spread of the city during the 12th century is indicated, and thereafter there was not much growth till after the 18th century. Around 1880 the semicircle of walls (R.R.) and moats was removed, and the area given over to Boulevards known as the 'Ring' (Fig. 62). A second line of defence about half a mile further from the centre of the city was erected in 1881.

To-day Cologne extends far beyond the early walls, forming suburbs such as Nippes, Ehrenfeld, etc., which are shown on the diagram. One interesting feature of the city development is given in the map, which is taken from the book by H. I. Triggs (*Town Planning*, London, 1909) which I have mentioned previously. It suggests the method adopted for coping with congestion in a great city. The whole area in and near the city is divided into Zones, where conditions vary with distance from the centre. Thus in the older portions labelled '1' about twenty-five per cent of all building-land must be left free. In zone '2' thirty-five per cent must be left free, but buildings may be of four storeys. In the outer belt, Zone '3', still more space must be left, while in the villa area there must be thirty-three feet between buildings, and fifty per cent of the area must be free.

CHAPTER IX

EARLY MEDIEVAL CITIES

The Dark Ages and the Revival of Towns

WITH the decline and fall of the West Roman Empire, which occurred between A.D. 400 and A.D. 476, we find a corresponding change in the whole way of life of western Europe. During the zenith of the Roman Empire city life flourished, primarily because the Pax Romana guaranteed freedom of travel and trade, and city life cannot flourish if these factors are unduly restricted. We have a remarkable record of travel, in the times of Constantine in the fourth century, as described by the 'Bordeaux Pilgrim'.[1] He travelled from Bordeaux through the famous Roman cities of Arles and Nîmes. He crossed the Genevra Pass, and so reached Milan and Verona. Thence he journeyed east up the Peartree Pass, and passed through Sirmium and Singidunum (Belgrade); after which he turned south to Nish, Sofia and Constantinople. He followed the military road through Ankara and the Cilician Gate to Antioch. Then by way of Tyre and Sidon, he reached his objective Jerusalem. Such a journey, carried out in security through many flourishing cities, would by no means have been possible for many ensuing centuries; perhaps not till the end of the Crusades.

With the dawn of the Dark Ages roads and cities rapidly deteriorated. Civilization itself received a great setback, as may be gathered from the small number of great teachers who date from this time. Between 400 and 800, when the relatively stable control of Charlemagne gave peace to western Europe, there are hardly half a dozen learned men. Boethius (died 524), Bede (733), and Alcuin (804) kept the light of civilization burning, for the most part in monasteries; but few others illuminate the Dark Ages. Pagan tribes invaded what had been flourishing communities. In England the Anglo-Saxons at first despised the civilized Romanized Briton; indeed the Angles ignored the towns, and developed many small settlements, largely on the loams and better soils of the eastern vales (see p. 147). London itself—a great town under Rome—was practically deserted for a generation or two. Not until 600 did the

[1]C. R. Beazley, *The Dawn of Modern Geography*, pp. 57-66, 1897.

Saxons become Christians, and begin to acquire a civilization equal to that of the earlier Britons.

Much the same evolution occurred in France and Spain. The Franks became civilized sooner than the Saxons, but the invasions of the Moslems in Spain—after 711—set back the development of northern Spain for centuries; though the Moors gave a remarkable culture to southern Spain. In the north, however, we find the Spanish devoting their lives to war against the Moslems, and the most characteristic feature of the times was the building of numerous castles for defence in what has ever since been known as Castile (see page 152). In the seventh century the former flourishing Roman provinces of Moesia and Illyria (now Jugo-Slavia) were overrun by pagan Slavs from the northern plains. Not for centuries did this part of Europe return to the conditions necessary for the steady development of an urban way of life.

As Bucher has pointed out in his *Industrial Evolution* (New York, 1901), the characteristic human agglomeration in medieval times was a walled city. Such cities were scattered at regular intervals over the country, and the increase in the number of the citizens continued only until the city was full. When once it had as many inhabitants as it needed, to man its walls and supply all the branches of industry, there was room for no more. Hence the Middle Ages developed no large cities; the medieval economic and commercial system forbade it.

The most illuminating study of the medieval city is that by Lewis Mumford,[1] in which he discusses at length the differences between the medieval and the modern industrial towns and cities. Since the large monasteries were often respected by the pagans of the Dark Ages, it was natural for them to form the nucleus of the cities of the future. Such towns as Wurzburg, Brandenburg, and Oldenburg (in Germany) seem to have originated as religious foundations. As more settled ways developed in western Europe, earlier invaders such as the Saxons and Franks found it necessary to defend themselves against their enemies such as the Danes and Normans. Under stresses like these it was natural for them to make use of the walls which had been erected by their predecessors the Romans.

With the development of the Feudal regime we find the castle of the local baron or knight acting as the nucleus of a new town. In England many of these earliest castles, just after the Norman Conquest, were built of wood, usually upon artificial mounds. Naturally to-day nothing remains of these except the earth mounds.

[1] *The Culture of Cities*, 1938, New York. The author has kindly permitted me to make considerable use of this striking and well illustrated volume.

During the worst period of the Dark Ages, which from the city point of view seems to have been in the ninth century, many of the former large cities were mere empty 'caves of masonry'. Such was the case with Mainz and Trier, though they later acquired as much importance as in Roman times. Indeed from the fourth to the eleventh century the chief value of a town lay in its walls. In times of peril the peasants abandoned their hovels in the vicinity, and fled within the walls until the enemy withdrew.

Village life changes relatively slowly, so that there is not a great deal of difference between a small community of the nineteenth century and one of the early medieval period. But with the growth of industry and commerce in the early middle ages we find a considerable hastening of the evolution of the towns and cities. The peasant had little sense of co-operation, and so we find none of the Peasant Wars or Jacqueries affecting national development to any great degree. But with the influx of the peasant into the walled town, there developed the important institution of the Guild. In Worms and Mainz this originated about 1100. The Guild was an early form of trade union, and in many parts of Europe, especially perhaps in Lombardy and Flanders the Guilds, and the middle class folk who controlled them, acquired much power. Ultimately they defied their feudal lords, so that the cities became strongholds of the trading class; and this change considerably modified the character of the towns themselves, as we shall see.

This revival of trade, and necessarily of towns, began in general in the eleventh century. Great Fairs were held at stated intervals, and these attracted traders from all over the continent, and in turn led to the growth of the towns themselves. Indeed, as we shall see, Universities, such as that at Cambridge, in part developed from these trade fairs. Byzantium, Augsburg, and Venice are examples of early growth due to such trading at the close of the Dark Ages. Meanwhile the monasteries continued their important role of civilizers, and in the twelfth century the Cistercians led the way in clearing the dense forests of the northern plains. Some 250 towns of North Germany date from these early centuries; so that this period of change has been compared to the similar evolution of the United States in the seventeenth and eighteenth centuries.

Better tools were being evolved, horses were becoming of much greater use for transport, windmills were constructed in large numbers, and as a result there was a notable increase in the area of croplands all through western Europe.

The thirteenth century saw the dawn of a rather flourishing civilization, which Mumford believes steadily deteriorated during the later medieval times. The cities were in general small, but though they were usually walled, they were not overcrowded. Gardens were numerous within the walled city, though as time passed these were gradually filled with inferior dwellings to the great disadvantage of the later city. However in this century both Venice and Milan had populations of 100,000; and were two of the largest cities in Europe. Their large size was of course a direct response to the growing trade with the East, and in part depended on their relation to the Crusades, which occupied so much of the attention of Europe from 1095 to 1291. [In this last year Acre was wrested from the Christians by the Moslems, and the hope of a revival of Christian rule in Palestine was abandoned.] By 1200 Paris had a population of 100,000, and in the next century it had grown to 240,000.

In early medieval times the shop and house were of course contained in the same building. Often the rooms were built around a court with the shop fronting on the street, and the living-rooms over the shop, while the bedrooms were at the back. In the fifteenth century glass became common in the windows, and inside fireplaces with well-built chimneys were the rule. Houses were usually half timber and stucco in western Europe, though stone was used in more important dwellings. It is interesting to realize that baths were obtained only at the public Baths, and much more cooking was done at public bakeries than in the home.

In these early walled towns it was unusual for the streets to be arranged in a regular plan. To some extent the various trades were segregated in special quarters, and the streets penetrated these quarters rather haphazardly. The chief meeting place of the town was the market square, which often faced the main church of the town. There were no 'realtors' in those days, so that there was not that slavish adherence to the rectangular plot of land, which marks the modern age of land speculation and absence of city design! In some ways this plan of well distributed little towns of about 10,000 people, each perhaps about a day's journey from its neighbours, reminds one of the utopia of the modern geographical planner. 'No self-suffocating giant cities, and no community without urban amenities'.

City Types across Europe

H. Fleure has given us an illuminating discussion of the main features of the early towns in northern Europe.[1] He points out that as cities de-

[1] *Geographical Review*, Dec. 1920, New York.

pend on trade there must necessarily be a progressive change in the city types as we proceed from the west—say in France—to the east in central Russia; because trade conditions became relatively stable first in the west, and did not develop to the same extent in Russia till several centuries later.

This change, as we proceed from west to east, is illustrated in a somewhat generalized form in Fig. 63. In the south of France many of the chief cities date from the days of Republican Rome. At this time the Forum was the central feature of the Roman city and the streets led thereto, while the main buildings of the town were grouped around the Forum, as we have seen in the last chapter. Towns like Arles, Nîmes, and Orange, illustrate this type of Roman foundation. Fleure makes special reference to Toulouse, and a simplified diagram of the city plan appears in Fig. 64

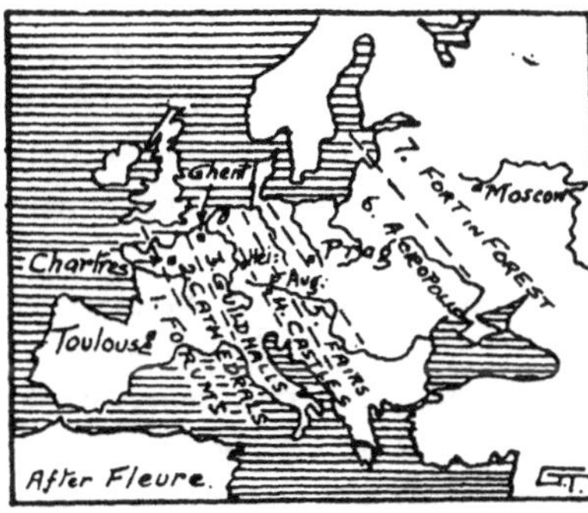

FIG. 63. Early types of towns in Northern Europe.

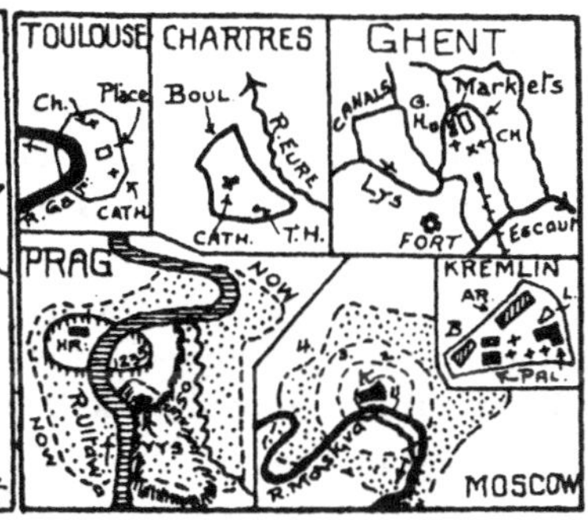

FIG. 64. Simplified plans of Toulouse, Chartres, Ghent, Prag and Moscow.

Tolosa was an important place before the Romans captured it (106 B.C.). The town was founded on the east bank of the Garonne, on the concave side of a broad meander (Fig. 64). It has always been an important centre on the road from the Mediterranean to the Atlantic. In the centre of the city the Romans placed the Forum or Market place, and the Town Hall is still in operation on the 'Place' in this same site. Later when Christianity reached Toulouse in the second century, churches were built in the city; but they necessarily occupied sites some little distance from the natural centre of the town. Our diagram shows the Cathedral well to the south of the Place, and another large church about the same distance to the north.

Northern France was settled in the days of Imperial Rome, when Christianity was a major factor in the development of the Roman

Empire. Accordingly we find that the Cathedral is usually the nucleus of the town, often enough with a 'Place' or open square fronting it. Chartres may be taken as an example of this type of city foundation (Fig. 64). The town is about fifty miles to the south-west of Paris on the River Eure, which flows north to join the Seine near Rouen. Here the magnificent Cathedral is the central point of the little town, while the Town Hall has a position considerably to the south. The town has not grown very much beyond its area in medieval times, so that the broad boulevards, which as usual have taken the place of the ancient walls, still contain the bulk of the city. As we shall see, Paris also developed to a considerable degree around the Cathedral of Notre Dame, and the same is true of Amiens, and of a score of famous cities in northern France.

As we move to the east beyond the major influence of the Roman type of settlement we find that trade developed there later, and another type of town grew up. Fleure uses Ghent to illustrate his third zone (Fig. 64), in which commerce was more important than Roman tradition or the early church, when the towns were becoming of note. Ghent is situated in the lowlands where the winding Lys joins the equally erratic Scheldt (Escaut). Various other creeks have been deepened to form canals; and Ghent may be divided into twenty-four insular districts separated by these narrow waterways.

Ghent has had several periods of marked prosperity, separated by times of stagnation. In the 14th century it was closely linked with England by means of the wool trade, and was very prosperous in consequence. Its second zenith was around 1520 under the control of Charles V, who was born there. But its waterways silted up, and it fell into disfavour with the later Spanish rulers. However one of its outstanding features is the number of markets, and the prominence of the old Guild Hall (G.H. in Fig. 64).

Western Germany, bordering on the Roman Empire, and later forming the margin of Charlemagne's empire, has been the scene of warfare throughout the centuries. Partly for this reason many of the older towns have developed around the castle of some border baron. Heidelberg is one such site, where the castle dominated the narrow valley where the Neckar enters the broad graben of the middle Rhine. All along the Rhine valley to the north almost every village and town is protected by a medieval castle, and these relics of antiquity add much that is picturesque to this part of Germany.

Further to the east there are a number of towns where one of the chief influences determining their growth has been the location there of one of the famous Fairs. Augsburg may be cited as an example. It is situated on the river Lech, some distance above its junction with the Danube. Augsburg is therefore in the centre of the structural downfold which flanks the northern Alps, and this downfold (extending from Basel to Vienna) has naturally been one of the most important routes of trade in Europe. Even in Roman times roads led from Augsburg (another 'Augusta' town) to the salt mines near Salzburg, and over the lesser Alps to the Inn valley. It was famous in medieval times for the power of its traders, who defeated the patrician party as far back as 1368. A few years later they repulsed a Bavarian army; and under the family of the Fuggers exercised much influence in Germany. Early in the 16th century their merchants had independent dealings with places as far away as India and Venezuela. Augsburg lost its independence shortly after this time, and its mercantile pre-eminence soon withered away. Leipzig may be instanced as another German city whose growth has been largely influenced by its famous Fairs.

It is an axiom in cultural geography that as we move further away from a centre of evolution the types involved in such evolution shall appear more and more primitive. In our present study, which deals with the evolution of towns, we find the same general principle illustrated. While the cities of the west had evolved far along the road towards peace and prosperity, those in the east were still largely concerned with defence against enemies. Eastern Germany, Bohemia, and Poland, were all menaced by barbarian invaders long after the west of Europe had become stabilized. Thus Fleure's two eastern zones (Fig. 63) deal with towns which have developed around the *acropolis* type of nucleus, or with veritable marginal forts such as arose in the limitless forests of northern Russia in medieval times.

Prag—the capital of Bohemia—has had a very interesting evolution (Fig. 64). The river Ultawa here flows north through the undulating lands of Bohemia to join the Elbe. Just south of a marked bend in the river was a striking crag on the east bank. This was the Vysehrad Hill, and here a castle was built to protect the village which grew up on the low plain just to the north. A German colony settled here as far back as 1100, but the real prosperity of the town dates from the time of Charles IV (1346–78). About this time the city spread to the west bank, where the Hradschin Fortress was erected about 1333. Here also was built the palace, and later the Parliament House. The two walls of Prag, the inner built in 1235 round the Old Town, and the outer wall built in 1350, are shown on the Plan (Fig. 64). One of the earliest universities in central Europe was founded here in 1348.

The last in Fleure's series lies far to the east, in what was throughout much of medieval times the vast coniferous forest of northern Russia. The first settlers in the vicinity of Moscow were probably Finnish tribes, whose way of life was not far removed from that of the Amerind who roamed through similar regions in North America before the advent of the white men. We have no accurate data as to the arrival of the Slavs in this region of Russia. No doubt they cleared a portion of the forest, and built a log fort there much as they did in relatively modern times right across the Siberian forests.

Moscow lies on a meander of a relatively small river, the Moskva, which flows to the east into the Oka, a tributary of the Volga. The Moskva flows in broad meanders through a topography which reminds every geographer of the site of Paris, though the river is flowing the other way (Fig. 64). Here also is a long narrow island very like the Isle de Cité; and the evolution of the town in a series of concentric circles also resembles that of the French capital. A log fort was built on the north bank, and this in time developed into the great Kremlin Fortress (K. in the diagram). Alongside was a small district called the 'Kitai-Gorod', the City of Refuge. Outside this area developed the 'Belya-Gorod' (White City) where the better classes live, and the outer annulus (between '2' and '3' in the diagram) constituted the 'Zemlanoi-Gorod' or Peasant City. The dotted area in the diagram represents the limits of the city about the end of the 19th century, but to-day the houses have extended far beyond these limits.

The Kremlin is the acropolis of Moscow, occupying an area of more than eighty acres. (It is shown enlarged in the inset in Fig. 64.) The walls are forty feet high, and within them are a number of barracks (B), arsenals (AR), palaces (Pal), and churches, as well as other buildings where to-day the central government of U.S.S.R. carries on its business. The Kitai-Gorod is also surrounded by walls which were built in 1534, and they enclose an area of some 100 acres.

In its early days Moscow was an outlying settlement attached to the older centre of Vladimir. Its real importance began about 1325, when the leading Russian churchman took up his residence in Moscow under the aegis of Ivan I. About 1480 it became the capital of the Slav nation, but in 1711 Peter the Great moved his capital to his new creation of St. Petersburg. Since the revolution it has again become the capital, and has left the city founded by Peter far in the rear as regards population and influence.

Medieval Town Patterns in Germany

Probably German geographers have given more attention to the plan of their early towns than have the geographers of any other nation. We owe to R. E. Dickenson a comprehensive study of such towns, from which the maps given in Fig. 64A have been borrowed.[1] He points out that the Germans brought to their eastern lands not only Christianity and commerce, but also the town itself as a distinct habitat form. Thus in a zone from the Baltic to the Danube the medieval town stands in marked contrast with the Slavic towns to the east and the Magyar towns to the south.

At the end of the 11th century there were about one hundred settlements of embryonic urban character in the west German lands. These lay along the three main corridors, the Rhine Way, the southern edge of the Plains, and along the Danube as far as Vienna. Regensburg is one of the most interesting towns, and is situated at the most northern point of the Danube Downfold. Here the walls and gates of the Roman Castrum survived the barbarian invasions; and for a thousand years the bishop's settlement lay huddled within the walls of the Castrum (Fig. 64A at 1). The merchants' quarters became a town, and were walled in A.D. 920, while later extensions were walled by 1320. At Innsbruck we notice that the market (M. on map 9) is outside the original 'burg' or strong point, which is shown in black in map 9. The position of the market-place is an important feature in classifying the later towns. Sometimes it resembles the external position in Innsbruck, as at Bremen, Munster and Treves. In other towns it is included in the stronghold, as at Osnabruck and Xanten (near Cleve).

Hoxter is an interesting example showing an irregular radial plan centred on the church (Fig. 64A, Map 4). It was a small village on the Weser in 833, where a bridge was built in 1113. Earthen defences were built in the 12th century, and the stronger walls about 1250 (Dickenson). The main phase of the growth of 'Burg and Market' settlements, along the road from the Rhine across to Magdeburg, was between 1150 and 1250. Braunschweig had a remarkable evolution, since here seven adjacent settlements fused between the 10th and 13th centuries. They were walled separately (Fig. 64A, Map 2), and were not merged to form one town until 1697.

Hannover developed on a ridge near the river Leine. A small stronghold commanded the crossing at first (1 in Map 3, Fig. 64A). The market quarter (2) was founded in 1189, and here the market is in an expansion

[1]Morphology of Medieval German Towns, *Geographical Review*, Jan. 1945.

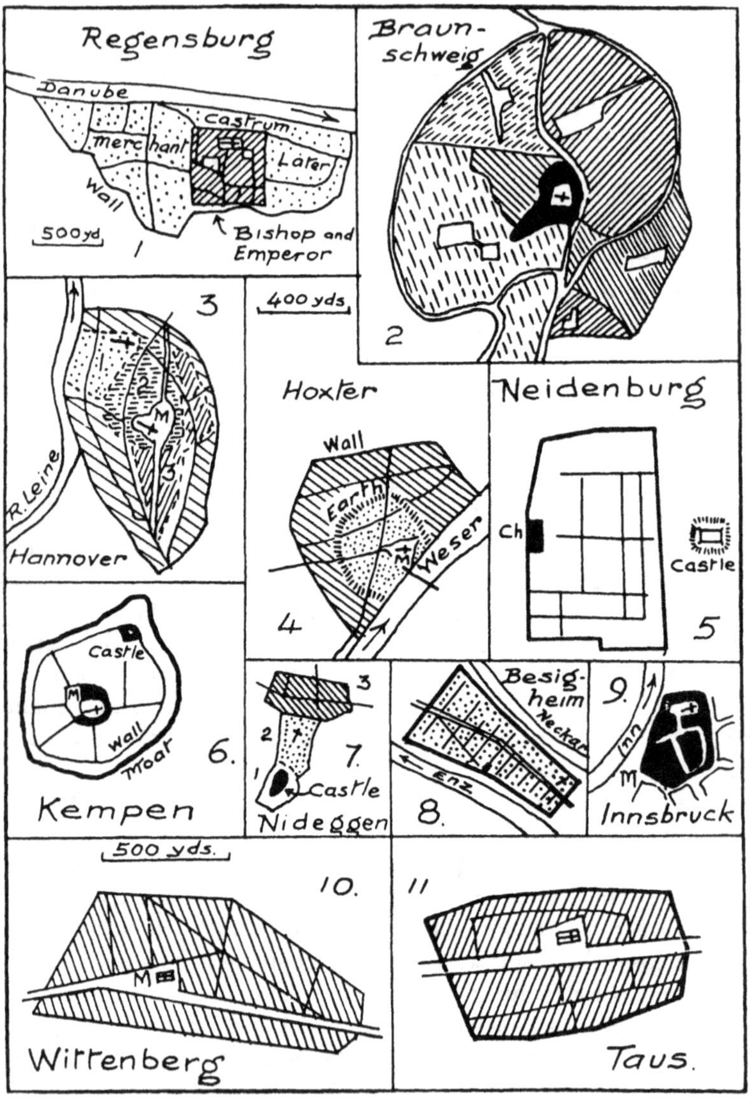

FIG. 64A. Simplified plans of various Medieval German towns. *1*. Roman castrum; *2*. Seven united settlements; *3*. Market in street; *4*. Old earth rampart; *5*. Church on wall: *6*. Central market. *7*. Three stages; *8*. Rib streets; *9*. External market; *10*. and *11*. Market in centre near main street. (*Based on R. E. Dickinson.*)

of one of the two main streets. The towns of the lower Rhine Plateau were founded in the 13th and 14th centuries, for example Nideggen in the Eifel. Here the castle was built on a high spur above the river in 1177. The parasitic settlement with the church was walled in the 13th century, while the market settlement at the end of the spur was founded in the next century (Map 7 in Fig. 64A). Kempen (near Cleve) grew up around the central church, with the market alongside. The streets radiate to the walls, which are almost circular in plan (Map 6). Many Westphalian towns used this plan, where defence was obviously one of the chief factors.

The south German towns are usually built around one through-route, which extends from gate to gate, and is usually far wider than any other street in the town. Taus in Bohemia (Map 11 in Fig. 64A) is an example, while near Heidelberg, Besigheim (8) shows the 'rib' streets crossing the main route, which runs along the ridge between two rivers.

In the belt of colonized towns in the northern plains, we find many which were founded about 1300. Among these is Wittenberg on the Elbe, where Luther was priest later on. Here two main streets converge to give a spindle form, and the church and adjacent market lie between the two in enlarged spaces (Map 10). In the far east, where the Teutonic Order built many towns, we see a rectangular plan in the walls. Often the castle is independent of the town as at Neidenburg (Map 5 in Fig 64A), which is in East Prussia. The position of the church alongside the wall is characteristic of these foundations. There was a rectangular street-net, and other features which it is thought may be due to their experience of fortifications in the Orient.

Some Norman and later towns in Britain

No great period of town development occurred in Britain, after the withdrawal of the Romans, for many centuries. The Saxons soon became farming peoples; and it is not difficult to see, as S. W. Wooldridge has pointed out, that their settlements can be rather closely related to the structural and soil characters of the east of England.[1] These are shown somewhat diagrammatically in Fig. 65. Unlike the early British folk they did not concentrate on the open lands of the summits of the cuestas but settled rather closely on the lighter soils of the slopes, and areas of drift (till) in the vales. They had no fondness for the heavy clay soils which are prevalent in considerable areas of the vales. We learn from their burial places that they did not penetrate to the west

[1] *Historical Geography of England, op. cit.*

of a line from the Tees to Dorset. Beyond this area Welsh place-names and British settlements (much affected of course by Rome in the east) were dominant.

Wooldridge shows that the earlier (mostly pagan) Saxon burial places are found chiefly in eight regions; and we may assume that these mark out fairly closely the areas which appealed to them as settlers, where the villages were most numerous. Along the coast the first district is along the South Downs, and the second is to the north of the North Downs in Kent. The third is to the south of London, and the fourth is far away in Norfolk, where light soils occur north of a great expanse of clay and sand in Essex. The Fen country was not attractive, but other Saxon settlements were made in east Yorkshire. Inland we see a group in Hampshire, another in the vale of Ouse and near Oxford, while the last was in the Midland Plain between Trent and Avon.

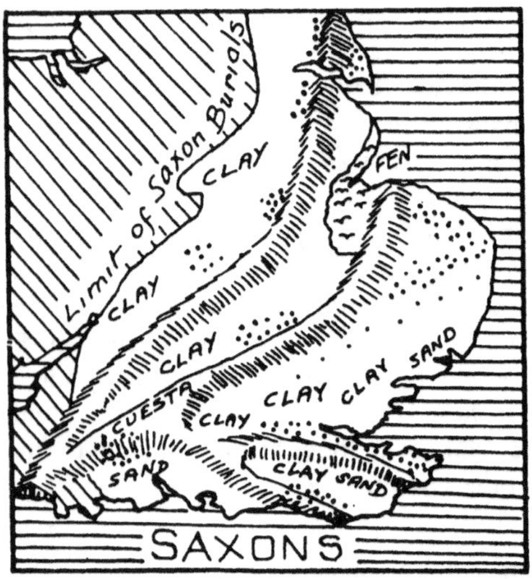

FIG. 65. The control of Saxon settlement by soils. (*After S. W. Wooldridge.*)

After the Norman Conquest there ensued in England a period of invasion something like that during the expansion of the Roman Empire. For the same reasons the invaders built strong posts, and later established outlying colonies of veterans who might be trusted to stabilize the control of marginal areas. Various large castles were built in England, especially in the north and west fronting the rebellious Welsh and Scots. Small villages soon grew up in the vicinity of such castles, as at Tutbury (near Derby), and Dudley (west of Birmingham). Okehampton (in Devon) and Arundel (near Brighton) offer other examples of castles dating from about this time.

It is inteiesting to note that the Roman towns were not in general expanded to become the cities of the future, though we have already noted some important exceptions to this statement. But as Haverfield has pointed out, their streets generally vanished, and even if the site was used, the medieval streets were irregular and often winding as opposed to the checkerboard of Rome. We see something of this change in the map of Colchester (Fig. 58). In Lincoln very little similarity exists between the Roman and medieval plan; and in Rome itself the term 'Eternal City' certainly does not refer to the ancient street-plan, though it may express the religious aspect of the famous city.

In this brief survey of medieval towns we must pass on to the 13th century, when town development received a great impetus. Edward the First of England was a great administrator, not only in England but also in his vast possessions in the south-west of France. Here no less than fifty towns were laid out, and the new citizens became direct tenants

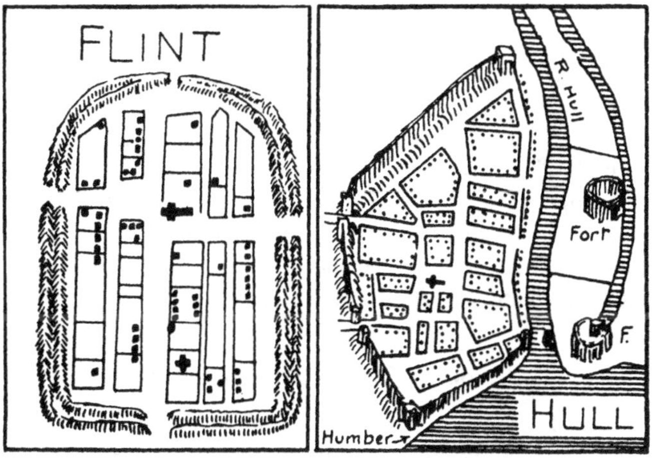

FIG. 66. 'Bastide' towns founded by Edward I about 1290 at Flint (near Chester) and Hull upon the Humber. (*Based on T. Adams.*)

of the Crown, and so were free from the burdensome claims of barons and bishops. These regular towns, planned often for veterans, were known as *bastides*, and the best examples occur in the vicinity of Bordeaux, such as Montpazier. Libourne and Sauveterre were other similar French settlements.

In England Flint and Winchelsea are of somewhat similar origin, while Conway and Carnarvon castles are well known relics from the same period. In Fig. 66 is given a diagram showing the original *bastide* settlement at Flint. This was placed about 1290 near the estuary of the Dee on a little crag which commanded the main coastal road into north-east Wales. To some extent it was the Norman counterpart of the former adjacent post of the Romans at Chester. To-day Flint is an important manufacturing town connected with the chemical industry. But the chessboard pattern of Edward I, as shown in the diagram, can still be made out. The main highways led to the castle, and as Adams points out in his book on Town Planning, there is a great similarity between these new Welsh towns and the French bastides, not only in general plan, but also in the character of the plots assigned to individual settlers.

Another foundation of about the same date has given rise to the modern flourishing port of Hull. Originally there was a small village of Wyke on the river Hull, which enters the Humber at this spot. At this time the crumbling of the cliffs at Ravenspur, under the attack of the North Sea, showed the need for a better port in this part of England. About 1290 Edward decided to establish one of his military settlements on a site near Wyke, and he gave it the name of Kingston-upon-Hull. Kingston was laid out somewhat after the bastide pattern (Fig. 66), though it is not as regular as Flint or the French examples. A central market square containing the chief church was provided in the plan. It is of considerable interest that Hull was the first town where brick was the chief building material, though it had of course been used to a considerable degree in an earlier age by the Romans. The town and harbour were completed by 1299. Edward II in 1321 granted the burgesses the right to enclose the town with a ditch and a wall of stone and lime.

Carcassonne and Castles in Spain

The town of Carcassonne in the south-east of France, about forty miles from the Mediterranean, is one of the best places to investigate the development of medieval and modern cities; and the author visited this district in 1938. The site was occupied long before the Roman Conquest, about 100 years before the time of Christ. It became one of the chief towns (Carcaso) of the province of Narbonne, and was fortified by the Romans. Some portions of the Roman wall can still be identified at the base of the medieval walls. The Visigoths fortified it in the sixth century, and much of the northern portion of the Inner Rampart dates from this time. The Citadel (or Château) was built near the Western Gate in the 11th and 12th centuries. The most impressive part of the defences is the Outer Wall, which was finished about 1300. There are some fifty round towers on these walls, and many of them are shown in the sketch appearing in Fig. 67.

Carcassonne was a stronghold of the Albigenses (early Protestants), and they suffered greatly in the wars of the 13th century. In 1262 the citizens were expelled from the old town (Cité), and started the Ville Basse across the River Aude. This new town was protected by walls in 1347, the plan being a hexagon. These walls, as in almost all other examples, were demolished in the 19th century; and their sites occupied by wide boulevards. Modern suburbs radiate along the main highways and are indicated on the map. Hence we have in one site Roman and medieval walls, razed medieval walls, modern boulevards and the radiating suburbs familiar to all modern citizens.

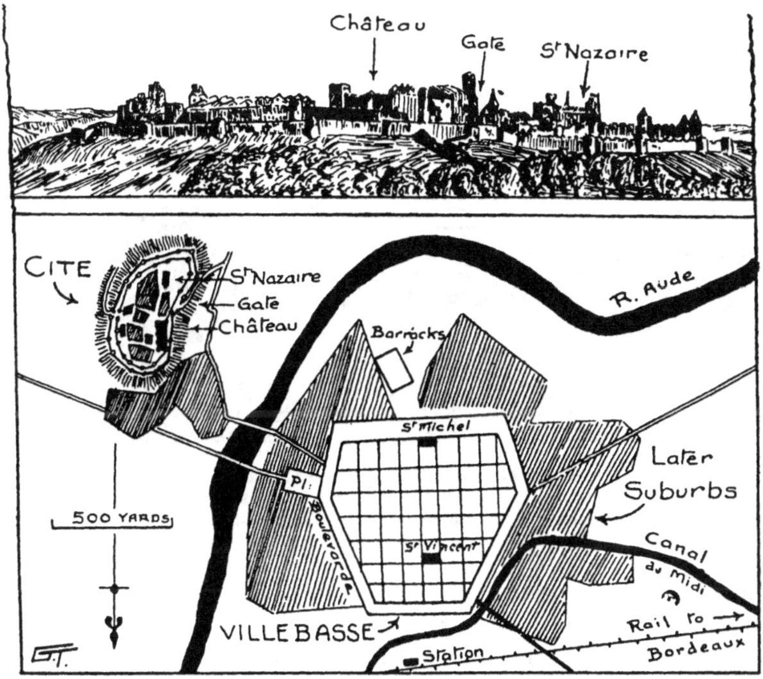

FIG. 67. Sketch of the Walled 'Cité' of Carcassonne in South France, looking to the south-east. The map also shows the newer 'Ville Basse' across the River Aude, where the walls here have been replaced by Boulevards.

It would be rather a serious omission—in dealing with conditions in the early Middle Ages, where so much depended on defence against invasions—if some small space were not given to the type of town which developed in the north of Spain. Here, from the eighth to the end of the fifteenth century, the Christian Spaniards were trying to drive out the Moslem Moors. As mentioned earlier this period of constant warfare is suggested by the name of the northern Provinces, Old and New

'Castile'. These names indicate the prevalence of *castles* of defence throughout the large region between the mountains of Asturias on the north and Madrid in the centre of Spain. Segovia and Toledo are two of these towns where castles played a large part in the expulsion of the Moors.

The author's acquaintance, however, is confined to Avila and Burgos, both of which are in the province of Castile (Fig. 68). On the northern slopes of the mountains near Madrid the rivers flow north to join the Douro. A promontory approaches close to the little river Adaja, and this formed the site of the Roman town of 'Avela'. On three sides this promontory was rather steep, and was only easily approached on the east. Its position to the south of the Christian part of Spain, meant that it was under attack by Moors or Christians for several centuries. In 1090 Alphonso VI surrounded it with the remarkable walls, which remain

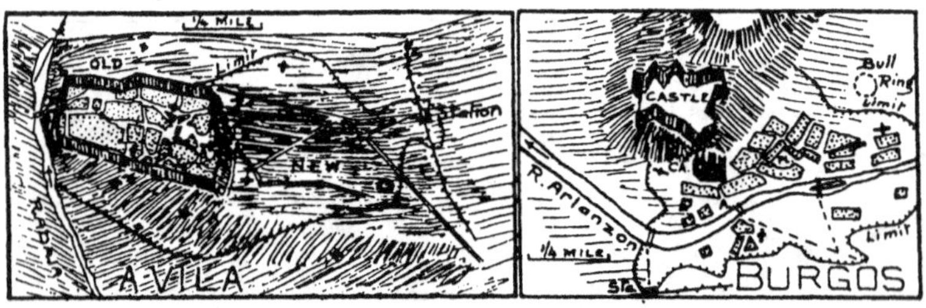

FIG. 68. Site-Views of two famous fortified cities in Castile, Spain. Avila is a walled city, whose walls are well-preserved; Burgos has a huge castle and Gothic cathedral. Dotted areas are blocks of old buildings. (*Ca.*=Cathedral; *Pl.*=Plaza; *A.*=Arch.)

without much alteration to the present day. It flourished during the Middle Ages; but, like many other towns of Spain, its prosperity waned with the expulsion of the Moriscoes about 1610.

The site-view given in Fig. 68 shows the general situation of the older and newer sections of Avila. The early portion is naturally the more defensible, overhanging the rather shallow stream in the west. The newer portions of the town fill up much of the area between the walls and the railway station to the east of the walled town. A prominent feature in the walls, which are about thirty feet high, is the number of towers, of which there are eighty-six. The ancient gates are also in a very good state of preservation. The old cathedral on the east of the city is actually built into the wall, so that it forms part of the defence system. My visit occurred in the height of summer, when the interior of Spain is very dry and dusty. The river was naturally very low, and the reverse of sparkling in appearance. Some irrigation on a small scale is practised in this rather treeless arid region.

Within the old city are rather narrow irregular streets leading to small plazas. The famous Roman Catholic, Saint Theresa, lived in Avila, and there are a number of shrines in her honour. Considering the small size of the town (about 4,000 people), there are a great number of large churches and other religious buildings. Some of these are denoted by the heavy crosses in the sketch. To the student of history the numerous relics of the Keltiberians (the predecessors of the Romans) in the local museum will be found of much interest.

Burgos which lies about 120 miles to the north of Avila is a much larger and more modern type of city. Its site was determined by a crag on the north bank of the Arlanzon, a river which flows into the Douro from the north. Here about A.D. 900 a castle was built, which became of great importance during the early struggles against the Moors, as well as in later wars. However it is much dilapidated now. During the 11th century Burgos was the capital of the Counts of Castile, and here about this time flourished the famous condottiere Ruy Diaz (the 'Cid').

The town grew up under the protection of the huge castle, and just below it in 1221 the foundations of one of the finest of Gothic cathedrals were laid down. Along the river near the Cathedral is the famous promenade, the Espolon, with a remarkable Arch (A. in Fig. 68) erected about 1550. There seems to be no specific plan about the city of Burgos, which consists of the usual rather narrow streets alternating with irregular plazas, common in most Spanish cities. The castle, though comparable in size with the Old Town at Avila, has been abandoned as a dwelling place.

CHAPTER X

THE TRANSITION FROM MEDIEVAL TO MODERN

The Origins of the University and Town of Cambridge

PROBABLY the associations of medieval times which have changed less than most, are those connected with the two great Universities of Oxford and Cambridge. Many of the colleges at these great institutions still preserve the buildings, and, to a small degree, the way of life of later medieval times. We may well devote some time to a consideration of one of these ancient University towns. Since the author spent four years at Cambridge it is natural that he should choose the eastern example; especially as Cambridge is in a sense the more typical University town, since it is smaller and much less commercialized than Oxford.

Cambridge has a very interesting situation in the east centre of England. The great chalk cuesta runs immediately to the east of the town, and here is called the Gog Magog Hills (Fig. 69). Indeed some outliers of chalk build up the little hills shown on the map immediately to the west of the town, though most of the town of Cambridge is built on the softer Gault—and so is in lowland country. In ancient days thick woods grew on the eastern slopes of the chalk ridge, though the summit was for the most part pasture. Hence one of the great corridors of early travel ran along the top of this chalk cuesta from Royston to the Norfolk area. This famous route was called the 'Icknield Way', from the tribe of Iceni who lived to the east in East Anglia.

Just to the west of this chalk cuesta the land was very low, and probably in Neolithic times was invaded by the sea. In early historic times it formed a very extensive area of swamps, which were known as the Fens. The Ouse flowed sluggishly to the north-east through the Fens, and finally reached the sea at Wisbeach (i.e., Ouse-beach). Another great route across England linked Colchester to Chester in Roman times, and this was called the *Via Devana* (i.e., road to the Dee). This road naturally kept to the west of the great Fens, and it came nearest to them at the site of Cambridge. Here it is believed the Romans had a strong post, perhaps at Grantchester, but no certain relics of it remain.

Of these early days the most remarkable remains are the five great dykes or ramparts which defend the slopes between the cuesta ridge and the fens below. These are shown on Fig. 69, and are named (from east to west) Devil's Dyke, Fleam Dyke, Worsted Dyke, Brent Dyke and Brant Dyke. From the position of the ditches, always on the west of the rampart, we may be sure that they were constructed by the British tribe of Iceni to defend their country from invaders coming from the west. Fleam Dyke reaches the river close to Cambridge at a little village called Fen Ditton, i.e., 'The end of the Ditch at the Fen'.

To the north of Cambridge an island projected above the general level of the fen, known as Ely. In this solitude a religious community settled in the seventh century, and later Edgar built an Abbey there. It was linked to the dry land by the two causeways (c.) shown in the

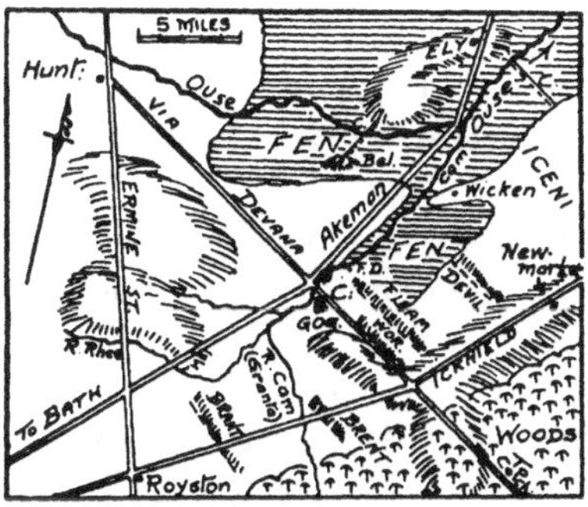

FIG. 69. The topography around Cambridge (black triangle), which is placed at the margin of the Fenlands where ancient roads cross.

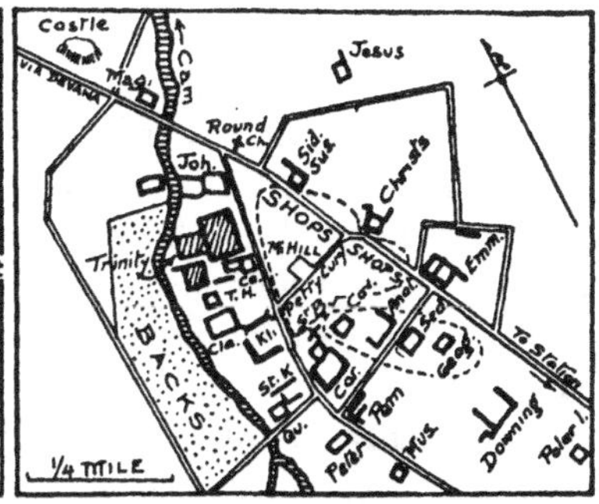

FIG. 70. Plan of Cambridge showing the seventeen colleges. The commercial area and the science area are enclosed in broken lines.

sketch map. In modern times almost all the fenland has been drained and turned into prosperous farms or pastures. However near Wicken a small area of the original fen has been preserved as a sanctuary for the interesting fenland flora and fauna. Well to the west of Cambridge, through Royston and Huntingdon, runs another of the great Roman roads, which was known as Ermine Street (Fig. 69).

It is worth noting that the original name of the first English settlement here seems to have been 'Grantabrygge', meaning the 'quay on the Granta'. It seems likely that 'Cambridge' is a softening of this word, perhaps by Norman tongues. If this be so it does not mean the 'bridge over the Cam', as is usually supposed.

The history of the town of Cambridge goes far back into antiquity. Four circular camps tell us that Britons frequented the vicinity. Detached villas were characteristic of the shire in Roman times, and pavements from these are preserved in the local museums. The Anglo-Saxon invaders entered England in large numbers by way of the Wash and the Ouse, but they did not occupy the earlier villages and manors, but in many cases destroyed them. Anna, the king of East Anglia, was a Christian, and he died fighting the Mercians at the Dykes near Cambridge. His daughter, Etheldreda, founded an abbey on the Isle of Ely in 673. In 905 Alfred defeated the Danes at the Dykes, but later Canute, in 1010, won a victory here for the Danes. Finally, mention must be made of the last stand of the Saxons under Hereward the Wake, which took place on the Isle of Ely against William the Conqueror.

The University of Cambridge had a somewhat complex origin. Various religious foundations, some linked with the larger institution at Ely, were established in Cambridge, such as Barnwell Priory (1092), and St. Rhadegund's Nunnery (1150). The latter is now the site of Jesus College. About this time there started one of the great annual Fairs of England which was held at Stourbridge nearby; and travelling scholars gave 'lectures' at these Fairs.

Oxford antedated Cambridge, and students from the older University moved to the fenland centre early in the 13th century; and were attached to one or other of the institutions on the Cam. Various hostels were founded towards the close of the century, partly with the help of Hugh Balsham of Ely. Finally, in 1284, this munificent churchman founded Peterhouse (Fig. 70), which is the earliest of the seventeen colleges of Cambridge.

The colleges of Cambridge were founded at various periods throughout the centuries since 1284. However they fall into five groups as indicated in the following table.

FOUNDATION OF CAMBRIDGE COLLEGES

I	II	III	IV	V
1284 Peterhouse	1326 Clare	1441 King's	1542 Magdalen	1800 Downing
	1346 Pembroke	1447 Queens'	1546 Trinity	1873 Girton
	1350 Trinity H.	1475 St. Cath.	1546 Caius	1873 Newnham
	1352 Corpus	1479 Jesus	1583 Emmanuel	
		1505 Christ's	1589 Sid. Sussex	
		1510 John's		

In a useful little book by my old professor, T. McKenny Hughes, we find a very good description of the typical college buildings.[1] The College plan and that of the medieval country house were essentially

[1]*Cambridgeshire*, 1909, Cambridge University Press.

the same; both having the great hall as the nucleus, generally on one side of an open court, with the gate house on the opposite side; and in the case of the college with the chapel on the third side. The hall, which occupied the whole height of the building, was entered through the court by a low passage called the 'screens', with the hall opening on one side of it; the gallery open to the hall over it, and the kitchen and buttery opening on the other side. At the further end of the hall was the dais, raised a foot or so above the rest of the floor: it was lighted at one end, or both, by a bay window; and on this dais in private houses, the family took their meals, as do the Dons (i.e., tutors, etc.) in College now.

Let us now study the way in which the colleges are arranged in the town to-day. There is no definite plan, such as we find, for instance, in a modern University like that in the south of Chicago. If we approach Cambridge from the north-west along the Via Devana (now called the Huntingdon Road) we see, as we reach the Cam, the great mound, on which the Normans erected a Castle about 1080. Nothing of this remains, but a short distance to the east is the outlying college of Magdalen, the sole college with most of its offices on the west bank of the Cam. The entrance gateway opens into a single quadrangle, containing the hall, chapel and library. In the latter is the famous diary of Samuel Pepys.

We now cross the Cam, which is so narrow a stream that the boat races are determined by 'bumps'. (One boat follows the other at a definite interval, and strives to bump, i.e., touch its leader). A few hundred yards further on we see one of the few Round Temple churches to be found in England, which was built as early as 1101. Here the roads branch, Sidney Street leading to the station and King's Parade paralleling the river Cam. All the colleges but one are found on these two roads. The more interesting are to be seen along the western of the two roads. First comes John's, which is one of the larger colleges; and consists of four courts, entered from the street by the finest gateway in Cambridge. Next comes Trinity, by far the largest of all, with five courts. Caius (*pron.* Keys) adjoins Trinity, and behind it is tucked away the small college of Trinity Hall.

Just to the east is the centre of the city, a flat open square which has the curious name of 'Market *Hill*'. From it a narrow street leads east to Sidney Street with the name of Petty Cury. This is said to be a corruption of 'Petit Quatre'; and contains some of the chief shops of the town. Close by is the Senate House, and to the south the ancient Saxon Church of St. Benets, one of the few surviving in England. Clare College is set back near Trinity Hall, and contains but a single court; and is said to be the most elegant building in Cambridge. However the best known is the lovely chapel attached to King's College, which is just to the south on King's Parade. It was built between 1446 and 1515. It has been described as 'one of the rarest fabrics in Christendom'.

St. Catherine's, Queens' and Peterhouse are rather small colleges on the west side of the street. All of these so far named have the Cam flowing at their backs; and across the narrow stream are gardens and lawns—one attached to each college, and each about 200 yards wide—which constitute the famous 'Backs'. Corpus Christi College and Pembroke on the east side of King's Parade complete the series of colleges along this road. A little further to the south, however, are two important buildings, the Museum and the University Press.

We may now pass down Downing Street to the east and reach the other main street, On the way we traverse the main scientific quarter of the University. Here is the famous 'Cavendish Lab.' where Thomson and Rutherford worked on the electron and the atom. Across the way is the Sedgwick Museum, one of the greatest centres in Britain of geological research. In the vicinity is the largest geographical institute in the Empire. We see the sole Puritan college, as we enter the main street, founded by Mildmay in the time of the Commonwealth, and named Emmanuel College. Some of the buildings formed part of a Dominican institution. Farther to the north-west are Christ's College, and Sidney Sussex, the latter occupying the site of a Franciscan monastery. One quarter of a mile to the east is Jesus College, where many of the buildings of the ancient nunnery are incorporated in the present institution. Isolated from the rest in the south-east is the last of the men's colleges, which was founded by Sir George Downing two hundred years later than any of the others. Girton and Newnham, the two colleges for women, were founded in 1873.

We may sum up this survey by stating that no definite plan controls the buildings dealing with any of the functions of the University. The colleges grew up often on lands owned by religious orders in quite a haphazard arrangement. The University offices are scattered along King's Parade for the most part. The scientific laboratories are clustered along Downing Street; or on the large areas formerly belonging to Downing College. The shops are mostly in the triangle surrounding Market Hill, as shown on the plan in Fig. 70.

Later Medieval Towns

In the period of the Renaissance many social changes developed which had a marked effect on the plans of the towns of the time. In early medieval times the church was very powerful, but as time elapsed this factor—to use Mumford's phrase—became recessive, and later was present merely as a survival in the evolution of cities. Two other factors also operated; the Court and the improvement in artillery. Both of these played a part in the changing pattern of the towns.

In the earlier medieval towns as we have seen, the Cathedral often occupied the central position, usually alongside the market-place. The town walls were so strong that the feeble artillery of the day made little impression on them. But as the artillery became stronger, and after cast-iron shot was invented in 1480, the old straight walls offered little pro-

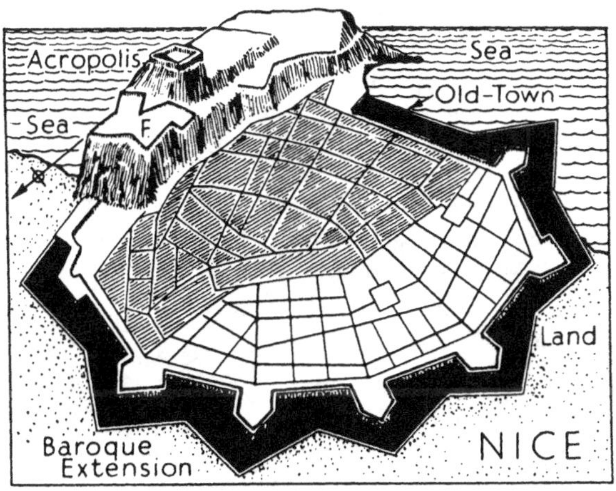

FIG. 71. A plan for converting Nice into a fortified 'Baroque' town, which was not completed. The modern city extends far to the west. (*After Mumford.*)

tection to powerful cannon, and the military engineers began to develop the elaborate star-shaped walls of the 16th century. With these projecting bastions it was possible to fire at besieging artillery; while the latter, though they could attack the folk in the bastion were too far from the centre of the city to harm the bulk of the inhabitants. Later ravelins were built in front of the main walls, and ultimately the plan of the fortress became a geometrical puzzle, each portion of which could for a time resist the attacking force. Such a design, as applied to the 'acropolis town' of Nice in Southern France, is illustrated in Fig. 71. The modern town covers four times the area shown in the diagram, but little indication that the 'baroque plan' was carried out can be observed to-day.

A medieval city in Italy which has changed less than most during later centuries is Padua. This interesting city is situated on the Lombard plain about twenty-five miles to the west of Venice, and it was therefore very near the coast in early Roman times. The surrounding country consists of some of the latest deposits brought down to the sea by the rivers Brenta and the Bacchiglione. It is therefore very level and low, and settlements could rather readily be protected by surrounding them with moats. The early town of Padua (in Roman days Patavium) developed in a meander of the Bacchigilione, which acted as a moat on all sides but the south. This oval outline of the early city is clearly indicated in Fig. 72.

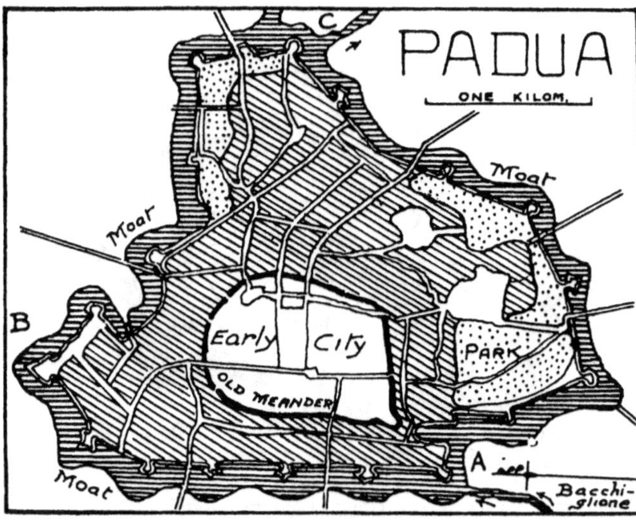

FIG. 72. A plan of Padua in the Lombard Plain, showing the inner early city and the later medieval city surrounded in 1623 by a wall with bastions and a moat. To-day the moat is partly filled, and suburbs extend in strips along the external roads.

In late medieval times the city had grown beyond the narrow confines of the early meander, and a much larger *triangular* plan (of walls, bastions, and moats) was adopted in 1623 for the protection of the city. The river entered the moat surrounding these new walls at A in the south-west corner, and was led due north to B and then to the south-east to join the normal course of the river where it finally left the larger settlement at C. Another arm of the river supplied the south-east section of the new moat. To-day water still flows round much of the older city and also along the north-east of the later city; but elsewhere the moat has been filled in, as is usually the case. In our present study the triangular outline of the later walls, and the score of bastions placed about half a kilometre apart, are the chief features of interest. They show an intermediate stage between the old straight walls, and the stellate pattern with numerous bastions as favoured by the school of Vauban. To-day Padua has a population of 139,000.

An excellent example of a fortified town of the 16th century, with typical star-shaped walls, is to be found at Mazagan. This is a port about sixty miles to the south-west of Casablanca in Morocco. A photograph will be found in the very illuminating paper by Jules Bache on page 487 in the *Geographic Review* for October, 1921.

The Baroque type of City

Mumford has introduced the useful term 'Baroque' to define the change in planning (and social organization) which developed towards the close of medieval times. This was an age of great scientific and mechanical discovery, which was reflected in the towns by the adoption of formal geometric designs. These to some extent replaced the crowded plans of the walled cities of the past. At the same time it was the age of absolute monarchs, whose general policy was to concentrate their power in one centre. Here, as an emblem of their authority, they built large and imposing palaces, and laid out extensive grounds according to the new formal designs. Possibly in some cases the town-plans followed the pattern of the royal gardens, rather than the other way round.

With the concentration of power at the capitals, the latter rapidly grew in population. London, Paris, Naples and Milan became cities of 200,000 or more. It was the age of Vauban and his bastions, which revolutionized military engineering. The monarchs collected large standing armies, and in Mumford's words 'the barracks replaced the abbey'. In Berlin it is stated that in 1740 one quarter of the inhabitants were soldiers.

One of the main features of the Baroque city is the dominance of the avenue. Wider streets became a necessity, as vehicles became larger and more numerous. A marked separation of the classes into rich and poor developed about this time, and the segregation into mansions and slums became noticeable. Straight streets were admired, and the old cul-de-sacs of medieval towns were often swept away.

Many new towns illustrate this type of design. Karlsruhe is one such city, developed about the year 1750 as the residence of the Grand Duke of Baden. The streets are laid out in the form of a fan radiating from the ducal castle (Fig. 73). The town blocks occupy rather less than one half of this Renaissance or, as Mumford prefers to call it, Baroque plan, which indeed copies a wheel rather than a fan. There are no less than thirty-six streets radiating from the palace area, which, however, do not meet at a centre, so that there is not the usual difficulty of crossings.

In the Baroque plan for the ducal town of Karlsruhe, the administrative quarters for the control of the state of Baden were housed in fine buildings around the plaza fronting the Palace. Here also were Museums and Art Galleries, as well as the Departments of Finance, Justice, and Mining. In modern times Karlsruhe has spread far beyond the area shown in our diagram. As usual the buildings tend to cluster in suburbs along the main outlets of the original town. Muhlberg lies due west of the main plaza,

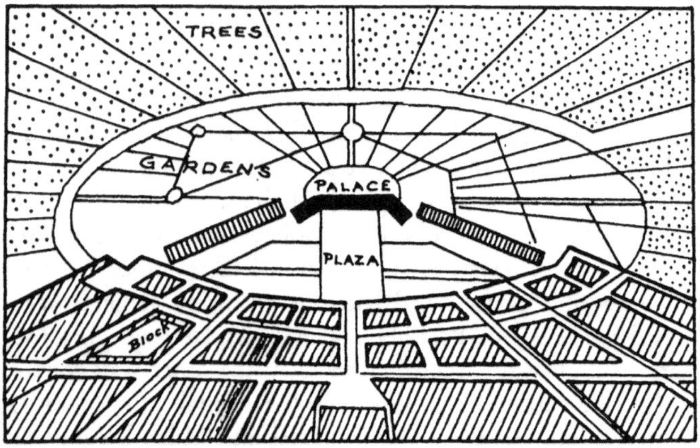

FIG. 73. The baroque design of the 'court' town of Karlsruhe in S.W. Germany. Thirty-six avenues radiate from the royal palace.

while Beiertheim is to the south-west. Also as usual, the radial pattern is forgotten, and the streets are arranged in the usual checkerboard. However it must be remembered that there is no centre in a *suburb* of such immediate importance to every citizen as is the case in the centre of the *whole city*. Hence this absence of radial streets in the suburbs is not such a disadvantage as one might assume. To quote Lanchester 'In the park the radial lay-out stretches for a considerable distance, but as this extent would become monotonous in a town it is cut through by a broad main street, beyond which, though the radials continue, the circular roads are omitted and straight ones at varying lengths are substituted'. It is obvious that a wheel plan of this complexity can only be developed on a level surface, and indeed it carries the geometrical idea too far to be practical, and has not been used in other places.

Bernini in Rome had previously used the plan of main avenues leading to a central plaza, often on a somewhat radiating pattern. This technique was followed in France, especially under Louis XIV, where Le Notre laid out the gardens of Versailles; and where later the Place de la Concorde embodied the same ideas as part of the improvement of a great city. In Germany many other towns adopted such patterns, as at Potsdam, Mannheim, Munich, etc. Indeed as we shall see it is still often adopted, and is a salient feature of such American cities as the Washington of L'Enfant (1791), and in modern times of the improved plan of Detroit.

Versailles and Washington D.C.

These two settlements, so important from the point of view of city evolution, were both designed long before town-planning was a recognized profession, and yet they exhibit many of the ideas which are fundamental in any modern system of town planning, and which many readers may have thought were the creation of the last fifty years. Versailles goes back to 1668, while the plan of Washington was charted in 1791.

In both cases the position of the main buildings was decided before the rest of the plan was drawn up. The concepts of fine vistas and of axes of design are exemplified in both. The rather formal geometric pattern—which is characteristic of the transition period between the medieval and modern periods—is very obvious. Circles where the radiating roads intersect are inherent in both designs; though it must be remembered that Versailles was essentially a garden plan, while Washington was laid out with a view to housing thousands of prosperous citizens at the nation's capital.

Versailles is situated about ten miles to the south-west of Paris (Fig. 77 at A) on the north slope of the Bievre stream. A hunting château was built here in 1624, but it was due to Louis XIV that the magnificent palace and gardens were laid out, mostly between the years 1668 and 1682. Most of the palace was built to the design of Mansart, and could house 10,000 persons, and the main façade is 635 yards long. The gardens were laid out by Le Notre, and are two miles long and nearly as wide. The map given in Fig. 74 shows that the main axis of the design runs nearly west-east, and is at right angles to the façade of the palace. A large cross-shaped pool—a mile long—occupies much of the axis, and near the western end is a circle, where eight routes meet, which is called the *Etoile*.

Other pools and avenues cross the marvellous gardens, and further details can be gathered from the plan. It is stated that twenty million pounds were spent during the construction of Versailles. At first there were only a few houses in the vicinity; but soon a town sprang up, mainly to the east of the palace. In 1802 it gave its name to a bishopric. The chief built-in area of to-day is indicated by the broken diagonal lines in Fig 74.

It is clear that the aims of Le Notre at Versailles and of L'Enfant at Washington were very different; yet it is unlikely that the later plan owes nothing to the gardens of Versailles. We are told, in the description by T. Adams, that Jefferson had studied European cities; and brought back to President Washington plans of Karlsruhe, Paris, Turin, Milan, and many others which had interested him on his travels.

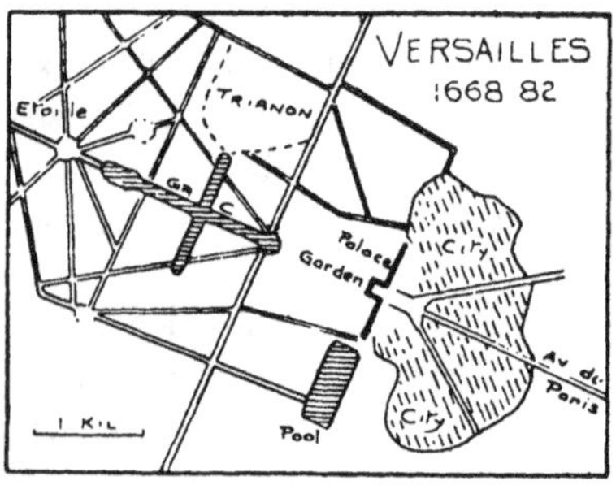

FIG. 74. The baroque plan of the palace at Versailles (1682).

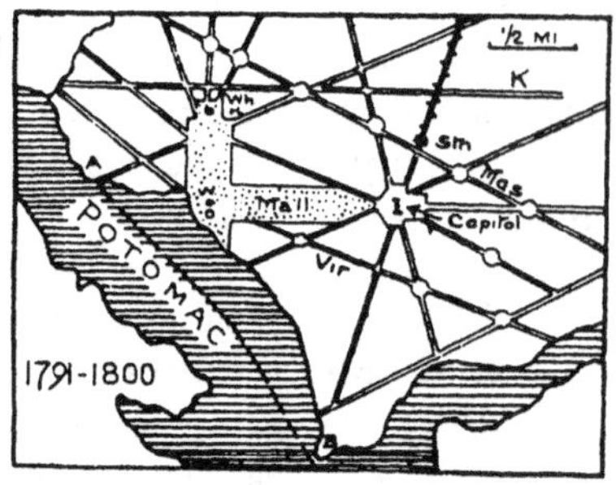

FIG. 75. The somewhat similar plan used by L'Enfant for Washington, D.C. (1800).

It is believed that Major Pierre L'Enfant first laid out the rectangular streets (which to-day fill every portion of the major blocks), and on these superimposed his diagonals, and skilfully adjusted them to the configuration of the ground. He developed them as approaches to strategic sites already selected for monumental buildings. The effect—as Adams points out—was to subordinate completely the rectangular to the radial part of the plan.

Major Ellicot completed the plan, altering some details of L'Enfant's masterpiece. The grand avenues were planned 160 feet wide, with lesser widths for less important streets. The main axis extends west from the Capitol along the Mall through the Washington Obelisk. At right angles to this vista is another, also passing through the Obelisk, which shows the White House at its northern end. Many of the finest buildings in the United States have been erected along the Mall, but they are not shown in the diagram. (Fig. 75)

Of course some modifications have been made in the last 150 years. For instance, a good deal of the Potomac river has been filled in—as far as the line marked A, B, in our diagram. The coming of railways did not occasion much alteration, and the rails were laid along one of the diagonals coming into the capital from the north-east. Automobile traffic has been a much more serious problem, and leads to some congestion at the numerous 'circles'. However no plan created over a century ago could have made provision for the vast city traffic of to-day; and all authorities acclaim the main features of L'Enfant's design.[1] We shall see elements of his plan used in most attempts to improve the lay-out of the cities of to-day.

Edinburgh—an epitome of Town Evolution

Few cities in the British Empire illustrate the various stages of city development better than Edinburgh, the capital of Scotland. Its position is shown in the first map in Fig. 76. Here we see the Firth of Forth cutting across the narrow 'waist' of Scotland, with a short crossing possible at Queensferry. The lowland road reached the coast at the mouth of the Esk, where there was a camp in Roman times. About four miles from the coast was a prominent crag rising 450 feet above sea level, which was obviously destined to be the acropolis of a primitive settlement. It is a volcanic plug of Carboniferous age. Probably there was a British stronghold on this crag, which later was crowned by the famous Castle of Edinburgh. To the north-west of this crag flowed a small river called Water of Leith, which reached the sea at Leith. This port has always served Edinburgh, and to-day the two places form one 'conurbation' as suggested by the dotted area in the first map. A much higher (822 feet) and more massive mount lies a mile or so to the east of the Castle Crag, which later received the name of Arthur's Seat. Its steep crags on the west are called the Salisbury Crags, and these appear in several of our maps. It was not so well suited for an acropolis, nor were the Calton Hills (355 feet) used for this purpose either.

In glacial times the ice seems to have moved from the west in this district, and as it passed around or over the Castle Crag it tore off much debris, and dumped it in the lee, i.e., to the east. Possibly it slightly eroded two shallow valleys, running alongside the ridge of debris (Fig. 76 at B). This ridge determined the site of High Street, and gave its

[1]For interesting maps of U.S.A. cities about 1800, consult *Mirror for Americans*, R. H. Brown, Amer. Geog. Soc., N. York, 1943; also Plates 156–9 in *Atlas of Historical Geography of U.S.*, C. O. Paullin, Amer. Geog. Soc., N. York, 1932.

main axis to Edinburgh. The northern shallow valley (150 feet) contained the Norloch until modern times, while the southern one was a grassy hollow, still known as the Grassmarket. Another parallel gully to the south was the swampy Borough-Loch (Fig. 76 at D).

Before the Romans arrived this region was controlled by a British tribe, the Brigantes. They seem to have had settlements in the vicinity of the Arthur's Seat, and to the north-west of the castle crag was a small chapel in Christian times in honour of St. Cuthbert. The Romans do not seem to have made this site one of their main settlements. The history really begins with Edwin of Northumbria, who built a fort about

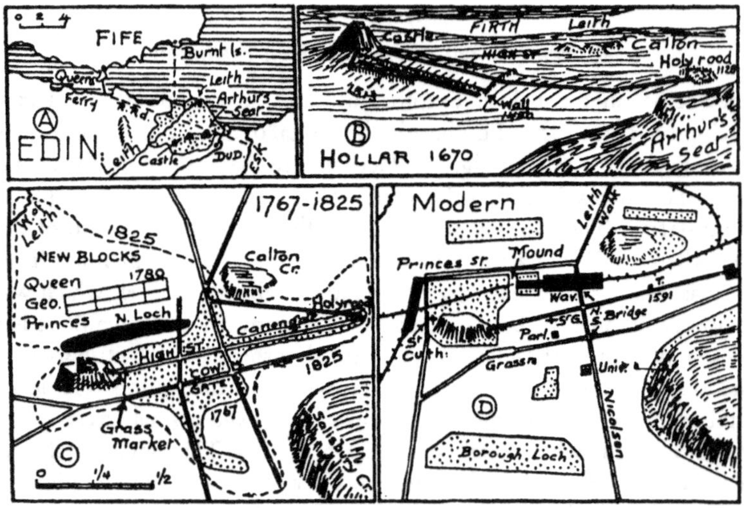

FIG. 76. Diagrams illustrating the evolution of Edinburgh. In *A* the present city is shown dotted. *B* is from a contemporary sketch. In *C* the relation of old town (dotted) to the new plan is shown. In *D* are the railway approaches of to-day, and parks are shown dotted.

620 on the crag, which was known as Edwin's Burg, or in the Welsh Dun Edin. A palace was in use on the crag during the times of Malcolm Canmore about 1070, and a settlement of Angles and other southerners grew up along the road to the castle. This naturally ran along the ridge mentioned previously, and has been known as High Street ever since. In 1128 an abbey was built at Holyrood by David, about a mile from the castle to the east (Fig. 76 at B), and around it grew up the suburb of Canongate.

In 1450 the important part of the settlement was enclosed in a defensive wall. This included much of High Street and the adjacent slopes to the south. Owing to the troubled times folk flocked within the shelter of the walls, and space was at a premium. This led to the building of high narrow houses, and the cross streets or 'wynds' were very narrow lanes. This feature is still characteristic of the older parts of Edinburgh. After the disaster of Flodden in 1513 an outer wall, including the Cowgate, was built to extend the south-west defences of the city. In 1436 Edinburgh was made the capital of Scotland (replacing Perth), and the city prospered for two centuries. But in 1603 James left Scotland to ascend the throne of England, and much of Edinburgh's importance disappeared.

Not much change occurred during the 17th century. The Holyrood of to-day dates mostly from about 1670, when the old abbey was replaced by the modern palace buildings. By this time the city extended as far as Holyrood and so included Canongate. The Tolbooth, built in 1591, still survives. Towards the end of the next century, a great change began in the city plan. The limits of the town are shown by the dotted area in Fig. 76 at C. A prominent feature at this time was the Norloch, a small lake about half a mile long to the north of High Street.

The Old Town includes the streets already described formerly included within the walls, as well as Canongate to the east. About 1768 a new period of planning set in, and the fine streets to the north (Princes, George and Queens) were laid out beyond the Norloch. Many fine squares and blocks of buildings were planned for this section of the city, extending north of Queen Street and surrounding Calton Hill. The coming of the railways greatly affected these æsthetic plans, and in some cases—as in the area north of Calton Hill—spoilt the fine plans of the architects of the day.

However, the railway was brought into the heart of the city by draining the Norloch, giving a sort of sunken valley down which the railway runs to the Waverley Station near Calton Hill (Fig. 76 at D). By 1829 a great mound filled much of the east end of the former Norloch, and on this has been erected the beautiful National Gallery. Most of the Norloch has been converted into the Princes Street Gardens, and high-level bridges (shown in Fig. 76 at D) carry Nicolson Street across the one-time gullies of Cowgate and the Norloch.

All students of town evolution should study Geddes' pamphlet *The Civic Survey of Edinburgh* (Edinburgh, 1911), especially the map of the city as it was in 1829. The pattern of the 'New Town' between the sunken gardens of the former Norloch and the gully of Water of Leith, is an interesting mixture of open spaces, circular gardens, rectangular and diagonal streets, which might have been planned to-day, after a century of town-planning progress. But there were some faults in this early praiseworthy design, which Geddes proceeds to enumerate. First the planners omitted adequate consideration of relief and contour, and so their schemes broke down where crags and marshes intervened. They had made no social survey, and 'assumed a practically indefinitely increasing population of the well-to-do . . . and forgot to provide for the workmen's homes. Thirdly they omitted any provision for anything so vulgar as workshops; and consequently the beauty for which they laboured was in many places destroyed by the filling up of every vacant space with irregular factory and workshop.' Two of the views in Geddes' survey are of special interest. They show the close resemblance between the site and the plan of classical Athens and those of modern Edinburgh. The acropolis, the distant port, the surrounding hills, the conflict between well-planned town and non-planned town are present in both cases.

To sum up our survey of Edinburgh, we see its beginnings before Roman times. It does not offer us a Roman castrum, as do some other of our examples. But the control of the medieval site by the crag and the ridge is very definite. The building of the walls, and their passing, is also characteristic of most medieval cities. The chief interest is in the development of a Renaissance plan well before 1800; which has little of the baroque, but more of the modern spirit than is exhibited by most of the towns so far considered. Finally Edinburgh offers good and bad illustrations of the way in which modern railways can be brought into the heart of a modern city. The approach along sunken gardens is good, the spoiling of the renaissance design to the north of Calton Hill (not clearly shown in our sketch map) is not to be commended, though perhaps it was almost unavoidable.

Paris: a study of Bulwarks and Boulevards

A glance at the structural map of north-west Europe shows us that the dominant feature is the hour-glass shaped Paris-London Basin; composed of 'saucers' of rather late geological formations. These have two independent centres, one in England near to London, and the other in north France near Paris. Lying just to the south-west of the London Basin is the incomplete Southampton Basin. The centre of this structural feature is the great port of Southampton. So also there is a similarly incomplete basin to the south-west of Paris, at whose centre has developed the rival French centre of Bordeaux. There can be no doubt that geological structure has played a great part in the evolution of these four cities: Paris, London, Southampton and Bordeaux. (See Fig. 89 in the companion volume *Environment and Nation*.)

The uppermost and youngest of the 'saucers' near Paris consists of Eocene (Early Tertiary) strata, which are necessarily almost level-bedded.

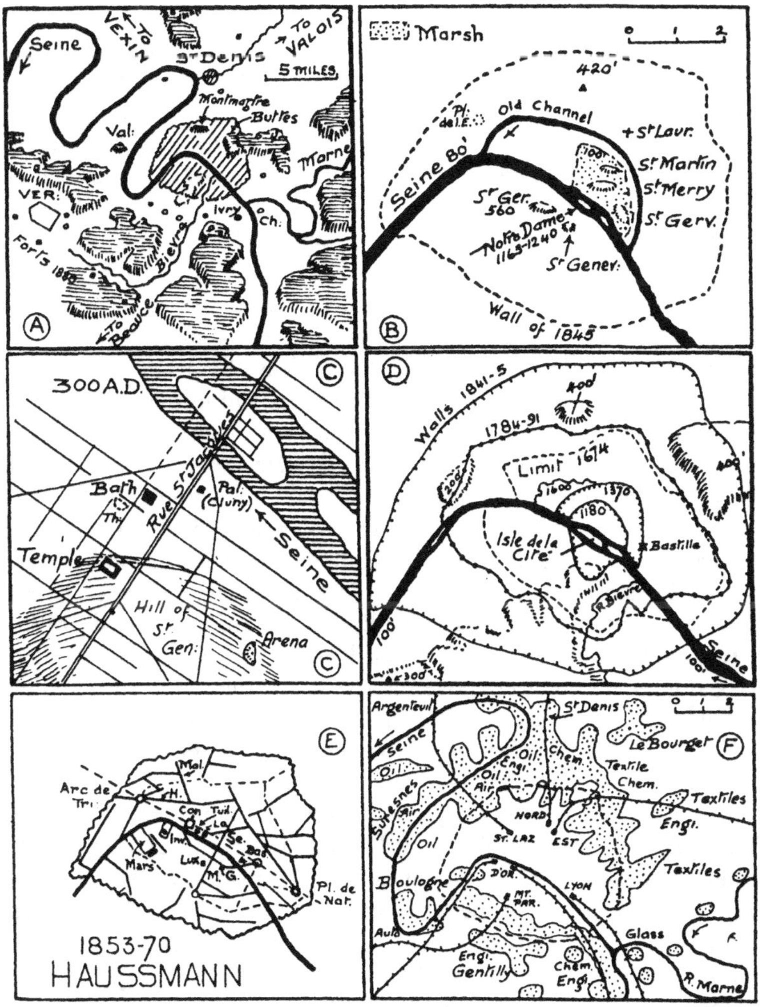

FIG. 77. The evolution of Paris. At *A* the low buttes are mapped, at *B* the old meander and early churches, at *C* Roman Paris, at *D* the five concentric walls, at *E* the new Boulevards due to Haussmann, at *F* the modern industries on the margins of Paris.

Into these Eocene beds the Seine has cut its valley to a depth of two or three hundred feet in the vicinity of Paris. The Seine hereabouts is just 100 feet above sea level, and still has to flow over 200 miles to reach the sea. The Marne joins the Seine just to the east of Paris; and to the south the great bend of the Loire brings the waters of the latter river only fifty miles away. The two rivers are linked by the valley of the Essonne, which reaches the Seine at Corbeil. (Fig. 77 at A).

These rivers break the upper 'saucer' into several low plateaux; such as Brie to the east, Valois to the north-east, Vexin to the north-west, and Beauce to the south. The upper layers are usually travertine or limestone, covering layers of clay and gypsum-bearing beds. Underlying the Eocene are the upper beds of the Cretaceous, which bound the Eocene 'saucer' about fifty miles to the east. The Seine, after flowing past Brie, commences to follow those remarkable 'incised meanders' which are the dominant feature of the lower Seine.

Just where the little stream, the Bievre, enters the Seine from the south, several elongated islands developed in the winding river. One of these became the Isle de la Cité, and the nucleus of Paris. To the north the Seine formerly made a much wider sweep, and in the early days of the city an abandoned meander flanked by wide marshes on the west, was a feature of this district. These marshes are clearly shown (at B in Fig. 77) by the dotted area in the map. The alluvial deposits and the clays and sands of the basin gave large crops of grain. The limestone and travertines were quarried extensively from the earliest days, and furnished the building stone for much of Paris. In Map A, Fig. 77, these plateau-relics of the hard upper layers of the Eocene 'saucer' are charted. Montmartre and the Buttes on the north bank are actually included in modern Paris, rising over 300 feet above the river. On the south bank the hills are lower, and that named after St. Genevieve is only about fifty feet above the Seine. However these low hills played a very important part in the early lay-out of the city.

There was a small village in pre-Roman times on or near the Isle de la Cité (Map D), but the local tribe of Parisii followed the leadership of the Sennones at Sens. They fought against Roman domination for a time, and were decisively defeated at a battle near Paris. The Roman city, however, was known as Lutetia, and was not of great importance in the early years of Imperial Rome. A large temple to Jupiter was erected on the Isle de la Cité, and gradually a town developed, mostly on the south bank.[1] The Roman settlement is sketched at Map C, and grew up on the low Hill of St. Genevieve and on the slopes north of this hill. Christianity is supposed to have reached Paris about 250, and later Emperors, such as Constantius and Julian, favoured the French city as a place of residence. The Roman baths near the Hotel de Cluny probably

[1] *Manuel d'Histoire de Paris*, H. Lemoine, Paris, n.d.

date from Julian's time. The size of the amphitheatre suggests a Roman population of about 20,000. During the closing years of the Empire Syagrius made Soissons his capital; and it was not till 497, when Clovis the Frank conquered the country, that Paris surrendered to him (on the advice of St. Genevieve) after he had become a Christian.

In Merovingian times many new churches were built near the banks of the Seine. The monks naturally chose the small hills which rise above the marshes here, and some of these early churches are indicated on the map at B. St. Martin, St. Merry, St. Gervaise and St. Germaine were all built in the early days of Christian Paris. In 1163 the most famous building in Paris was begun, when the Cathedral of Notre Dame was erected, according to the new Gothic principles now beginning to dominate architecture.

Around 1200 the city progressed greatly under the rule of Philip Augustus. He laid out the first buildings of the Louvre (Lo. in map E), and also inaugurated the famous University of Paris. This formed a group of buildings on the south bank on the slopes of the Hill of St. Genevieve, where there were already a number of monastic institutions. Commerce tended to keep more to the north bank, and here markets and river ports developed in the vicinity of the (later) Bastille. Philip erected the first massive walls around 1180 (Map D), and he inaugurated street-paving, and the lighting of the town at night. It is to be noticed that the walls included much of the town on the south of the Seine, but many of the monasteries were left outside.

The second wall was built around 1370 by Charles V, and he erected the Bastille as a special fort to guard the gate on the east side of the new city wall (Map D). He did not add to the defences of the south portion of the city. The Louvre was greatly extended after 1550, and Henry IV built the New Bridge across the Seine near the Isle de la Cité, and this is now the oldest of the thirty-one famous bridges of Paris. In the time of Louis XIII the Quarter of the Marais (Marsh) was drained and occupied by houses, and the island east of Isle de la Cité was also covered with dwellings.

In the time of Louis XIV the old fortifications were in part removed, and the space used for boulevards connecting the Bastille with the north-west of the old city. Many of the open spaces, such as the Place Vendôme, date from his reign, and the Hotel des Invalides was erected about 1670. The Place de la Concorde belongs to the reign of Louis XV, and around 1785 many important changes took place. Thus a 'Customs Wall' was erected (1784-91), which is clearly shown in Map D. It extended from the Place de la Nation in the east to the famous Champs-

Élysées avenue in the west. The new wall enclosed many new suburbs on the north bank; and, extending across the Seine, took in considerable areas of what was at that time open country. Once more, older walls were torn down, and fine boulevards and buildings—such as the Opera House and many theatres—took their place.

During the 19th century, with the coming of the Industrial Revolution, the city approached the million mark. In 1841 the last great walls were built, and they enclosed outlying satellites such as Belleville and Montmartre in the north. This increased the population of the city to 1,700,000. Various canals were constructed, and soon the railway era began. The main lines of railway entering Paris are shown on Map F.

Perhaps the feature in the development of Paris which is best known is the improvement due to the planning of Haussmann between 1853 and 1870. However, it is incorrect to suppose that all the fine boulevards and 'Places' are due to his efforts; and indeed the present account should prevent that error. Already there were certain significant axes of development, as for instance the Champs-Élysées; and many fine boulevards had already replaced old walls. Some of Haussmann's achievements were the Boulevard de Sebastopol (Se. in Map E), as well as St. Michel, Malesherbes (Mal.) and Haussmann (H.). Boulevard Henri Quatre linked the Bastille to the Sorbonne (4 in Map E): and the beautiful Place de l'Etoile (with the earlier Arc de Triomphe, 1805-36) is part of his work.

Haussmann spent some twenty million dollars a year on these improvements, and unfortunately added greatly to the city indebtedness. It has been stated that the disaster of the Commune at the close of his regime in part was due to this enormous expenditure. Moreover the radiating streets meeting at the various 'Places' lead to many problems of congestion. The necessary demolishing of hundreds of buildings to make way for each boulevard could hardly be carried out on such an extensive scale under our form of government. In the map at E, the heavy lines show some, but not by any means all, of the new streets which Haussmann opened across the congested slums of Paris.

The last of the series of maps in Fig. 77 at F, shows the latest developments in the city of Paris. This deals with the changes brought about by the industrial revolution, and will be considered in some detail in the next chapter. The replacement of walls by isolated outlying forts will also be discussed.

CHAPTER XI

THE MODERN CITY

THERE is a very voluminous literature dealing with modern industrial cities, but a survey of it will show that the contents are mainly concerned with sociological or architectural aspects of city life. Problems of environmental control, of functional patterns and of distribution in general do not seem to interest the authors to any large degree. These aspects however are precisely those with which the geographer concerns himself; and in the present chapter I have tried to place before the reader the findings of geographers in these fields; and to draw some conclusions which will help him to acquire some idea of the general geographical pattern of the modern industrial city.

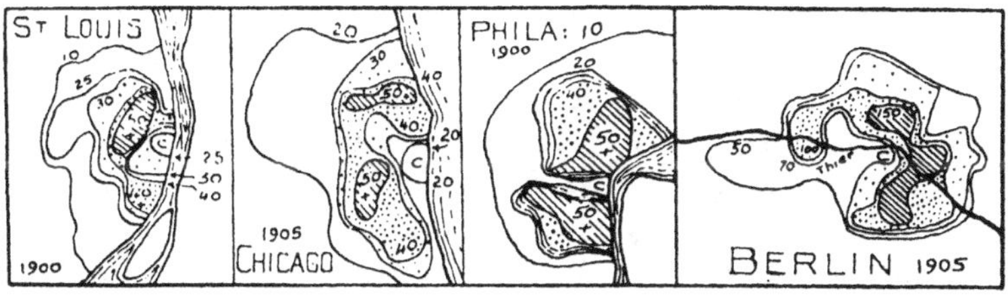

FIG. 78. Most large cities show a core (C) which is relatively empty of actual residents, with adjacent wards showing much higher densities (thousands per square mile). (*Diagrams based on an early memoir by Mark Jefferson.*)

Perhaps we owe more to the veteran geographer, Mark Jefferson, than to any other in this field. As long ago as 1909 he wrote a brochure on the Anthropogeography of some great cities,[1] which was a pioneer among *general* urban studies. It dealt largely with variations in the densities of the wards of the great cities. Many great cities show a relatively empty nucleus surrounded by a zone of wards of high density (Fig. 78). Outside this again is a zone of larger wards with much lower densities. Various great American and European cities were charted by him to show this feature, and four are redrawn in Fig. 78. In each case 'C' is the relatively empty 'core' of the city.

[1]Bulletin of the American Geographical Society, 30 pp., with 10 maps. New York, 1909.

In this chapter are presented a number of the correlations which will be needed in a complete synthesis of the features of a modern industrial city. The first sections deal with the definition of a city. Then follows a section showing some climatic controls for the greatest world cities. In order to disabuse the reader of the idea that only one or two factors need to be considered, I have summarized a fine piece of research by E. N. Torbert on the evolution of the small town of Lebanon in New Hampshire. This study .was selected because it seems complementary to Goldthwait's well-known study of a dwindling town in an adjacent township (Lyme), which is presented in brief herewith; with specially drawn block diagrams of Lebanon and Lyme, since the topographic factor is dominant in both instances.

The growth of cities and their relative importance vary a good deal from century to century, and some of the reasons for this 'change of rank' are discussed in the next section. Here again Jefferson's deductions are illuminating; as is his concept of the Primate City, and its effect on the young folk of a nation. I have summarized the latter memoir, though it may fairly be claimed as sociology as much as geography.

The pattern of the great city seems to follow certain general rules, which have been examined by Burgess of Chicago. I have described his 'Five zones' as applied to the ecology of Chicago, and summarized his stimulating discussion. As a check on the rather simple pattern offered by Burgess for Chicago, I have discussed the more complex series of zones which seems to be apparent in Toronto.

The relation of the under-privileged sections of a city to the commercial core is of some interest; especially with regard to foreign immigrants, who—in America—so often form definite 'islands' in such sections. Here again it is profitable to contrast Chicago and Toronto, especially as the writer has lived over seven years in each city.

Later chapters will of course deal with details of the growth of the modern industrial city, as modified by the dominant environmental controls. But it is felt that this introductory chapter will prepare the way for the more detailed chapters, and will give the reader an idea of the questions with which we are primarily concerned.

What constitutes a city

In the introduction to this volume it was pointed out that there is no general agreement as to the number of folk necessary to justify the name 'city'. In Melbourne, Australia, the writer found it somewhat confusing to find each of the main suburbs officially known as a city. Thus as part and parcel of the main agglomeration even in 1900 we find the City of Prahran (1879), the City of Richmond and six other 'suburban' cities; the name being given to a borough which has an annual revenue of not less than £20,000. (It is declared a 'town' if the sum is £10,000.)

Some sociologists use the density of population as a standard. Wilcox in *The Urban Community* proposes that all territory settled to a density of less than 100 per square mile be rated as *rural*; from 100 to a 1,000 be considered a *village*; and that territory having more than 1,000 persons per square mile be rated as *urban*. Mark Jefferson in his study of Great Cities suggests that over 10,000 per square mile entitles the place to be called a *City*.

In Germany there are grades among the towns which are worth noting. Thus from 2,000 to 5,000 the cluster is called a 'Landstadt'; from 5,000 to 20,000 is a 'Kleinstadt'; from 20,000 to 100,000 is a 'Mittelstadt', while over 100,000 is a 'Grossestadt'. Most of these definitions do not specify where a town turns into a city. Weichel quoted by Jefferson[1] gives an interesting table which links all the types of human industry in terms of density, as follows:

Density	Economy	Density	Economy
0–8 p. sq. mile	Hunting and fishing	256–381	Industry predominates
8–26	Grazing and forestry	381–512	Industrial suburbs
26–64	Agriculture starts	512–2560	Industrial towns
64–192	Agriculture	2560–5120	Centres of small cities
192–256	Industry starts	5120–12800	,, moderate ,,

Perhaps the best definition is based on the practice in U.S.A. cited by Gist and Halberg[2] that the larger towns are not incorporated as cities until they reach a population of 10 or 15,000.

Primary Controls determining City Sites

If we glance at any population map of the world it is clear that there are certain regions where dense populations—and therefore cities—occur, while in polar regions and in real deserts there are naturally no cities. No geographer believes that the site of a city can be explained by one or two factors only. Temperature and rainfall are quite important;

[1] *The Anthropogeography of some great cities*, Am. Geo. Soc., 41; 542.
[2] *Urban Society*, New York, 1938.

but, as we have already indicated, political importance, presence of coal or oil, facilities for ports, and many other factors must be considered to arrive at anything like a real picture of the factors which have led to the growth of a large city at a given place. However it is of considerable interest to test a few of the more obvious controls to see if any useful conclusions can be arrived at. Such a first approximation in our study is illustrated in the chart given in Fig. 79.

In this chart only two controls are considered, those of average temperature and annual rainfall. The graph is of the *isopract* type, since for any

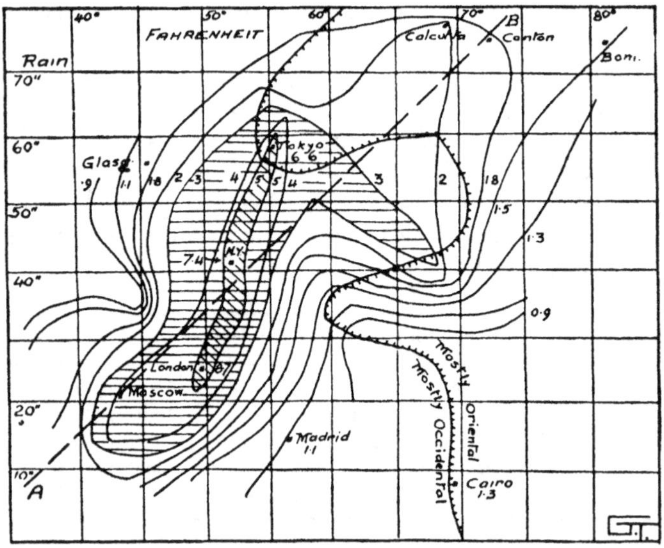

FIG. 79. CITIES, TEMPERATURE AND RAINFALL. A tentative isopract graph suggesting how great cities are controlled by temperature and rainfall. Figures on the isopleths indicate millions of city population.

place—say London—the temperature and rainfall are considered, and the position labelled with the total population. For London the temperature is 49 degrees F., and the rainfall about twenty-eight inches. Here London will be found on the chart. Attached to its position is the figure 8.7; which give the population in millions. The same thing is done for all the other cities with populations over 900,000. There are forty-six of these, with Toronto holding the last place.

Among these forty-six cities fifteen are inhabited by folk with an oriental or at any rate non-European type of culture. On the graph Canton, Calcutta, Bombay and Cairo are indicated; though most names must necessarily be omitted through lack of space. Nearly all these lie in the right half of the chart, i.e., where the temperatures are high and the rainfall (except for Cairo) rather heavy. If we ignore this aspect of the problem we see that the three greatest cities of to-day, i.e., London (8.7), New York (7.4), and Tokyo (6.6) lie in a narrow strip which is shaded in the graph. I have of course been compelled to generalize somewhat in this complicated problem, but the isopleths for four millions, three millions and two millions run approximately through the cities with these population.

It would serve no good purpose to discuss all the generalized curves exhibited by the isopleths in Fig. 79; but it seems reasonable to state that all the great cities of the world have developed very largely between temperatures of 45°F. and 70°F.; while the rainfall limits are twenty inches and seventy-five inches. Moreover there seems to be a definite *slant* to the isopleths, so that the cooler cities are distinctly drier than those in the warmer sites. (As stated earlier the latter naturally are more usual among *oriental* centres of large population.)

I have added to the graph a line AB which seems to lie about the central axis of these isopleths. There seems to be a tendency for the cities to cluster near this line AB. The straight-line equation for this line is ($T = 30+.56\ R$); where T is temperature and R rainfall. Assuming that it has some application to our data, we are in a position to state what is the normal rainfall for any city with a given temperature. For instance Toronto has a temperature of 46; and substituting in the equation we find that the 'normal' rainfall would be about thirty inches. For Calcutta, with a temperature about 70, the corresponding rainfall would be about seventy inches. In both cases these are near the actual amounts. Most cities, however, would not agree so closely with what we might call the normal relation.

The Shift from Farm to Factory: Lebanon, New Hampshire

A number of careful studies have been made of the changing factors, chiefly industrial, which determine the growth of a town or city, and one of the best is that by E. N. Torbert dealing with Lebanon, New Hampshire.[1] It should be read in conjunction with the similar study by J. W. Goldthwait on the adjacent town of Lyme.[2] (This latter paper is considered in the next section of this book.) Migration to the Lebanon region

[1] "The Evolution of Lebanon," *Geographical Review*, April, 1935.
[2] "A Town that has gone Downhill," *Geographical Review*, October, 1927.

(Fig. 82) was discouraged until about 1760, when the Indian wars were over. In the next few years many townships were thrown open for settlement along the Connecticut valley, and by 1762 five families occupied Lebanon. A crude survey of the lands seemed to show that the 'warm rich loamy soil' which supported beech and maple trees was the most advantageous for cultivation.

At first poor subsistence farming was the rule, and in 1789 the average area of 118 farms was only about fifteen acres. By 1803 there was enough grassland cleared for their cattle and sheep, and heavier crops of maize and barley were raised on the better tilled fields (Fig. 80). The pioneer stage was now over, and frame houses had supplanted log cabins. There were five crude stores, nine taverns, and three grist mills, about this time. (See Fig. 81 for the population graph.) The chief asset in the township was the Mascoma River, which gave power to the small mills. The chief mill was at Mascoma Lake at a site now called East Lebanon. Below the lake were two 'intervales' with good soils, and here other clusters developed. The middle village gave rise to the town of Lebanon of to-day.

During the first decade of the century boats navigated the Connecticut River as far north as Lebanon; and about 1800 a good turnpike road was constructed over the hills to the Merrimack valley in the east.

Dairying developed about this time, and several hundred fat cattle were driven annually to Boston or the vicinity. The change over to factories however started by 1830, when thirty folk were employed in a textile mill at a site (near the lake) which at first was called 'Lebanon City'. The village further down the river, known as 'Lebanon Center' became dominant after a fire in 1840 destroyed the mills at the 'City'.

Before 1830 the 'Center' contained a Masonic Hall and a Library, while there were eleven residents engaged in crafts or professions. After this date there was a great expansion in wool-growing, which rose to a peak about 1840, but declined rapidly soon after that date. During the forties much of the tilled land was converted into pasture for the use of the sheep. There was a temporary revival during the Civil War, but the period of successful farming was over in many cases; for there were thirty abandoned farms in 1860, and by 1892 this figure had risen to fifty-two.

In 1848 the railway reached Lebanon, and there was an important junction just across the river in Vermont, which led to the growth of West Lebanon in New Hampshire (Fig. 80). This was the period when the town definitely became industrial rather than agricultural. From 1850 to 1870 the chief factories produced iron products, such as small engines, machinery and nails. But Lebanon was too far from coal and iron ore to succeed in heavy industry; and a shift to wood-working factories occurred about 1870. Now furniture and organs were manufactured, but this section of industry also declined as the difficulty of obtaining special timbers increased. In 1886 the emphasis now lay in the textiles, for four mills of this type employed 250 workers. These were situated at Lebanon Center.

Agriculture, however, persisted, but dairying was the best paying occupation, so that there were twice as many cows in 1908 as at any earlier date. Special care was needed to make a profit, and there was a

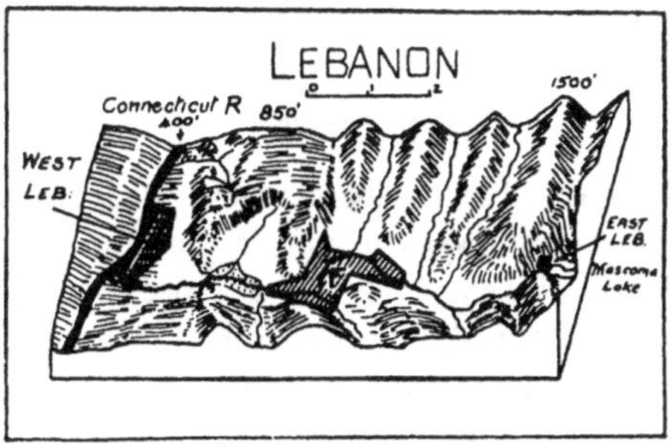

FIG. 80. Block diagram of the vicinity of Lebanon (N.H.) to illustrate the shift from farm to factory. Modern factories are mostly at 'F'. (*Based on E. N. Torbert.*)

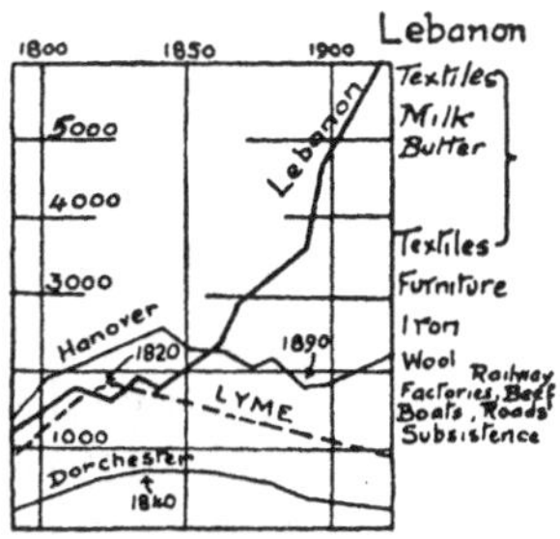

FIG. 81. The decline of Lyme and Dorchester—and rise of Hanover and Lebanon. (*After Goldthwait and Torbert.*)

great increase in the use of ensilage. In many areas, however, the woodlands are spreading over abandoned farms. Market milk rather than butter is now the aim of the dairyman, and it is sold in the villages or to distant creameries.

In 1935 Lebanon had a population of 4,700, West Lebanon (which is in a sense part of White River) had 1,300, while East Lebanon had dwindled to twenty houses. There were 1,700 industrial workers, of whom 1,200 were engaged in the textile factories. Clothing, furniture, bricks and special tools engaged the attention of the other 500 workers. Nearly all these factories are near the letter 'F' in Fig. 80, where the Mascoma River flows down some small rapids. There is a good water-supply from the hills so that there are three power stations on the Mascoma River between the lake and the big river.

These changes in the functions of Lebanon during the past eighty years are summarized in the list at the right hand side of Fig. 81. They could be paralleled in almost any district in the east of the United States or in adjacent parts of Canada. To summarize them, we find that the pioneer stage endured from 1760 to about 1800. Small factories began about 1830, and accompanied an important pastoral development in the township. With the coming of the railway the shift to factories was assured in those areas where good water power was available. (Compare with Lyme, p. 181.) The factories passed through various stages of their own; i.e., iron goods, wooden products, and now textiles. Dairying, at first for butter with local creameries, and now for milk, represents a survival of agriculture, which has, however, lost its importance in competition with factory products. As elsewhere in the hill country of New England, tourist traffic and summer residents help to bring prosperity to many a small town and village like Lebanon.

The Decline of Modern Towns

The causes of the decline of towns has not been investigated nearly as fully as the reasons for their growth; yet declining towns and cities are very abundant in the Old World, and equally so in the pioneer portions of the New World—as the many so-called 'ghost towns' indicate. There are, however, some interesting examples in parts of the go-ahead eastern United States, and one such example has been fully investigated by J. W. Goldthwait in New Hampshire near the well-known Dartmouth College at Hanover (Fig. 82).[1]

In the middle Connecticut valley are a number of small towns all of which were settled about the same time, in the same type of environment, but their evolutions have been rather different, as is explained in Goldthwait's paper. This portion of New England was settled mainly by farmers from Connecticut at the close of the French wars, i.e., about 1760. Crops sprang up on every sunny hillside, for they preferred the better drained hill slopes. Log huts, sawmills, gristmills, and later, frame houses soon spread all over the region. Altitude and distance from the village seem to have mattered little with these settlers, and each neighbourhood was practically self-supporting.

Early in the 19th century the population began to decrease. It was found that the soil soon lost its fertility, though at first very good yields of wheat and maize were obtained. But there were many other factors at work, notably the rise of factories in the seaboard areas of New England. Inland, if a good water-supply was available, the village grew, as in the case of Lebanon and Dorchester (Fig. 81), while adjacent Westmoreland and Lyme with no water-power steadily dwindled in population. By

[1] "A Town that has gone Downhill," *Geographical Review*, October, 1927.

1850 most of New Hampshire was linked to the seaboard by railways, and these did not so much open up the farmlands, as drain away the young men to the newly established factories at places like Manchester and Nashua. (See inset in Fig. 82.)

A glance at the graphs given in Fig. 81 will make these changes quite clear. The four little towns all progressed favourably in much the same

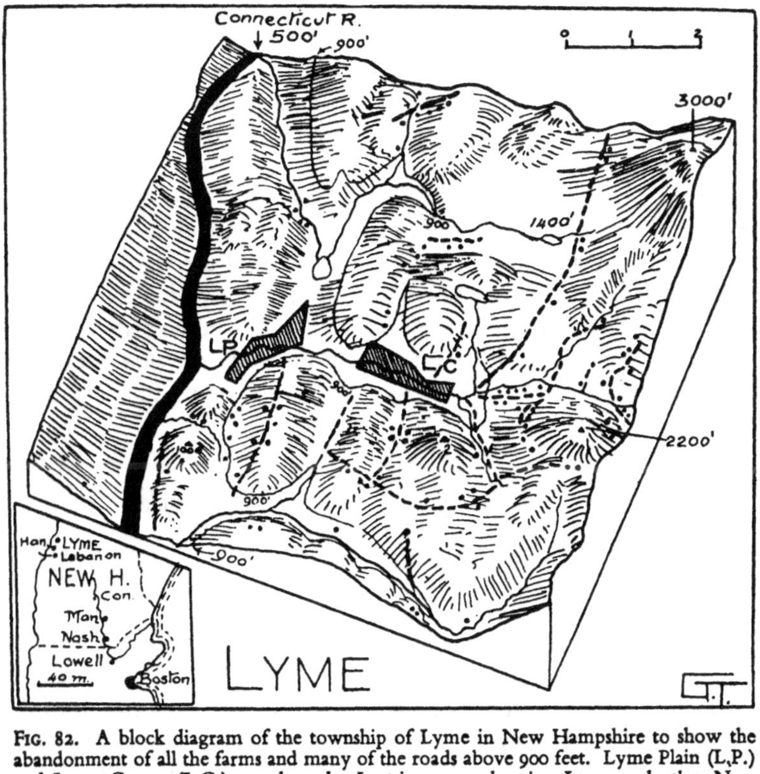

FIG. 82. A block diagram of the township of Lyme in New Hampshire to show the abandonment of all the farms and many of the roads above 900 feet. Lyme Plain (L.P.) and Lyme Center (L.C.) are charted. *Inset* is a map showing Lyme and other New England towns. (*Based on J. W. Goldthwait.*)

fashion from 1770 to 1820; though Hanover was larger than Lebanon or Lyme, and Dorchester nearby was somewhat smaller. But within the next decade or so, all four became stagnant or worse. Lyme and Dorchester never recovered, and have been slowly decreasing from that time; the first dropping from about 1,800 folk to 800, and the latter from 750 to 200 to-day. Hanover has been saved by the prosperity of the well-known Dartmouth College, as is seen by the upward trend of the graph after 1890. Lebanon has had another way of life altogether, which has already been described.

Goldthwait made a special study of the way in which the countryside around Lyme has changed in the last century. He traced all the relics of abandoned farms, the overgrown and fallen-in cellars being perhaps the best clues to a former farm. I have tried in Fig. 82 to show how the topography has been a major factor in the decay of Lyme. To-day there are two villages, Lyme Plain and Lyme Center (L.P. and L.C. in the block diagram), situated along the main valley. A number of surrounding farms still survive, but the map shows primarily the abandoned roads (by broken lines) and the abandoned farms (by black dots) which were flourishing during earlier decades. To-day Lyme Center contains twenty houses, a church, school, and one store. Lyme Plain is larger with two or three stores and a hotel. The vanishing of family names in the cemeteries indicates another major factor in this decline; the migration of whole families to the rich soils of the western prairies. A survey of active and abandoned farms to-day shows that the 900 contour line marks the upper level of farms in 1925. Almost all the abandoned farms, shown in Fig. 82, belong to the more elevated portion of the township of Lyme. Well has Goldthwait chosen his title *The Town that has gone downhill.*

The Evolution of the Megalopolis of London

It has been shown in an earlier section that London was for a century or two the largest city in the world. Indeed even to-day it is running almost even with New York, for the relative totals of population depend to a great degree on the somewhat arbitrary area which is assumed to comprise Greater London or Greater New York. In both regions the human agglomeration, clustered around the nucleus of the city, is of the order of ten million people. Obviously a large volume would be necessary to describe either of these immense groups of dwellings, and in this chapter only the salient points will be discussed. The writer has chosen to describe London more fully than New York, because he is much more familiar with the former, which indeed happens to be his birthplace. In the next few paragraphs I have borrowed freely from my *Environment and Nation*, where London is described at some length in the chapter dealing with the evolution of the English nation.

In the first place London's pre-eminence depends on her setting in the centre of the great London geological basin. As is often the case the centre of such a basin is somewhat lower, and almost always consists

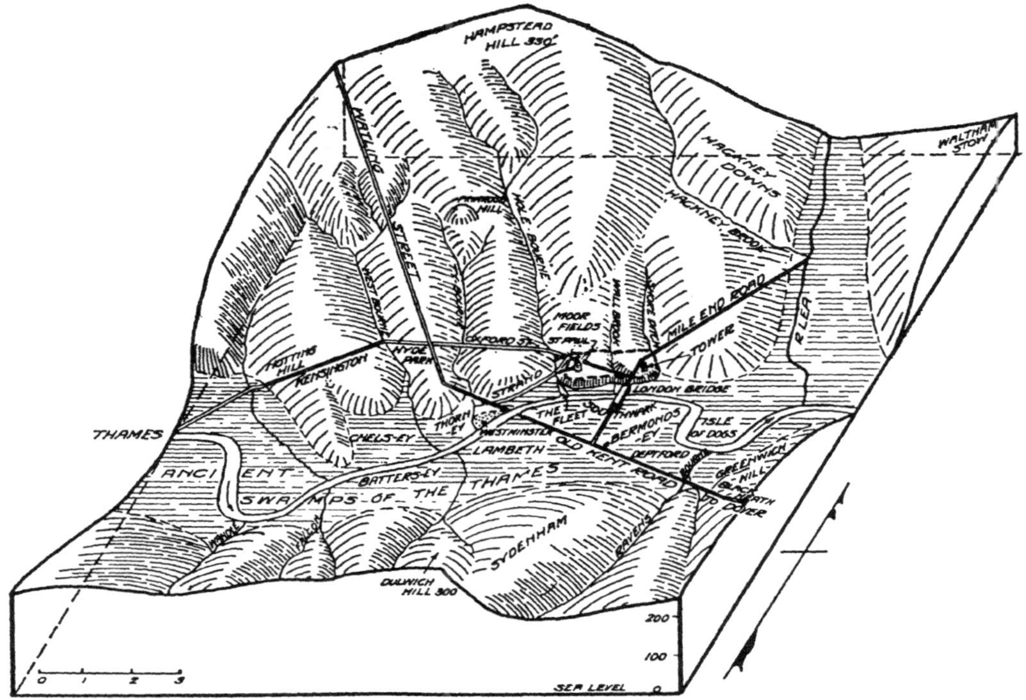

FIG. 83. A block-diagram of London, showing the control of settlement by topography. Note that Watling Street originally crossed the Thames by a Ford near Thorn-ey (Westminster). The Port developed near the Fleet Stream. The heavy broken line around the 'Twin Hills' shows the site of the early London Wall.

of less resistant beds than the older outer 'saucers' which build up the basin. Hence the centre of such a basin contains the main river of the area, and in the case of London, the drowning of the main river valley has produced the fine estuary leading to the Port of London. It is a great advantage that this estuary leads fairly directly to the estuary of the main river of the adjacent portion of the continent of Europe, i.e., the Rhine. The shortest distance to the continent—across the narrow seas—is not so far away but that the trade of Thanet and Dover inevitably led to London.

It seems clear, however, that the early Gaulish tribes, who settled near London, preferred the open woodlands and pastures of St. Albans (Verulam, p. 130) and Colchester to the swampy flats of the Thames. The North Downs—open ridges of chalk—led the traveller from Rutupiae (near Dover) towards Brentford and Westminster, where he would cross the Thames and so reach St. Albans, some twenty miles to the north-west. Probably London was not in existence when Caesar made his raids in 55 and 54 B.C. The name 'London' is supposed to mean the 'dark pool', and is of Welsh (i.e., British) origin. It refers to the great sheet of water, some two miles wide, which at high tides covered much of the flats near the site of modern London (Fig. 83). Even to-day some of the wide marshes of the River Lea, as well as the lower suburbs of London are drowned at specially high tides in the Thames.

A dense forest covered the heavy clays between the Thames and St. Albans, and there was only a narrow belt of sandy heath near to the river, which was of little use for primitive agriculture. The name 'Blackheath' reminds us of one such heath on the south side of the river. In the marshes of the Thames there were gravelly islands called by the Saxons 'eys', and the names of to-day, Chelsea (Chel's or Chelis isle), Battersea and Bermondsey, refer to these early drier areas among the swamps. One island, formerly known as Thorney, is of special interest. Here was the lowest ford on the Thames, and here we may be sure the first nucleus of a village started in British times. Later Saxon kings including Edward the Confessor built monasteries here, and to-day the spot is known as Westminster.

From Thorney a rough road—later to become Watling Street—led to the north-west to St. Albans. It ran along the ridge between West Burn and Ty Burn, which are the names of London areas to this day, though the streams have long ago vanished underground as sewers and storm-water channels. South of the ford, the 'Old Kent Road' climbed Blackheath and thence proceeded to Canterbury and Dover.

Probably sometime after the invasion of Claudius (A.D. 43) London became of some importance as a fishing port. The river swings somewhat to the north below Westminster, and flows under fairly steep gravelly hills. These were trenched deeply by three streams, which were kept flowing strongly by the springs in the gravels. Hole Bourne, Wall Brook and Shore Ditch have carved the gravels into a pair of somewhat isolated hills (Fig. 84), between which lay snug little coves suited for the small fishing boats of Roman days. Roman London was erected on these 'Twin Hills'; and to-day St. Paul's crowns one while the other is just behind the Tower. The lower end of Hole-bourne was called the Fleet (i.e., flowing brook), and this was perhaps the first port of London.

The Romans built a fort in the vicinity of the Tower, and about 100 constructed a wall to enclose the twin hills. The valleys of Hole-bourne and of Shoreditch lay just outside the wall, and acted as moats to strengthen the defences of the city. It is not clear when the first bridge was constructed, but it naturally was placed where the river had a high gravelly approach, and this occurred somewhat downstream from the Fleet, and near the Tower. The north road was now deviated from the Westminster ford along the north bank, entered the Wall near (the later) Newgate, and then crossed the bridge to the southern side of the Thames, where a protected suburb (South Wark) soon developed.

During Saxon times London was almost deserted, until Alfred restored the walls and encouraged citizens to occupy the city. Around 1060 Edward the Confessor restored some of the importance of the early settlement at Thorney, and later monarchs by building various ecclesiastical and royal palaces established there a rival settlement to the growing commercial town further down the river near the bridge. William the Conqueror constructed the Tower so as to overawe the citizens, as well as to control the passage of the bridge, which lay close to the great keep.

During the early Middle Ages there was great rivalry between the court at Westminster and the city two miles to the east. It was not till Tudor times that the gap began to close, and even by the middle of the 18th century the close-set houses did not extend much beyond Westminster. The street along the bank of the river was called the Strand—as it is to this day. But the whole of the swampy land (formerly two miles wide) excluding about one quarter of a mile occupied by the embanked Thames, is now closely covered with buildings.

The maps given in Fig. 85 show very clearly the slow growth of London until 1800, and then the sudden increase as the result of the industrial revolution. We have only to compare the map of 1800 with that for 1933 (which is based on Unstead, and drawn to the same scale) to see the development of a 'megalopolis'; as Mumford terms this

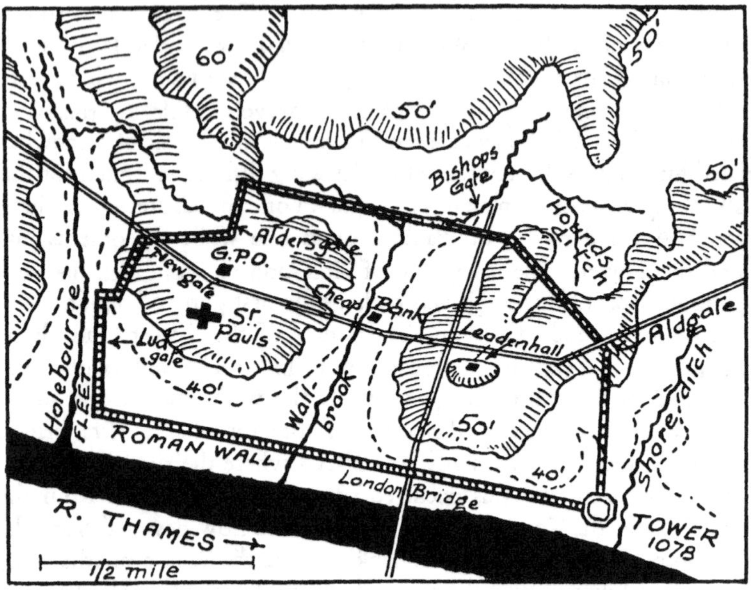

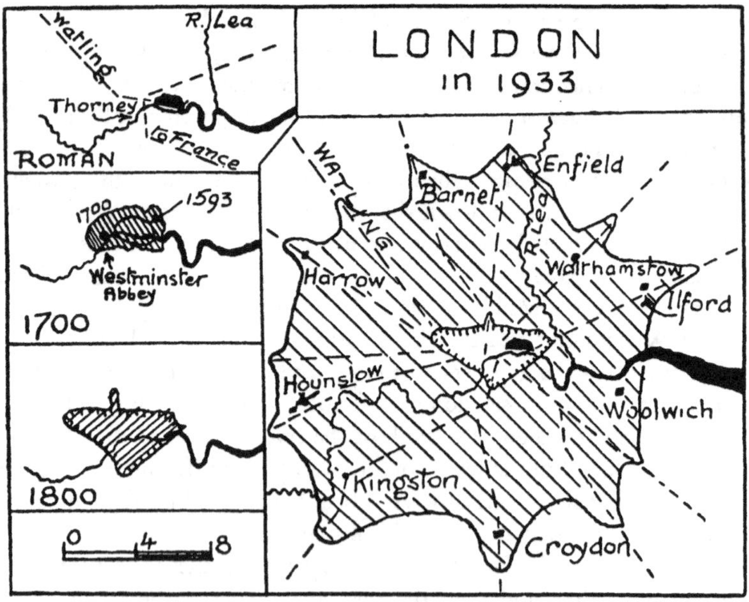

FIG. 84. (*Above*) The topography in the heart of the city of London—showing the Roman Wall.

FIG. 85. (*Below*) The growth of London from Roman times until 1933. All to same scale. (*Based on Unstead.*)

widespread agglomeration of bricks and mortar. Until several decades after 1800 there was a series of isolated villages surrounding London, and these are shown on the map. Croydon, Hounslow, Barnet, and Walthamstow, were several miles away, but they have all been absorbed by the vast growth of London. It is indeed something like a gigantic *amœba*—where its pseudopodia are represented by the extensions of the houses along the main roads—reaching out to swallow all it touches. To-day London is twenty-two miles across, and it includes an area of about 270 square miles. Thus it is more than 300 times the size of the important city of Roman times.

One of the best of the briefer studies of the growth of London is the chapter by O. H. K. Spate in the *Historical Geography of England* (Cam-

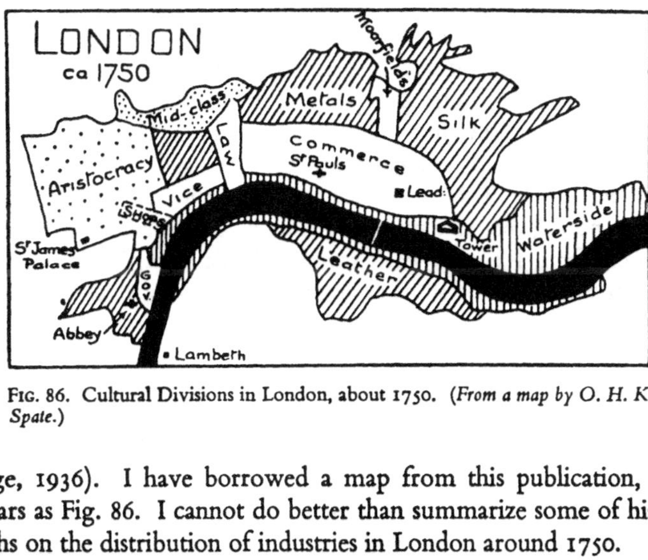

FIG. 86. Cultural Divisions in London, about 1750. (*From a map by O. H. K. Spate.*)

bridge, 1936). I have borrowed a map from this publication, which appears as Fig. 86. I cannot do better than summarize some of his paragraphs on the distribution of industries in London around 1750.

The City sheltered a host of small tradesmen, and their banks, markets and exchanges. Around the City were grouped three industrial sectors. In the vicinity of the Tower was a district of innumerable twisting lanes and alleys, where were many artisans employed in making clocks and jewellery. North of this was a district of weavers (Fig. 86) around Spitalfields. To the south-east were the taverns frequented by mariners; and here were many 'noxious trades' such as oil and soap works. There were very definite markets: such as the Haymarket (near the area marked 'shops' in the map) for hay. Smithfield was the great meat market, and was situated in the midst of the metal workers' homes in the north.

Leadenhall, which crowned the eastern 'twin hill' of early London, was a great mart for meat, leather, and provisions. Covent Garden was surrounded by the district where amusement and vice were dominant. Here fruit and vegetables were sold, then as now. Just above the bridge on the north side of the river was Queenhithe, where the great supplies of wheat for London's food were landed. About half a mile down river near the Tower was where the 'stone coal' from the north was unloaded on the wharves at Roomland.

Westminster and Blackfriars Bridges were constructed between 1750 and 1770, and they opened a vast area of land to the east of Lambeth (Fig. 86). The 'New Road' linked Marylebone and Moorgate in the north-west of the city of that date. The Fleet brook still flowed south near the area marked 'Law', though by 1760 it was mostly arched over. However, it was strong enough to give power to flour mills at this date. Around 1800 the fine buildings near the British Museum were constructed. Regent's Street and Regent's Park were developed by Nash before 1823. The most attractive portion of London was the 'west end'; partly because the prevailing winds prevented the fog and odours of London from greatly affecting this region patronized by the aristocracy. Here the parks near St. James Palace made this the best part of the city. It is worth noting that to-day there is an almost continuous belt of such open spaces from Regent's Park through Hyde Park, Richmond Park and Hampton Court for some fourteen miles to the south-west of the heart of London.

The varying rank of great cities

The rise and fall of various cities in the last four centuries can be grasped from the graph given in Fig. 87. In 1500 the chief cities were Constantinople, Paris, Naples, Venice, Milan, and Lisbon. To-day, only Paris among these is one of the front rank cities of the world. What has led to the decline of the others? Their environment is clearly the chief cause of their relative decay. All but Paris have one feature in common, they are centres of Mediterranean trade; and four of these are ports on that sea, mainly dependent on the traffic in goods brought from the East. When routes round the Cape were discovered, and routes to America began to be of major importance, their dominance gradually disappeared. Both these great events in human evolution happened in the last decade of the 15th century. Constantinople, under the Turks after 1463, declined rather rapidly. Milan and Naples kept their relative position for a century, and then fell off quickly in importance. Naples, not only a port but the capital of a large kingdom, persisted longer, until the dawn of the industrial revolution, when lacking fuels it gave way to the coal towns of other lands.

From 1600 to 1800 we see another group of cities coming to the front, as indicated by the upward trend of their graphs. Of these the most striking are London, Lisbon and Amsterdam. All of them are favourably situated to take part in the rapid growth of Atlantic trade; and by 1800 London was the largest city in the world, Lisbon fourth and Amsterdam ninth. Moscow and Vienna also grew rapidly during these centuries;

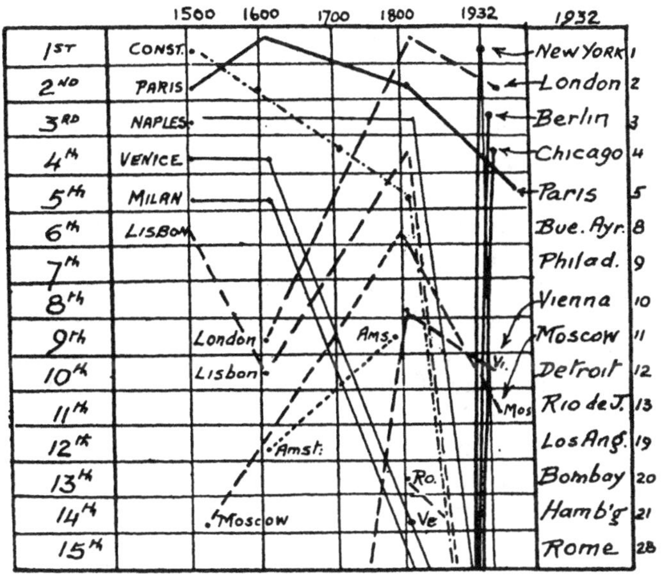

FIG. 87. Change in rank in cities. (*From a table by A. F. Weber.*)

suggesting, as is true, that other factors besides Atlantic trade determined the rapid growth of cities in this period. In both cities we may ascribe much of their growth to political importance, for they were capitals of great empires, as well as most advantageously situated for inland commercial intercourse.

The most obvious feature of the chart is, however, the growth of the large cities of to-day, which were of no importance only 150 years ago. At the beginning of the 17th century Berlin only had a population of about 20,000 and by 1800 it was below 200,000. It doubled its population by 1845, i.e., in forty-five years; but the striking change came in the next thirty years, for it reached one million by 1875, thus adding half a million people in twelve years (Reclus). In 1900 Greater Berlin included two-and-a-half millions, while to-day it amounts to more than four millions.

Of course it is in the new lands of the world that we see the most rapid increases in population. The greatest agglomeration of folk to-day is in New York and its vicinity, which has no less than twenty-seven satellite cities (each over 25,000) included in its metropolitan region. The total population of this district about 1932 was eleven millions; and its distribution along a strip from Perth Amboy to Norwalk is charted in Fig. 187. Los Angeles is usually cited as the most rapidly growing large city in the world. In 1860 it was a country town of less than 5,000 persons. Forty years later it was still rather an unimposing city of 100,000. But from 1900 to 1932 it had experienced a meteoric rise to the ranks of a metropolis of the first magnitude. An increase of more than 1,000 per cent had placed it above the million mark. (Gist and Halbert.)

The next graph, charted in Fig. 88, indicates the chief reason for the remarkable change in the character of the great cities which occurred around 1800. It was of course the Industrial Revolution, which may be roughly defined as the 'crowding of factory operatives made possible by the import of food from far beyond their homes'. The evolution of crowds of workmen clustered in one city was due to Watt's invention of the practical steam engine, about 1763, more than to any other single factor. There were plenty of small factories long before 1763, not only in Europe but in North America. Every little town had its gristmill and its saw mill, usually harnessed to a source of water power. But such sources of power were strictly limited in any district; moreover it was not feasible to collect workers from long distances since they could not readily be fed on local food products.

However, the environment played a part as important as human invention in the rise of the great cities. I have discussed in other volumes (notably in *Environment and Nation*, pp. 229-235) the shift of populations in Britain as various types of environment appealed to folk at various stages of their social evolution. Open downs appealed to Neolithic shepherds; loamy slopes to the Saxon; fertile vales, when cleared of dense forest, to the medieval farmer. Finally in the early 19th century there was a marked shift to the sterile grit-stone areas lying above the ancient rocks of the Oldermass, and beneath the softer more fertile strata of the Youngermass. This is a long way of stating that the coal deposits have determined population density in England almost completely during the period since 1850.

We are now able to interpret Fig. 88 which shows critical periods in the growth of four great European cities. All four grew but slowly until 1750; but London began to increase rapidly first of the four, to be followed by Paris, Vienna, and Berlin, at shorter and shorter intervals. This rapid start of London in the race is due to the especial advantages of the English capital. It naturally utilized a British invention before its rivals on the continent, and England in general had more coal, and coal which was easier to work than any of the continental countries. Moreover the island position of Britain saved it from the destruction of war which marked the days of the French Revolution and the imperialistic

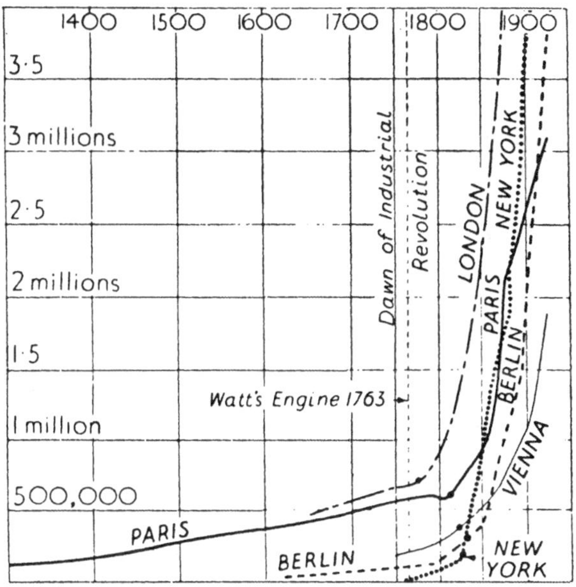

FIG. 88. Population graphs of five great cities showing the sharp rise at the dawn of the Industrial Revolution, which appears first at London.

designs of Bonaparte. The critical angle in the graph of Paris is seen to be about 1810, to be followed by that of Berlin in 1830, and Vienna in 1840.

Something of the same sort is shown by the earliest cities in America; and the graph for New York appears in Fig. 88. The critical angle in the curve of population seems to occur about 1820, i.e., as soon as in Paris, and before the industrial revolution greatly affected Berlin and Vienna. This is what we should expect, for engineers such as Fulton in America soon benefited by the inventions of Watt and others in England. Indeed

Fulton's steamer on the Hudson in 1807 was perhaps the first successful application of the steam engine to navigation.

But we must not imagine that the Industrial Revolution about 1800 was the sole great stimulus which changed the countryside into a land dotted with large towns and cities. Such stages in human development have occurred several times since. For instance the invention of the practical locomotive by Stephenson in England (about 1830) made a very great difference to the distribution of populations, and nowhere was the change so marked as in the new lands of North America. Canals were of similar, though less, importance perhaps a decade or two earlier. The automobile has changed the map of the world, in regions of occidental civilization at any rate. The most spectacular rise in modern times among occidental cities has been that of Los Angeles—already referred to. What caused this city to slumber from 1827 to 1880, when the population was only 11,000; and then grow so rapidly in the early twentieth century that it has over one and one half millions of citizens in 1940?

Los Angeles offers a good example of *multiple* factors making for great populations. It was naturally not much affected by the early industrial revolution, though there was a Mission there as far back as 1781. Railways reached it in 1876, when it had been an incorporated city for twenty-six years, and they undoubtedly helped its rise. With the growth of a western population, the city naturally became a notable port for California and later for the outside world. The climate was perhaps the main factor. In another volume[1] I have shown that the *isoterps* (lines of comfortable climate) are unusually good in this part of the world, as indeed Californians have always boasted. This led to the growth of the great motion-picture industry, and it is of course one of the world's great fruit regions. The discovery of the remarkable oil field, practically in the suburbs of Los Angeles, also helped its rapid rise. It is informative to contrast the rise of Los Angeles with the crystallization (not to say stagnation) which has marked the evolution of Charlottetown, the provincial capital of Prince Edward Island, founded about the same time, which has long stopped growing with a population of 14,000 (See page 70).

Social Groups in Chicago

Few cities in the world have more complexity in their social groups than Chicago, where the author was a member of the staff of the University of Chicago for seven years. During his last few years in Chicago,

[1]*Our Evolving Civilisation* (Toronto), 1947.

he devoted a good deal of time to a partial survey of the southern side of the great city. Three well-known writers on social subjects, Merriam, Park and Burgess, were colleagues at that time, and the world owes much to their efforts to understand the anatomy of cities and city life. A discussion of the evolution of Chicago illustrated by maps of the city will be found in the writer's companion volume on *Civilization*, e.g., Fig. 68.

The map in Fig. 89 is from the book by Park and Burgess (*The City*, 1926), and illustrates the views of Burgess with regard to the social groups in the city. He pictures the three million citizens as living in five zones, centred on the 'Loop', i.e., the business centre of the community.[1] In New York the 'Downtown' section corresponds to the 'loop' in the Chicago river, now crammed with offices, etc.; in Pittsburg it is the 'Golden Triangle'. Here is the commercial core, together with the large theatres, skyscrapers, and department stores. This zone is crammed with business people and shoppers, but is often relatively empty during the night, since most folk sleep away from the offices.

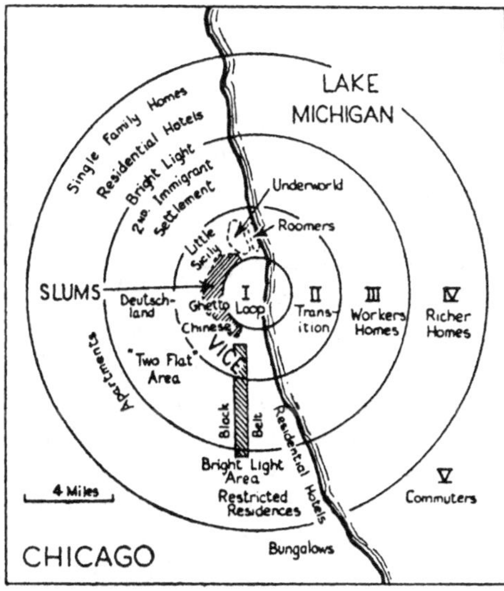

FIG. 89. Somewhat generalized zones showing culture-groups at Chicago. (*From a map by E. W. Burgess.*)

The second zone is a very interesting one, which Burgess calls the zone of Transition or Deterioration. The folk engaged in this belt are essentially unstable. The buildings are always being invaded by growths of the most progressive part of the city. Property is often dilapidated, although land values are high. 'It becomes a catch-all for social pariahs'; and is often the first place of residence of poor immigrants. Here then in Chicago are the Ghetto, Chinatown, Little Italy, etc., as shown in Fig. 89. Here also is usually found the 'black belt' where the negroes live, crowding out other culture groups. As these under-privileged folk improve their status they tend to move further from the centre. Thus

[1]For variants of this simple 'Zone' plan, see Harris and Ullman, *Nature of Cities*, Amer. Acad. Pol. Sci., Nov., 1945.

in the case of the 'black belt' it extends somewhat south into the next zone. The slums of a city are here, and in the north of Chicago here is the 'bohemian quarter', in the vicinity of the old Water Tower. For this reason Chicago has a 'Towertown' corresponding to 'Greenwich Village' in New York.

The third zone contains the 'Proletariat' homes, where dwell folk who have risen above the earlier settlements in Zone II. Here also are the thousands of workers who must be relatively near their employment in the shops and offices. Many of the homes are 'two-flat' dwellings built of wood, with the owner living on the lower floor and a tenant on the upper floor.

Zone IV contains the middle-class workers, professional people, clerks, etc. Apartments are perhaps the commonest dwellings, but there are still many houses with a tiny garden or yard separating each from its neighbour. Outside of this zone is the belt where live the commuters, who endure a long journey in and out for the benefits due to more space and better living conditions. This zone is naturally not so congested as the others, and on the periphery it merges into relatively open country. The outer edge of a city is usually 'stellate' in pattern, with the 'rays' extending some distance along the main railways and roads, while fields still occupy the sites not so readily reached. On the edge of this zone may be 'satellite towns' not yet engulfed by the megalopolis. In these the houses are almost empty of menfolk in the day; but at night serve as dormitories for the workers in the great city.

Often enough 'strings' of satellite towns develop along the main railways. For instance, round Chicago, one such string of contiguous incorporated towns extends along the north shore of the lake as far as Waukegan. Another fills up most of the land along the railway to Arlington Heights in the north-west. To the west is a similar string linking Wheaton to the great city. To the south-east there is a solid mass of buildings all the way to Gary; though the country is fairly open to the south-west of Chicago. For a description of the rural-urban fringe see the writer's "Our Evolving Civilization," pp. 234-237.

Zones and Cultural Minorities in Toronto

Burgess has produced a generalized pattern for a large city (i.e., Chicago) which is admittedly too simple to be accurate (See Fig. 89). As a check we may consider briefly the pattern that seems to have developed in Toronto, a city of 800,000 with a definite though small minority of foreign folk. These groups are inserted in the plan appearing as Fig. 90. The nucleus of the town, around 1800, was at 'A', near the head of the harbour; but the city spread to the north-west away from the deep gorge cut in the soft glacial till by the Don River. To-day the commercial core is situated where Yonge crosses Queen Street. As in the Chicago plan we find that there are zones of 1st class (Ha), 2nd class (Hb), 3rd class (Hc), and slums or rather unattractive dwellings (Hd). These zones are, however, determined by the site; for the houses are not quite free to spread equally in all directions. The main railway came in from the north-west along the Humber valley, and this determined, as ever, the site of the factories. So also a large area of swamp and 'made land' is now used for factories to the east of the harbour. Hence the

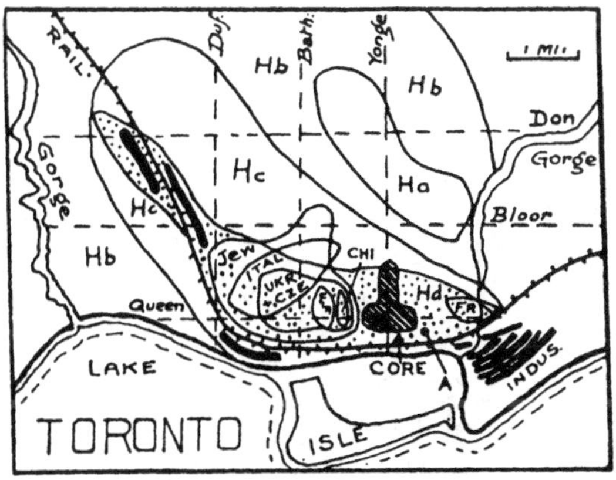

Fig. 90. Simplified cultural zones in Toronto, showing approximately where the minorities of foreign origin dwell. Poorest districts are dotted. (*Partly based on N. Deacon.*)

sites along the railway axis are unattractive, and are occupied by Hd or Hc dwellings. The better-class houses, Ha and Hb, form zones on the outer edge of these poorer belts. As a result the whole set of zones is twisted, and is not so symmetrical as Burgess would indicate.

With regard to the poor immigrants who have flocked into Toronto, about eight-and-a-half per cent have come from continental Europe,[1] while the Chinese add about half a per cent more. In Fig. 90 only the more concentrated groups are indicated. The Jews are densest in the Hd zone to the west of the 'core' and dwell in a number of blocks south of Bloor Street. Italians rank second in numbers and occupy the western portion of this same section of the city. Ukrainians and Czechs seem to

[1]For a discussion of the evolution of Toronto, see N. A. H. Deacon: *Can. Geog. Jour.*, August, 1944.

live in the same portion of the 'Foreigners' Quarter', while the Finns and Chinese are mainly in the portion nearest to the core. Only one minority group is not represented here; and these are the French, who favour a more eastern site by themselves. Notice that here as in Chicago, the minorities tend to spread into the adjacent better zone (Hc).

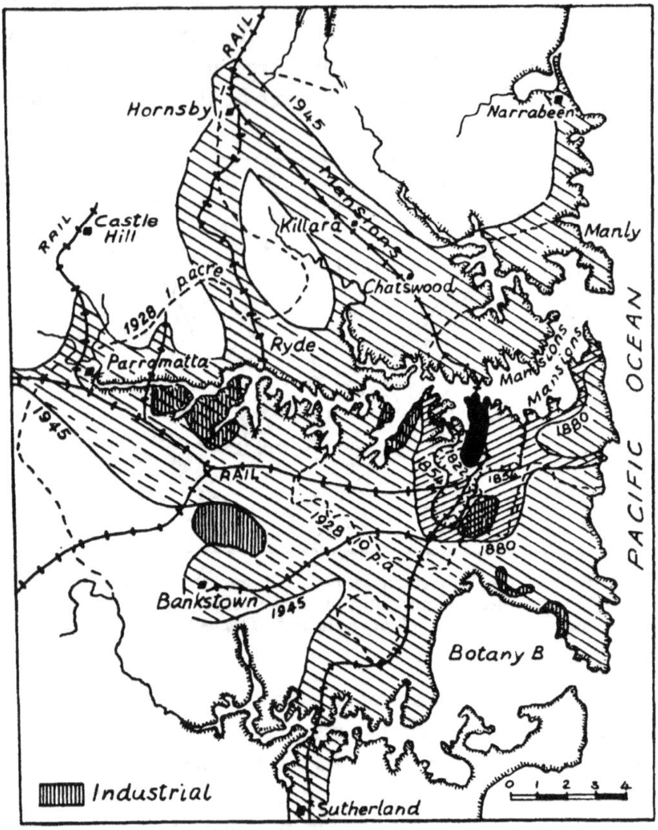

FIG. 90A. The growth of Sydney (Australia) from 1822 to 1945. Densities per acre in 1928 from J. M. Holmes. The commercial core occupies much the same area as the town of 1822 (black). (See also Fig. 94.)

PART III

TOPOGRAPHIC AND OTHER CONTROLS

The modern Town Planner
  Inscribes on his Banner:
"When shall we be rid
  Of the curse of the 'GRID'"?

M. NOY

CHAPTER XII

GEOLOGICAL CONTROL: TOWNS IN THE PLAINS

Topographic Control in general

IN the preceding chapters there have been illustrated many examples of topographic control of the sites of villages, towns, and cities; but so far no special attention has been given to this factor in the broadest sense. It is quite obvious that in general the cities of the world are most commonly found on low-lying plains, where agriculture is widespread and communications are easy. Temperature, however, alters rapidly with elevation, at the rate of about 1° F. for each 300 feet of ascent. Since most folk like a temperature around 65° F. it is clear that the optimum lies at different levels in different latitudes. This is suggested in the graph given in Fig. 91. At sea level cities in latitude 35° N. are favoured with the best climate, since the average here is 65° F. If one lives in latitudes warmer than this, it is well to live at higher elevations as suggested in the graph. At the Equator, where the sea-level temperature is about 80° F. it is necessary to climb 4,500 feet to reach the elevation where 65° is the average. The graph suggests (at B, C, D and E) the heights of the mountains which give optimum climates.

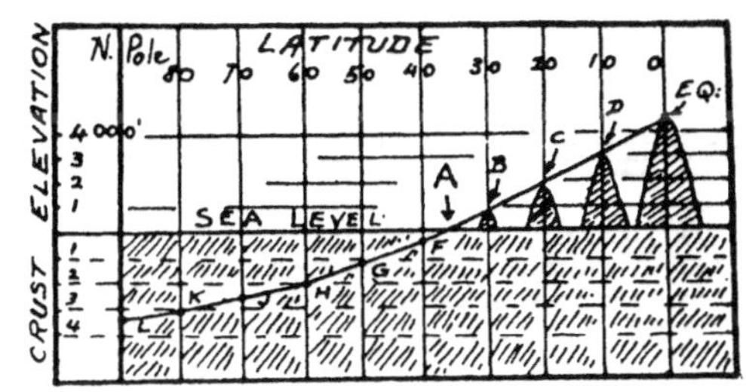

FIG. 91. Diagram showing in section the position of a sloping plane (from Equator to North Pole) which would have an average temperature of 65° F.

What about folk who live in cooler latitudes than 35° N.? They may in the future burrow down into the earth to take advantage of the increasing heat of the earth's interior, as we descend. The moderate depth of fifty feet or so takes us to a 'cellar' where the temperature is not affected by seasonal changes of temperature. In Toronto therefore this 'cellar' would be 45.5° F. For fifty or more feet further down there would be little change, but below that depth the temperatures rise about

1° for each 52 feet of descent. So that our cellar at Toronto with a constant temperature of 65° F. would be about 1,000 feet below sea level! The depths for similar 'comfortable cellars' at various latitudes are suggested by the letters H. J, K, L at the appropriate levels. Possibly in some far distant future the cities of the world may be built on an inclined plane. This would tower above the earth to the south of latitude 35° N. (at any rate as far as the Equator), but would descend into the bowels of the earth (to obtain underground dwellings free from the atomic bomb) for cities which lie too far north of the happy mean of 35° N. latitude. This is admittedly a very far-fetched prophecy, but it suggests some important truths with regard to the best situations for prosperous cities.

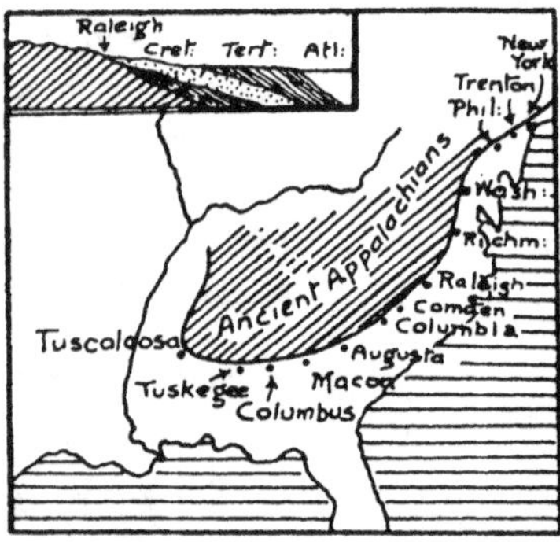

FIG. 92. City sites along the 'Fall Line', where younger rocks contact hard old rock (see section in *inset*), in S.E. United States.

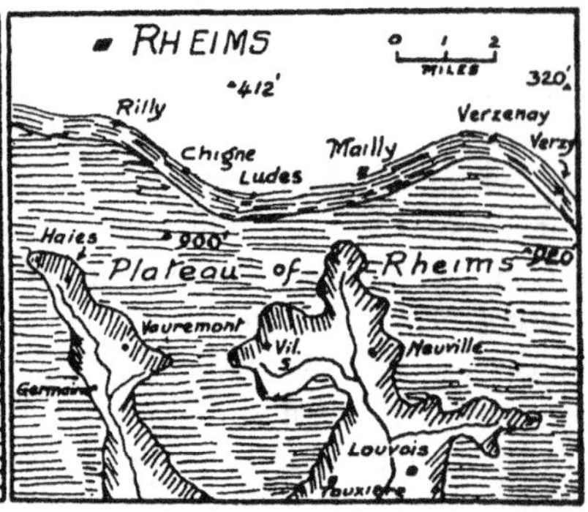

FIG. 93. Village sites in the Eocene limestone plateau near Rheims. They are determined by springs on the edge of the plateau. (*Aurousseau.*)

Leaving the temperature factor—which obviously depends on the topography to a considerable degree—we may examine other ways in which topography affects city development. Perhaps the best known of these correlations is the series of towns along the 'fall-line' in the United States. This is illustrated in Fig. 92 in the inset; where we see that the Appalachians to the west of Raleigh (or elsewhere on the eastern seaboard) consist of the hard granites, etc., and so naturally resist erosion better than the softer strata of the coastal plain. These latter are formed of Cretaceous and Tertiary rocks, which readily erode under the action of rain and rivers. Hence there is almost always a considerable drop where the rivers of the Appalachians reach the coastal plain, as at such a location as Raleigh. In the early days of settlement this was of prime importance as a site for the simple mills of the day, and indeed plenty of power is developed to this day in the cities along the fall-line. The main map in Fig. 92 shows that many of the chief inland cities of the United States from Tuscaloosa to New York have had their locations in large part determined by such fall-line topography.

A very similar illustration on a somewhat smaller scale is to be seen in southern Ontario at the southern edge of the Precambrian Shield. A most striking change in the landscape, and also in the human response to the environment, can be observed as we cross the line linking Midland on Georgian Bay with Kingston on Lake Ontario. To the south, on the fairly good soils derived from Ordovician strata, are closely set farms with numerous villages and towns; but to the north on the granites of the Shield are few farms, and no towns except near the widely spaced main highways to the north (Fig. 186).

Sydney—a Metropolis strongly affected by Geological Controls

Sydney, the largest city in Australia, and next only to London among the cities of the Empire inhabited by 'white' people, has perhaps the most unusual umland (i.e., surroundings) that the author has investigated. Another reason which makes its development of interest is that the first settler only arrived in 1788, so that all the factors affecting its growth are fairly well known.

The First Fleet under Captain Phillip first entered Botany Bay, acting on the advice of earlier explorers. But the low sandy shores of this 'drowned plain' did not seem suitable for a pioneer settlement, and Phillip proceeded north for twelve miles, where he found a much better harbour. Here is a fern-leaf shaped inlet (of the type known as a *ria*) with rocky cliffs on the north shore, and lower shaly slopes on the south shore —where the conditions for settlement seemed much better. Accordingly he set up his tents at Circular Quay, a minor inlet on the south shore about five miles within the entrance to the Harbour (Fig. 90A, p. 196).

The most interesting feature of the city is emphasized in the inset map in Fig. 94. Here we see a black patch, which represents the built-in streets of Sydney on both shores of Sydney Harbour. To the west is a settled area of shales about thirty miles across, and to the west, north and south of this patch of shales is an area (left white in the map), which to-day is not much more settled than in the days of Captain Phillip. When we have crossed this belt of empty country—on the average about thirty miles wide—we again reach lands which have become fairly closely settled.

What is the explanation of this unusual belt of empty lands around one of the largest cities in the world? It is almost wholly due to the topography and geology of the region. A section through the region appears on the front edge of the main diagram (Fig. 94). Here we see that a thin layer of shale caps a very thick layer of Hawkesbury Sandstone of Triassic age. This is peculiarly sterile, and its soil grows very little except rather drought-resistant shrubs and trees. Under the sandstones, which are about 3,000 feet thick, comes a thick series of Permo-Carboniferous rocks. These carry valuable coal seams near the top of the series, which indeed have been worked by deep coal-pits in the heart of Sydney itself. Only the shales above the Hawkesbury Sandstone weather to a fair soil, and accordingly the farmlands and orchards to the west of Sydney are almost wholly found on this shale formation.

The topography, however, plays an equally important role with the geology. The region to the west of Sydney has been bent upward—as a simple monocline or warp—along a hingeline running north through Picton and Penrith (Fig. 94). To the west of this hinge, nothing but sandstone occurs at the surface, and as the land rose up at the close of Tertiary times, the rivers cut down into the warp. Thus were produced the famous *bottle-neck valleys*—called in western U.S.A. 'box canyons' —several of which are sketched in the block diagram. Thus communications are very difficult to the west of Sydney, for the warp has lifted up the hard sterile sandstones to a height of 3,500 feet above the sea near Wentworth Falls (Fig. 166). Roads and railways to the west of the State have had to surmount this great elevation to cross the sterile belt of deeply eroded plateaux, which surrounds Sydney on most sides. As the small map in the inset shows, when these dissected plateaux have been traversed we reach the wheat and grazing lands (shown dotted) which have led to the development of Bathurst to the west, and Goulburn to the south-west.

The major warp has been in the west, but minor warps of the same type enclose the shales on north and south. The Nepean River rises in the southern warp, and enters the lowest portion of the basin near Camden. Here it flows through a broad area of silt, which is due to a temporary lake, which developed here while the western warp was rising. The Nepean actually cuts into the rising warp, at Bent's Basin and again at Wallacia, before finally flowing along the foot of the hinge near Penrith. There is another broad area of lake silts near Richmond, where the river Nepean enters the northern warp. Now it flows between rocky cliffs of sandstone, which become higher and higher as the ocean is approached at Broken Bay. Salt water now reaches nearly to Richmond, and this part of the Nepean system is known as the 'Hawkesbury River'.

The settlement of the hinterland of Sydney has been greatly affected by these variations in the river's history (Fig. 94). South of Douglas Park the river is cutting its way into the sterile sandstones of the southern warp. Here is poor forest country, now largely utilized as water-collecting ground for huge reservoirs. Then at Camden are the fertile lake-silts, and here were established almost the first farms outside of Sydney itself. The shale country is occupied by farms, though the crops are not as good as on the lake silts. At Richmond are other large areas of maize and sorghum which are grown on the lake silts there. North of

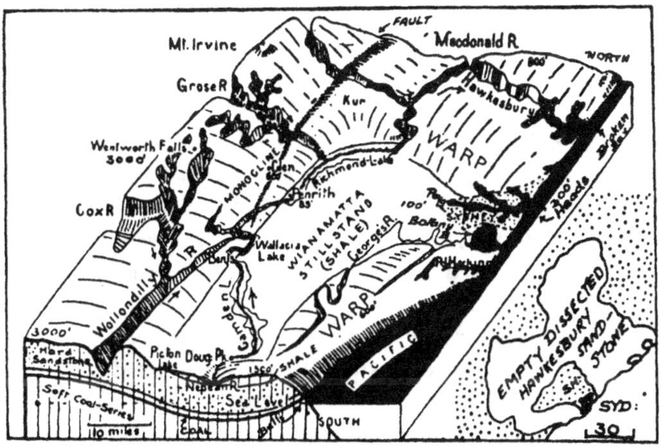

FIG. 94. Block-diagram of the Warped Littoral around Sydney. The three 'Silt Lakes', through which the Nepean River flows to the north, are indicated. Geological section shown in front. Dotted area at Sydney shows close-set houses. (*From* Taylor, *The Geographical Laboratory*, 1925.) Inset shows the empty hinterland.

this point the land is nearly empty, and is given over in large parts to the huge National Park south of the Hawkesbury River.

Clearly the city of Sydney itself has developed along a sort of minor east-west hinge of the northern warp. The older suburbs around Circular Quay and extending to Parramatta (P) on the *south* shore of the harbour, have been built on shales. But the *northern* suburbs have developed on upraised sandstones, and the little streams have cut quite deep gullies fringed with steep cliffs in many of the northern suburbs. The better suburbs are spreading along the railways to the north of Sydney, where the land is higher (770 to 900 feet) though the soils are very sandy, and it is hard to grow flowers, etc., in the gardens (Fig. 90A, p. 196).

Since railways and roads are much easier to construct on the south side of the harbour, the industrial areas are to be found here. Others have developed on the north shores of Botany Bay, which somewhat resembles in its topography the southern suburbs. The land rises towards the Heads, i.e., to the east from the city; showing that the sea has breached the eastern rim of the great shale-covered basin. In all probability this shale area is really a 'stillstand', i.e., a portion of the crust which has remained at rest, while all round it, even to the east, the crust has been upwarped to varying degrees above its original level. Accordingly the basin is sometimes referred to as the 'Wianamatta Stillstand', from the site where the Triassic (Wianamatta) shales are best developed. This basin is shown blank in the larger diagram, and is the smaller dotted area in the inset in Fig. 94.

Villages and Towns on Cuesta slopes: The Weald

Another example of control of town-sites by topography and geology is illustrated in Fig. 93. Anyone who has had occasion to travel in the plateau country just to the east of Paris, will be struck by the absence of villages on the plateau, whereas there are many clustered along the slopes of the scarps, as illustrated in the map. This is neither a question of soils or communications, but of water supply (Aurousseau). In earlier and more primitive times a ready supply of water was necessary in villages and towns; and the surface of the plateau was not satisfactory in this respect. But all round the scarps at the base of permeable limestones, where they lay over impermeable shales, there were numerous springs gushing out of the face of the scarp. These springs formed the nuclei of the villages in the region; and a string of them, Rilly, Chigne, Ludes, Mailly, Verzenay and Verzy, appear on the map to the south of Rheims; which itself is in the vale some 500 feet below the hard beds forming the Plateau of Rheims. On the southern side of this plateau much the same sites for the villages are apparent.

Another example of similar character is to be found in the region of the Weald, just to the south of London.[1] Here is a classic example of an 'eroded dome'—which in a model looks very like an apple pie with the crust sliced off the top horizontally! In the centre (corresponding to the heap of fruit) are the rather hard basal Hastings beds of the Cretaceous (Fig. 95). These stand up as the sandstone hills around Ashdown Forest—and rise nearly to 800 feet. They contain clay-ironstone, largely used in medieval times. Surrounding these harder beds are the soft Weald Clays, which are about 200 feet above sea level. As we move north (or

[1]This paragraph is based on the valuable study of *Great Britain* by 26 authors (edited by A. G. Ogilvie), Cambridge, 1928.

south) from the central hills we find alternating beds of hard and soft rock—of which the northern series is sketched somewhat diagrammatically in Fig. 95.

Thus the Weald offers a series of low vales where the rocks are easily weathered, and a series of ridges or cuestas where they are resistant. The Lower Greensand lies above the Weald Clay (Fig. 95), and this is moderately resistant, and so forms a ridge through which the main rivers draining from the centre have cut gaps. Above these harder beds come the stiff blue clays called the Gault, which forms a wide vale bordered on the outer edge by the high Chalk plateaux. In this valley of the Gault are the main towns of the Weald (as we shall see), and it has a special name, 'Holmesdale'. South of Rochester the Chalk is four

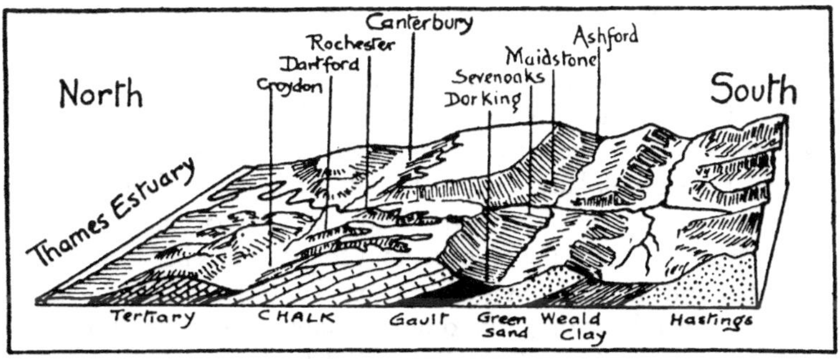

FIG. 95. Diagram of topographic and geological control of town sites in the Weald, south of London. The sketch represents the northern portion of an eroded dome. *(From a sketch by Fagg and Hutchings.)*

or five miles wide, though it narrows in other parts of the Weald. Continuing our traverse we find soft Eocene beds above the Cretaceous Chalk, which extend to the sea-coast, or in the north-west to the lower Thames valley.

The early roads followed the open lands near the top of the Chalk plateaux, but the line of springs at the base of the Chalk determined the sites of the villages. These clustered along Holmesdale especially, as the thick forests of the low dales were cleared away. In the early Middle Ages the road on the plateau was given up in favour of the roads which soon linked these villages in the vales. Where the Holmesdale village happened to command one of the main river gaps through the Chalk, as at Sevenoaks on the Darent, Dorking on the Mole, or Guildford on the Wey, here quite large market towns developed. (Their positions in Fig. 95 are somewhat diagrammatic.)

Another line of villages and towns developed on the outer side of the Chalk plateaux, below the famous 'Pilgrim Way' which followed for the most part the northern slope of the Chalk. Here we find Canterbury, Rochester, Dartford, and Croydon (Fig. 95). In later English history Watling Street was built from Canterbury to London along the same route through these towns north of the chalk 'Downs' (i.e., *Dun* or hill).

Pioneer Settlements on the open Plains

It may be worth while before proceeding further with our study of topographic control to consider what happens in an area such as the Canadian prairies, where there is no topographic control. What decided the pioneers for instance, in the region between Winnipeg and Wainwright—some 675 miles to the north-west—where they should build their huts, develop a village, and later establish a town? The answer is of considerable interest and explains many of the first sites, though naturally there are a number of exceptions. This great agricultural area—as everyone knows—was developed by the railways. The Canadian Pacific first about 1883 and the Grand Trunk later, carried the rails across the level plains, which to some extent represent the elevated floor of a Cretaceous Sea.

It is instructive to read the list of stations for 600 miles between Portage (Manitoba) and Wainwright (Alberta) on the (former) Grand Trunk Railway. This is how they run, omitting only a few names in the long list. Arona, Bloom, Caye, Deer, Exira, Firdale, Gregg, Harte, Ingelow, Justice, Knox, Levine, Myra, Norman, Oakner, Pope, Quadra. This covers 110 miles. Then eighty miles further west we find a similar list: Atwater, Bangor, Cana . . . Fenwood, Goodeve, Hubbard, Ituna, Jasmine, Kelliher, Leross . . . Punnichy, Quinton, Raymore, Semans, Tate, Undors, Venn, Watrous, Xena, Young, Zelma, Allan, Bradwell, Clavet, Duro—which brings us to Saskatoon. However the names go on Farley, Grandora, Hawoods . . . Juniata, Kinley, Leney, Mead . . . Oban, Palo . . . Reford, Scott, Tako, Unity, Vera, Winter, Yonker, Zumbro, Artland, Butze, Chauvin . . . Dunn, Edgerton . . . Greenshields which brings us to Wainwright well within Alberta.

Probably the reader will have gathered that these names are in *alphabetical order*; and the stations are placed about six or seven miles apart on the illimitable prairie, though occasionally the distances are a mile more or less than those quoted. The author has no idea as to what determined the *exact* location of the stations, which of course were nothing but simple platforms (for a 'signal stop')—as indeed most of them are to this day in the pioneer areas. But the first step of progress was the building of a small siding; so that wheat could be more easily collected in cars and linked into a train. By this time there would be a few farms scattered around the country, and later a village comes into being.

Some years ago (in July, 1936) I made a rough traverse of the wheat areas to the north of Edmonton between that city and Lac La Biche, 132 miles to the north. The little town of Bon Accord (at twenty-seven miles) has custom for four wheat elevators, which stand up eighty feet or so above the flat prairies, and are always the most prominent features in the landscape. Boyle (ninety-two miles) is smaller, with two elevators and some half dozen stores along the road by the railway. It also boasted a hotel. At Bondiss (ninety-seven miles north) there were twelve families, of whom four were on farms. Forty cows were pastured here, but at that time it had no elevator, though there was a station shed. However, at Noral, fifteen miles further on, I saw only two houses, and the railway station was an old box-car at the side of the railway! These notes will give some idea of the variations in the stations and settlement in the pioneer wheat belt of Canada, for Lac La Biche is at the northern edge of the wheat belt.

As an example of a small town in the level prairies we may consider Elbow in Saskatchewan. It is not quite typical, because its site was to some extent affected by the big South Saskatchewan River, which lies a mile or two to the west of the town. Here is a rather sharp bend in the submature valley of the big river, hence the name 'Elbow'. In the period at the close of the Ice Ages, there is little doubt that the river took a fairly direct path to the east here, and entered the present Qu'appele River, and so reached the Red River near Winnipeg. There is, however, not a very pronounced notch at Elbow where the postglacial waters once flowed to the east. (See Fig. 60 in the writer's *Canada*, 1947.)

Around 1900 homesteaders began to settle in the undulating district, and soon the railway surveyors laid out the branch line from Moosejaw to Rosetown, and ultimately to Edmonton. This was a land-grant railway and the promoters received alternate blocks along the new route. They laid out the usual square townships which developed into the conventional 'gridiron' towns of North America (Fig. 18). There was a pumping station on the river, from which water reached the railway, which was built about 1908. Here for a time was a notable ferry, in which the boat was attached to steel cables across the river. The current operated the ferry by driving it along the wires—a practice which is often adopted in Canada as well as in Europe. In 1926 a big bridge, carrying railway and road, replaced the ferry. This also helped to develope Elbow, though the river and these other features are quite a distance from the town itself. In 1913 its population rose to 600, but it has lost some of its dominance as the rails pushed further and further to the west. Much of the traffic runs right through the town (instead of halting there) to reach Moosejaw; so that to-day the population is only about 200.

Elbow is on the boundary of the wheat and grazing areas of Saskatchewan, and its rainfall is only about thirteen inches. To the north-west and south are large patches of sandhills, which are only of use for sparse grazing in summer. Other railways and new towns across the river have taken much of the custom from Elbow, so that its umland is an oval, about thirty miles long, between Riverhurst and Davidson.

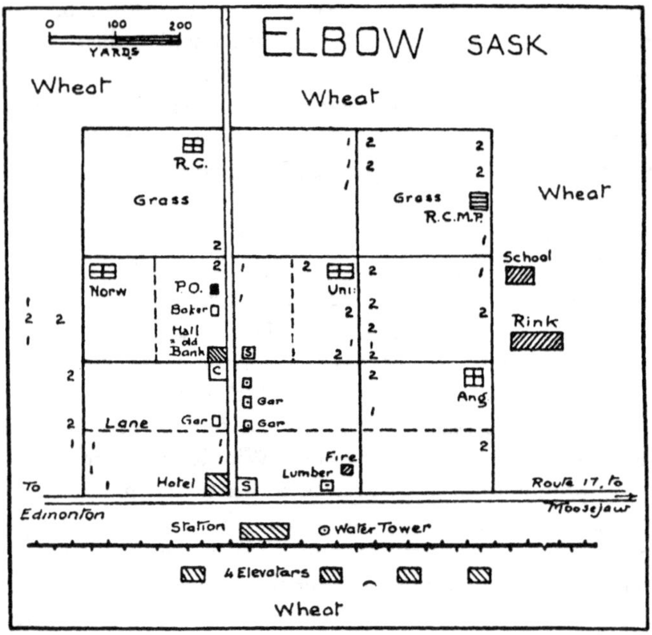

FIG. 96. Elbow, a small town in the drier portion of the great wheat belt in Saskatchewan.

Before 1908 there were two little stores some way from the new 'township', and they moved in to the town as it developed. Two of these buildings—some forty years old—still house the café and the drug store. Technical changes have also affected the population. For instance, there was a petrol 'combine' at work cutting the wheat during my visit in 1944, and it could do as much work as forty men in 1910; thereby lessening the need for labour in the vicinity. It is very hard to hire labour, and farms with six helpers at former harvests would have no one but the farmer himself to-day. Elbow is a 'four elevator' town, and these towers —some eighty feet high—are charted in the sketch survey (Fig. 96).

The functional plan of this little town shows that it has not advanced very far in forty years. The main road runs north from the station with a large hotel just at the station. There are two stores and a café, several garages, and a rather picturesque hall near the Bank. Since it is a farming district, there are four churches for the folk who live some distance out of Elbow. The Norwegian Church is perhaps the largest, but there are also Anglican and Nonconformist churches. It is not usually the case, but the Catholic Church was on a much smaller scale than the other three. A fine school, and the usual large Rink—for winter skating, etc.— were located to the east of the town. Wheat fields pressed hard on the town on all sides, as suggested in the sketch survey. The houses were all either one or two storey buildings, as indicated, and were all built of wood.

A second example of a prairie town is one which I surveyed in 1944 on my return from a traverse of the Alaskan Highway. At the southern end of this remarkable roadway we enter the very margin of the Peace River agricultural lands at Fort St. John. This old settlement is in 56° N. latitude, and was founded so far back as 1805 on the Pine River. However, the present post was only established in its present position in 1925; and it is some miles from the Peace River, in typical undulating poplar-scrub country, which is so abundant in that part of British Columbia (Fig. 97).

In 1928 there were three government offices in the young town, as well as the new Hudson's Bay Post. The first log hotel was just being built, but most of the private residents lived in tents. The Anglican Church was erected in 1930, and the hospital and flour mill in 1933. The development of the airway to Alaska, with a very important airport about four miles to the east of Fort St. John, together with the construction of the Alaskan Highway (which runs diagonally across the south-west corner of the town) have all greatly increased the growth of the town during the last two or three years.

The town plan is the conventional grid, based on the original township survey. In fact at the cross roads the four townships meet, their boundaries being the two main roads shown on the plan. The country is almost level, and there seems no obvious reason for choosing this particular crossing of boundaries as the site of a town. The chief shops are to be found on the west-east road between the bank and the chief hotel. There are two other hotels, a cinema theatre, and a police post, in the shopping centre. Three butchers, two fur dealers, several cafés, and a drug store, about completed the list in 1944. Further away are the large hospital, the new school, a rink, two auto camps, two garages and a flour mill. As usual the churches, of which there are four, are on the borders of the little town.

We may now consider a pioneer town under still more adverse con-

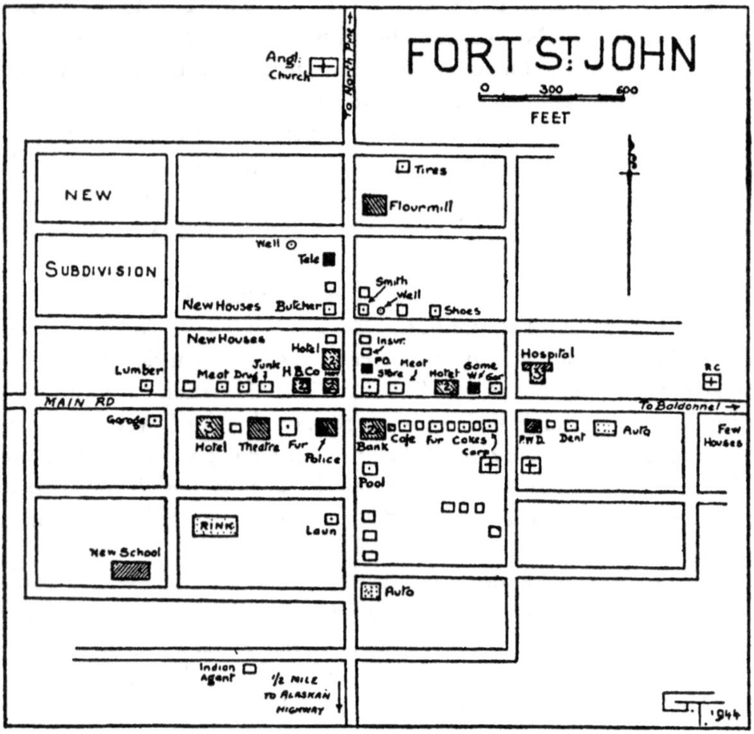

FIG. 97. Sketch survey of Fort St. John. Main buildings are black or ruled; shops have a single dot, churches a cross.

ditions, i.e., on the granites of the Canadian Shield. One of the most interesting types of settlement is that which has grown up along the railway, and which has no other contact with the outside world. These are still the rule in many parts of central and western Canada, as we have seen, but there are now not so many in eastern Canada. It is worth noting that it is only a year or two since it became possible to cross the Shield north of Lake Superior by car. Long after the Rockies had been crossed by several superb roads, the gap to the east of Lake Nipigon was still crossed only by rail, though now the last section from Nipigon to Hearst is open to the public. In this region of the Shield north of Lake Superior, but 100 miles to the south of the new transcontinental road, is the little town of White River. It is about 250 miles to the north west of Sudbury, and has the doubtful distinction of being the coldest place in southern Canada (Fig. 98).

A few years ago the writer made a sketch survey of this settlement, which dates from about 1884, when the Canadian Pacific Railway was built through the region. Two roads lead out of the town, to the north and east respectively, but they both peter out after half a dozen miles.

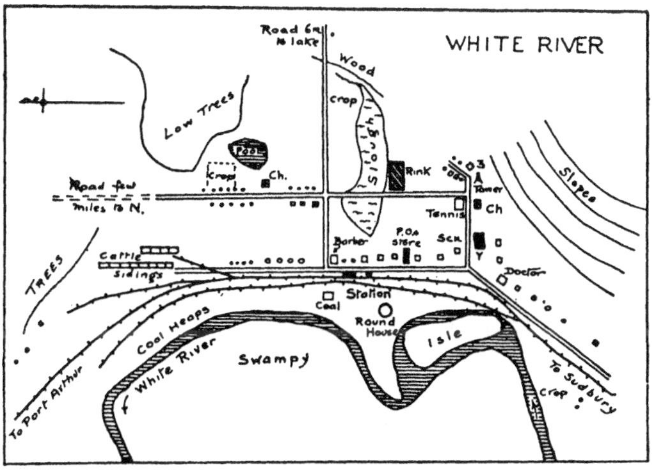

FIG. 98. A sketch survey of White River, a 'Railway Division' town 250 miles north-west of Sudbury. This settlement is not connected by roads to other towns. It extends half a mile from north to south.

The town is a 'Division Centre' on the railway; where the engines are fuelled, repaired, changed, etc., and where the drivers often end their day's run. Another important function of the town is that here cattle are watered and fed on their long run from the west to the stockyards of Toronto, Montreal, etc. How has this important local centre developed in the past fifty years? We must journey about 120 miles to the east to reach the similar Division town of Chapleau, while Schreiber has a like function to the west.

The White River here runs close to the railway and is flowing to the north-west in a somewhat meandering fashion, though rapids are common on its course nearby (Fig. 98). The Railway Yards are naturally extensive, with eight sets of rails near the station. Large heaps of coal and an elaborate coal shoot are situated to the north of the large Round House. The cattle sidings are of great length at the north end of the yards. The chief building in the settlement is the 'Y' (i.e., the large Y.M.C.A. boarding house) run primarily for the benefit of the railway employees, but also open to the public traveller. Similar useful institutions are found at Schreiber and Field, and no doubt at many other Division Stations.

What sort of town plan has developed during the existence of White River? It consists mainly of two long streets running north-south with a cross street and a few cross tracks. These are indicated on the sketch plan. There are about eighty houses in the town, all built of wood, and about half being two storeys in height. However, the house of the local meteorologist and one of the shops are of three storeys. As usual the smaller one-storey houses are on the margins, chiefly in the north of the little town. There are only two or three shops, the chief store being also the post office. A doctor, baker and barber live on the main road near the station, which is built of brick. There are two churches and a wooden two-storey schoolhouse, which are shown on the plan (Fig. 98).

Low rounded hills border the broad valley of the White River hereabouts, while several pools and sloughs occur in the vicinity. I was much interested in the agriculture in this part of the Shield. White River has no extensive lake silts, such as improve the character of the soils of the Clay Belt. Yet there were three ploughed fields close to the town as shown on the map, and several more ambitious areas further away. Oats and potatoes are the chief crops, but vegetables like lettuce and turnips do well. As is universal in such environments, the costs of transport prevent such crops from being of much value save to the few local inhabitants. The townsfolk felt that things would improve when they were linked by road with the rest of the world. Possibly tourist traffic will be of considerable help in the future. Since the town has developed no residential zones and no definite shopping centre, it is in the Infantile stage (p. 10) in spite of its life of fifty years.

Seven villages in the Corn Belt of Illinois

An interesting parallel environment, where topography is of quite minor importance in determining sites, has been studied by Clarence B. Odell in a paper with the above title dealing with the functional patterns of a number of settlements in central Illinois. It is published by the University of Chicago Libraries, 1939. Odell points out that fifty-two per cent of the folk in U.S.A. live on farms or in small towns, and that in the state of Illinois half a million folk live in villages with a population less than 1,000. Indeed there are probably 20,000 villages in the States with populations between 250 and 2,500; and Odell's purpose was to find out the way of life and the settlement patterns of this very important section of American life.

He chose seven villages in Maclean County, Illinois. This at one time was the leading county in the United States for the production of farm crops, and is still classed as one of the ten richest agricultural counties (Fig. 99). The landscape is level to undulating, with about four farms to the square mile. White frame houses and red barns, with a white school house each two miles, these are characteristic features in the county. In the 236 square miles of the portion specially studied there were 815 farms. The district is traversed by two railways, the Illinois railway runs from north-east to west, while the Nickel Plate traverses the southern edge of the district. As in the Canadian examples the railways are vital factors, but in U.S.A. very good roads traverse the region, and play a part of equal importance, which is not the case in the Dominion examples.

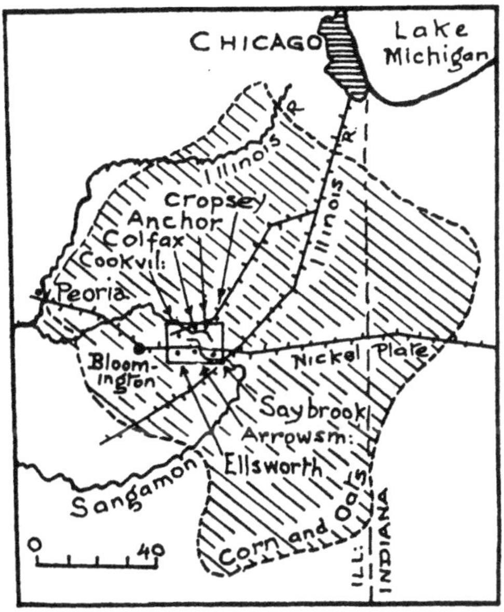

FIG. 99. Seven villages in the centre of the rich Corn and Oats belt of Illinois.

The origin of these typical little towns in the Corn Belt is of interest. The county on the edge of the Grand Prairie was the last part of Illinois to be developed for agricultural purposes, as it had little value until railways made available a market for agricultural products. The railways also helped to get the lumber of the region to its users. The settlements before the railway era were in the groves along the Sangamon River. 'Cheney Grove' was the first village, and it only dates back to 1856, though there had been some settlement since 1830. During the sixties a stage coach to Bloomington passed through the southern part of the county.

The Nickel Plate Railway built its line through the area in the early seventies. Odell states that the railway was diverted considerably south of its direct route to take in Cheney Grove (now Saybrook). The sites of Arrowsmith and Ellsworth along the railway were laid out on land in part at least belonging to directors of the new railroad. In the early eighties the branch of the Illinois Central Railway entered the area from the north-east. The elevators or sidings for grain were rather evenly spaced along the railroads, and villages have grown up with these as nuclei, as we saw in the Canadian Prairies.

The year 1900 saw the peak population for Saybrook and Colfax, the two largest of the settlements under consideration. This was the period of railway expansion, when the distance from Bloomington made these two villages important trading centres for that section of the county. On the whole the seven villages have declined since that date. The automobile has made a recent change in the economy of such regions. As Odell points out, the villages need the farm contacts now more than the farms need the villages. Any farmer can travel forty miles to buy goods, whereas in the early days he would not wish to travel four.

All these seven villages, and fifty more studied by Odell, have much in common. They are stations along railways; they are business, educational and social centres of small rural areas. They have stores, garages, schools, churches, elevators, cream stations, and residences like all the rest. The houses are either brick or frame, on streets bordered by trees, with concrete or brick footpaths. These seven villages have electricity but no gas, cleanliness but few bathrooms, baseball fields but no golf courses. There are, however, plenty of differences in the village patterns which we will now proceed to investigate. Bloomington is the main commercial centre, but the four northern villages are tied to Kankakee by the Illinois railroad, while the southern villages are on the Nickel Plate Railway. The three southern villages are some distance from the main paved highway, but the northern four (Fig. 100) are almost on paved routes.

Ellsworth was laid out in 1871 and incorporated in 1925, so as to provide funds to tile the streets for drainage, etc. The plan of the town is given in Fig. 100; and the town pattern was laid out symmetrically north and south of the railway. Even to-day Ellsworth has not spread much beyond the original lay-out. The chief business block is on the south side of the railway, and it is a 'two elevator 'town. Now there is only one train each way *per diem*, and the railway is not the centre of interest as it used to be. The better residential section is also south of the line, near to the two schools and two churches. Brick is used for most of the shops, but the houses are all made of wood, and belong mostly to the pattern in vogue in this area in the eighties.

There are eleven shops, etc., including three stores, lumber, tools, garages, radio, cafés, etc. The single doctor serves a community extending within a radius of about ten miles. Three out of four homes are owned by the residents, but the average family seems to consist of two adults and one child.

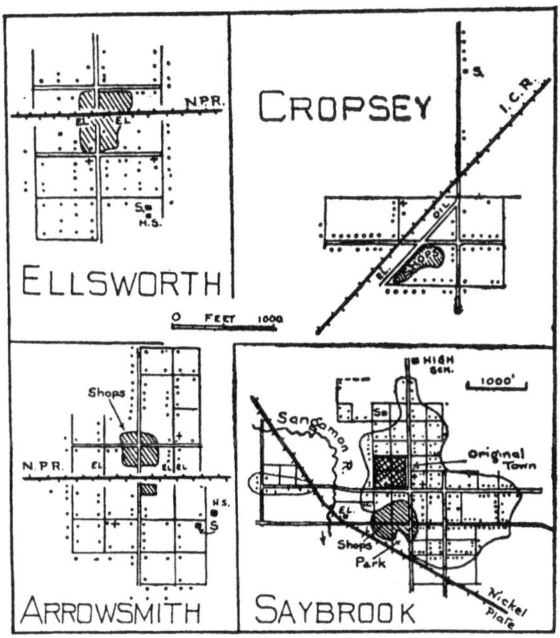

FIG. 100. Four villages in the Corn Belt of Illinois. Dots are houses. The diagonal ruling shows the shopping centre. (*After C. B. Odell.*)

Except for the transport of grain from the district, in which use is made of the railway, most of the transport goes by road. The village serves an area equal to one township, and the same sort of distribution of villages is found in adjacent townships.

Arrowsmith is much like the preceding village, but it has three elevators, and the shops are on the north side of the railway. The two schools are on the south side of the railway and are made of brick. Saybrook is the oldest, and the second largest of the seven (Fig. 100). Its pattern differs from the others, since the village was there before the railway which runs along the south-west corner of the village. It is over two miles from the main highway, and the main street runs west-east through the shopping area, which is located near the railway, but not in the centre of the village, There are three churches, and the whole settlement is built in a grove near a stream so that it is the most picturesque of the seven villages.

Cropsey (Fig. 100) is unlike the others since the shops directly face the railway, and it is one of the smallest villages. It has only one elevator; and the stockyards are now little used, since stock is taken out of the district by truck. There is, however, a large oil-distributing centre at Cropsey. The frame houses are painted white, and usually surrounded by pleasant gardens.

Summarizing the seven villages we note that they are about five miles apart, and each covers less than half a square mile. Colfax and Saybrook have about 1,000 inhabitants each, but five are under 500 in population. Most of the schools are of brick, but the churches are usually built of wood. There are three newspapers in the seven villages, and mail is delivered twice a day by truck (not railway). In conclusion, it may be pointed out that a number of villages in the pre-railway days have vanished, and to-day perhaps the highway takes precedence over the railway. This is the heart of the Cash Grain portion of the United States, but apart from this crop all the other farm produce makes use of roads rather than railways for transport.

Umlands of Towns and Cities

The concept of the Umland is an important one in our study of urban settlement. The umland of a town is that portion of the surrounding country which is linked culturally with the town as a centre. There are of course many such links, and in general the larger the town the more diverse the links. Stanley Dodge has worked out one of the simplest of such linkages in the region surrounding the little town of Howell in central Michigan. Howell is situated about twenty-five miles north of Ann Arbor, and has a population of about 3,000. It is surrounded by a farming population, and Dodge investigated the routes used by the farmers, and noted the character of the roads themselves. His paper on Howell follows one on Princeton, Illinois;[1] where the same ideas are carried out in a more general fashion.

Dodge and Wilson in their field investigations mapped each farmyard entrance to see the direction in which the traffic usually turns. Further-

[1] *Bureau and Princeton Community*, Annals Amer. Geog., Sept., 1932. *The Umland of Howell*, S. D. Dodge and L. S. Wilson, Mich. Acad. Sci., 1932.

more they found that a graded road generally drops to a lower state of improvement in the section between divergent turnings; or may disappear in a growth of bush. This is called the 'bad-road divide'.

In Fig. 101 based on their map, the general direction of the turnings is shown by the arrows on the roads. Within the heavy boundary line all the arrows turn toward Howell. Outside of this line the traffic leads to some other centre. On the north, minor villages such as Oak Grove confuse the pattern, and the authors conclude that they occupy the interstices where two or more umlands come together. The total extent

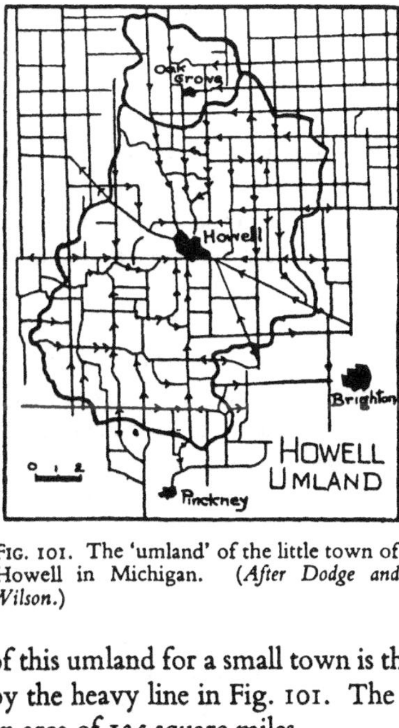

FIG. 101. The 'umland' of the little town of Howell in Michigan. (*After Dodge and Wilson.*)

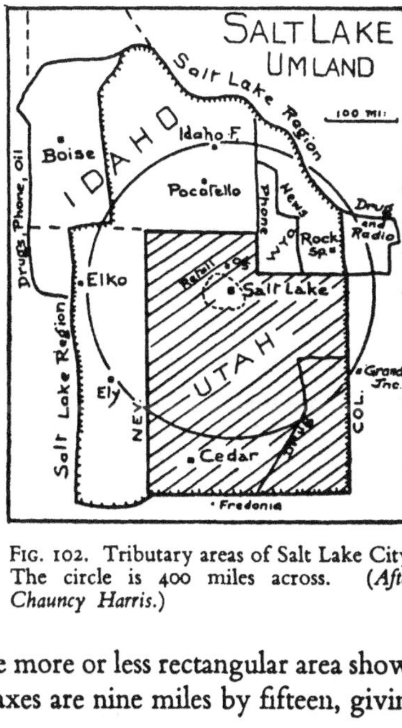

FIG. 102. Tributary areas of Salt Lake City. The circle is 400 miles across. (*After Chauncy Harris.*)

of this umland for a small town is the more or less rectangular area shown by the heavy line in Fig. 101. The axes are nine miles by fifteen, giving an area of 135 square miles.

A much more complete study of such 'tributary areas' has been carried out by Chauncy Harris with respect to Salt Lake City and its surroundings.[1] He points out that the city is the regional capital for an area called 'Salt Lake Region'; this region covers an area of 185,000 square miles and contains 790,000 people in Utah, south-western Wyoming, southern Idaho, and eastern Nevada (Fig. 102). The city has a population of 150,000, and is situated in the 'Wahsatch Oasis'. This latter is a transportation focus, since it lies on an axis marked by the Wyoming Basin

[1]*Salt Lake City, a Regional Capital*, Univ. of Chicago, 1940.

in the east and the Humboldt river in the west. This route is followed by the Union Pacific-Southern Pacific railway through Ogden.

Fig. 102 reproduces the essential data of one of the numerous charts in the memoir by Harris. Twelve selected services, performed by the city, have been investigated as the basis for this map. They are retail trade; wholesale grocery trade; wholesale drug trade; radio broadcasting; newspaper area; religious sphere of influence; telephone area; bakery distribution and oil distribution and a few less important services. On the basis of these twelve individual limitations the generalized boundary of the Salt Lake Region was drawn. It appears in Fig. 102 as the 'pecked' boundary, which extends far beyond the state boundary of Utah to the north and west.

The *retail* trade boundary includes a much smaller area than the others, It corresponds to some extent to the umland shown in Fig. 101 for Howell. It includes the areas from which people shop regularly in Salt Lake City. The wholesale *grocery* and the *bakery* limits do not extend far to the north of the city, although they agree with the general 'regional boundary' in other directions. These two northern boundaries, running near the state boundary, are not shown in Fig. 102.

The *religious* boundary limits the area in which the Mormons form more than twenty per cent of the total population. It agrees with the regional boundary (pecked line) on the south and east, but does not reach quite so far from the city on the north and west. The *newspaper* limit agrees very closely with the regional boundary except to the north-east, where its boundary is indicated in Fig. 102. Drugs, telephones and petrol are carried from the city some distance into Idaho beyond the regional boundary, as the map shows.

Thus the 'Regional Boundary' as drawn by Harris is based principally on the generalized trade area, the newspaper circulation, the religious hinterland, with lesser consideration of other tributary areas. On the south nearly all the services of the city stop near the state boundary. On the south side the barrier of the desert is reinforced by the Grand Canyon of the Colorado. On the north the services stop at the state boundary between Idaho and Montana. The south-east corner of Utah is (for some services) linked to the little town of Grand Junction in Colorado. In the west the region includes Elko and Ely in Nevada, but beyond these towns the state of Nevada looks to other trade centres than Salt Lake City. In the north-west Boise in Idaho is a town which is large enough to rival Salt Lake City, and accordingly we find here a transition area where the newspaper and other services grade off gradually.

CHAPTER XIII

TOWNS SITED ON RIVERS

Introduction

As pointed out in an early chapter it seems likely that the first villages grew up in the irrigable regions of south-west Asia. If we turn to the period of great migrations in historic times we observe that the rivers were often corridors of advance into the unoccupied territory. In Europe the Danube led the Alpine races into the centre and west of Europe. In North America the French advanced along the St. Lawrence and the great Lakes, and then swept to the south down the Mississippi as far as its mouth. Joliet and Marquette reached the Arkansas River in 1673, and La Salle reached the Gulf of Mexico in 1682. In South America the Amazon flows through hot steamy selvas, and is still not occupied in much of its course by progressive peoples. But such settlement as exists in these tropical forests is close to the rivers in every case. In more clement climes to the south we find the La Plata and its tributaries playing a dominant role here as in the northern hemisphere. In Australia it is a curious fact that the mouth of the Murray—the sole important river—was the last stretch of the whole coast to be charted (in 1802). But the country along the Murray and the Darling was occupied by pastoralists long before any one lived in much of the vast semi-arid regions nearer the coast of New South Wales.

In the third chapter of this volume, where the infantile beginnings of settlement are being discussed, a number of examples along the great Mackenzie River have been described. In all cases the settlements were in a pioneer stage, and often enough, though they had been occupied for a couple of generations, the population of whites was only about a dozen. There the *primitive* character of the settlement was emphasized, but certain features due to the site on the banks of a large river were brought out. We may summarize them here.

First of all it should be remarked that over a great portion of its course the Mackenzie is a 'mature' river. It flows for the most part in a *broad* valley, between 'cut-banks' of silt laid down by itself at earlier stages. There are present, in general, neither the cliffs nor the gorges of the 'juvenile' stage, nor the winding meanders with very low banks of the 'senile' stage. But of course every long river is bound to exhibit slight changes in the character of its valleys due to local interruptions of its even history. There is an amusing instance of this in Emil Ludwig's romantic book dealing with the Nile. Knowing nothing of the true *topographic* developments of the river he, perhaps naturally, describes the upper part of the Nile as 'young', and the lower part as 'old'. To the geographer the upper Nile flowing in the vast marshes of the Sudd in the Sudan is old (indeed 'senile'); while in the cataract area it is as young ('juvenile') as a river can well be (Ludwig, 1937, New York).

It will be well, therefore, to particularize the character of the river, when we discuss its relation to the town which has grown up on its banks. Fort Smith (240 folk) illustrates a settlement in a mature river incised some 100 feet into its flood plain. Aklavik (with rather fewer citizens) is in the delta, where, as ever, the river breaks into distributaries and these exhibit marked meanders, i.e., the river is here senile. In a few short sections of its course of a thousand miles there are examples of 'juvenile' topography, as in the five-mile canyon just south of Good Hope. Here needless to say there are no settlements, since in general the latter are fifty miles apart, and the pioneers choose the more convenient sites.

The data from the lengthy traverses made by the author in the Athabasca, Slave, Mackenzie, Lewes and Yukon Rivers—all in a state of pioneer settlement—lead to the conclusion that the site where a large *tributary* enters the main stream is the most likely one for a town to develop. This determined the position of MacMurray, Hay River, Simpson, Norman and Macpherson on the Mackenzie, and of Selkirk on the Yukon River. Next in number (and perhaps more obvious sites than the preceding) are the settlements which are determined by important rapids in the stream. Fitzgerald and Fort Smith on the Mackenzie, and Whitehorse on the Lewes are examples of this type. Special circumstances such as gold at Dawson City, oil at Norman Wells, end-of-rail at Waterways naturally dominate the topographic factor; since it is not difficult to plant a settlement, as far as topography is concerned, almost anywhere along the rivers cited.

Pierre Lavedan in his little book *Geographie des Villes* (Paris, 1936) gives a brief discussion of the role of rivers in the siting of towns. He points out that such towns rarely develop equally on both banks. Along the Seine between Paris and the sea we find Mantes and Vernon on the left bank; while Rouen and Caudebec are on the right. On the Loire Orleans and Blois are on the right and Tours and Saumur are on the left. In most cases this is due to a topographic difference, one bank being more subject to floods than the other. In other cases as at Cologne and

Maestricht the town grew up on the side protected from invasions by the broad river. Thus the Roman colony of Cologne is on the French side of the Rhine (i.e., nearest to Gaul), and as for Maestricht (Meuse 'trajectum') it was also built on the west side nearest to the main Roman territories.

If the town spreads across to the other side it is very rarely that both sections are of equal importance. We have already seen that in Paris and London, the north bank has the more important portion of the city. In both cases the chief wharves in the early days were on the north bank, though in the case of Paris (Fig. 77) this was the 'Marais', i.e., the marshy quarter. In rivers of a pronounced character like the Seine with its deep incised meanders towns may develop on the concave or convex side of a meander. We shall examine certain examples of this character shortly. Often enough a narrow incised meander is a good site for a town, since it forms a 'peninsula' with steep scarps, and only needs to be defended on the 'neck' of the meander. Bern is such a site and Cahors another; and many others have been noted in our study of early defensive positions.

Of course certain river sites are early marked out for the growth of cities, as for instance Vienna at the north-west gate of the Alföld plain of Hungary. Here the Danube has cut its way across the slowly rising Alpine ranges, and its valley is much restricted for some 100 miles to the west until it opens out again into the broad expanse of the Bavarian Downfold.

Finally, very special types of 'river towns' are those which have developed in direct response to water-falls. These furnish perpetual power, and such sites are in increasing demand as our sources of coal and oil are used up.

Some asymmetric river sites in Russia

Particularly interesting are some of the rivers of southern Russia, such as the Dnepr, Don and Volga. These flow to the south, and are cutting their beds into the western banks in all three cases. Thus they are encroaching on the rocky 'platforms' on the west, and leaving to the east wide swampy lowlands. Kiev, Stalingrad and many another town has had its site on the west bank determined by this encroachment to the west. It has been suggested that this is due to the 'Ferrel Effect'; whereby all moving bodies (including rivers) tend to turn to the right in the northern hemisphere, owing to the earth's rotation.

A brief description of some of the famous Russian towns charted in

Fig. 103 will be of interest, though lack of space does not permit of our giving detailed plans of them. The Dnepr (pron. *Din-yèpper*) is cutting into the Podolia Block, and the three chief cities on the lower Dnepr are Kiev, Cherkassov, and Dnepropetrovsk. Kiev is the most honoured city in the south of Russia, and was founded before A.D. 864. The city is 590 feet above sea level, and stands on a series of wooded heights nearly 300 feet above the Dnepr river, which is here about 500 yards wide. On the east bank the land is swampy, and there are practically no dwellings. Cherkassov has a somewhat similar position lower down the river. At Dnepropetrovsk the river swings to the south and cuts into the hard rocks of the Podolia Block. Here are a number of rapids, and the half

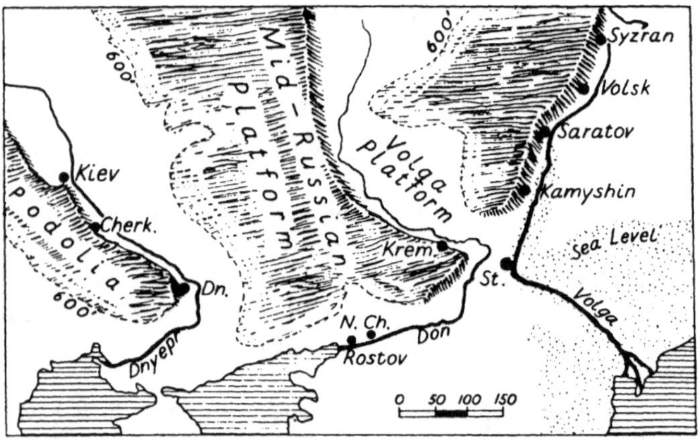

FIG. 103. Towns on the high western banks of South Russian rivers. Cherkassov, Dnepropetrovsk, Novo Cherkask, Kremenskaya and Stalingrad are indicated.

a million citizens are engaged in engineering and other industries linked to the tremendous power developed by the great dam across the Dnepr.

On the Don the towns are not so large, nor are they clustered so obviously on the west bank. However, Kremenskaya, and the two towns of Rostov and Novo Cherkask near the mouth are on the higher bank to the north-west, as in the case of the other rivers. The most striking examples of the sites we are here considering are to be found on the lower Volga. As the map shows there are five large towns on the west bank, and hardly one on the right bank. Syzren, Volsk, Saratov, Kamyshin and the famous city of Stalingrad are all perched on the high western bank facing the swampy country to the east. Below Volsk the steppes extend to the east, and here is a large colony of Germans, who have, however, founded no large cities. Stalingrad (formerly Tzaritsin) was founded about 1700 as a Cossack outpost. A small canal, to be replaced by a larger ship canal, now connects the Volga with the Don, some fifty miles to the west.

Vienna and the Danube

The three most important rivers in Europe are the Danube, the Volga, and the Rhine; and of these the former has probably had the greatest place in European history. Of the many cities and towns upon its banks Vienna holds first rank, and we may well commence our study of river

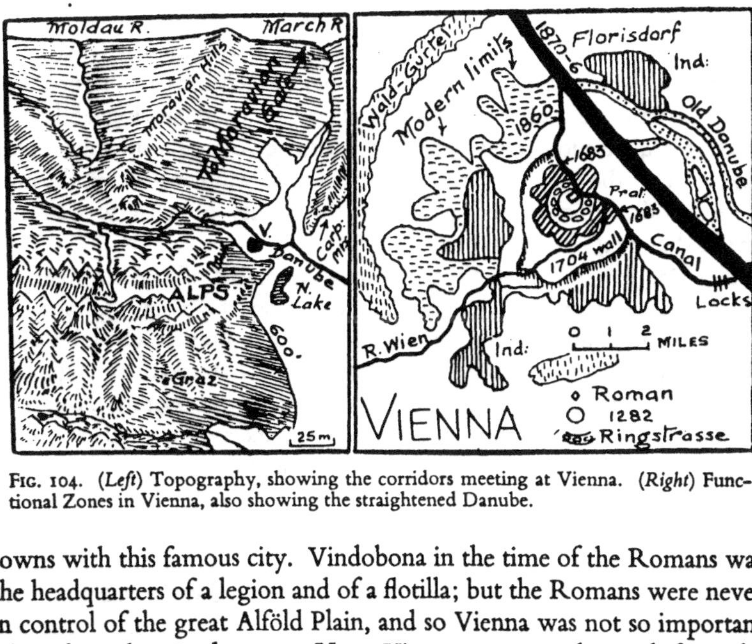

FIG. 104. (*Left*) Topography, showing the corridors meeting at Vienna. (*Right*) Functional Zones in Vienna, also showing the straightened Danube.

towns with this famous city. Vindobona in the time of the Romans was the headquarters of a legion and of a flotilla; but the Romans were never in control of the great Alföld Plain, and so Vienna was not so important a 'Gate' as it became later on. Upon Vienna converge the roads from the Adriatic, from Bohemia, and from Silesia; the latter by way of the Moravian Gate, the valley of the March, and the Oder Canal (Fig. 104). For centuries the city was the easternmost outpost of European civilization, as the word 'Austria' (i.e., 'eastern Reich') suggests.

The original town was established at the mouth of the little river Wien—hence the name Vienna. It was flanked on the west by the wooded heights of the Wienerwald, which in turn form foothills to the great Alpine masses of central Europe. To the north-west are the lower rounded summits of the Bohemian horsts; and the plains of the Alföld just reach to Vienna, and are bounded by the 600 foot contour (Fig. 104). Just across the river the outlying ridges of the Carpathians reach nearly to the Danube, so that the manner in which Vienna guards the 'Gate' into Germany can readily be realized from Fig. 104.

Vienna has passed through the usual stages of a continental city, and these may quickly be summarized. During the early middle ages it was little bigger than in Roman times. A strong circular wall defended the site, barely a mile across. As the city grew this was a great obstacle to ready access, but although Vienna soon extended beyond these walls the latter were not removed until 1860. As a result a series of parks and fine residential areas was obtained known as the *Ringstrasse*. New ramparts were laid out in 1704, and these in turn have been removed to form an outer boulevard known as the *Gurtel*. Just before the last Great War dislocated civic affairs in central Europe, the citizens of Vienna had acquired vast areas of woodland on the skirts of their city, and were creating a continuous *Wald Gurtel* at a distance of some six miles from the centre of the city. A grand panorama road was constructed through these woods, which added greatly to the beauty of what was already one of the finest cities in the world.

The Danube at Vienna already shows the marks of a senile river. It used to flow by many winding channels across the plain, and the city grew up a little away from the main stream on a branch which has long been canalized. About 1870 the large river was attacked by engineers, who rectified the whole river system in the vicinity of Vienna. A straight channel 980 feet wide and ten feet deep was excavated about this time to the east of Vienna, through which the Danube now takes its course, its old bed having been partly filled up. These changes are indicated in the map given in Fig. 104. To the south-east of the city, in the lowland between the canal and the rectified Danube, is a large park called the Prater. Across the broad river, a number of industrial suburbs have developed such as Florisdorf; though there were few residents in this area before the improvements of about a century ago.

In the early days of Vienna's development it is evident that the Danube did not play the part in the city's economy that might have been expected. The city was built over a mile from the larger channel of the Danube, and it was not till the Danube bed was straightened (around 1870) that river trade became important. The little branch was canalized about 1600, and about 1876 the so-called Canal was deepened and broadened so as to form an elongated dock. Many wharves have been constructed along the main river channel also. The little stream of the Wien has now been arched over, and carries an important railway. In 1900 the population consisted of 1,138,000 Germans, 133,000 Hungarians, and 103,000 Czechs. In 1939 the numbers had increased to 1,918,000 (See Fig. 88).

Dresden in Germany and Verona in north Italy, each built on the *convex* banks of a river meander, are similar cities which may be studied with profit.

New Orleans on the senile Mississippi

Of the two greatest rivers, the Amazon and the Mississippi, the latter is by far the most important, and the chief town is placed near its mouth at the base of the huge proliferating delta, as is apparent from Fig. 105. In the early days[1] all the region to the south of the Red River confluence was known as the Delta. New Orleans was the sole town for many decades in this large area. The delta was crossed by a maze of bayous, all of which could be reached fairly readily from the young settlement. A canal was cut (via the bayou) to the large lake Pontchartrain from the old French City (Fig. 107), and this opened up much country to the north-east. 'Down river' it was about 100 miles from the sea, and there was a large trade with Mexico and the West Indies in those early days. Tides had little effect in the crescent-shaped harbour; and the chief wharves, then as now, were on the northern concave side of the great bend of the Mississippi near the French city.

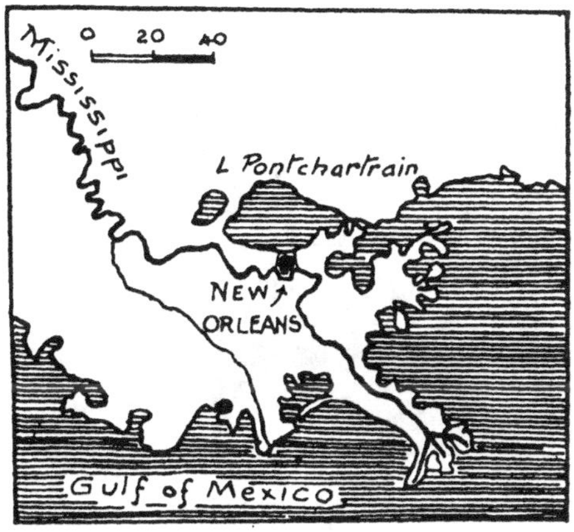

FIG. 105. The site of New Orleans between the Mississippi and Lake Pontchartrain.

There were three or four small settlements in the delta, such as Madisonville on Lake Pontchartrain, placed on a higher portion of the levee. At the time of the 'Purchase' (1803) there was one cotton mill in the city. About this time there were ten sugar refineries in the valley, and one at

[1]Two papers by E. F. Campbell in the *Geog. Review* of 1920 and 1921 may be consulted for a good account of the evolution of New Orleans.

16

New Orleans produced annually about 200,000 lbs. of loaf sugar. Lumber was important, and the planters cut mill races through the natural 'levees' (i.e., high natural ridges bordering the senile river), and erected saw mills which worked continuously during flood periods. In 1803 the port exported flour, tobacco, salt beef, and cotton to the amount of 40,000 tons. Until 1820 the downriver trade originated chiefly in the Ohio River, whence flour was sent to New Orleans. About 1817 the steamboat trade began to be important, and by 1821 287 barges came regularly to the city. In 1825 the commerce of New Orleans was worth seventeen million dollars, and by 1840 New Orleans was the fourth city in population in the United States.

The plan of the city about 1803 is given in Fig. 106 and has a number of points of interest. There were a number of disadvantages, the chief

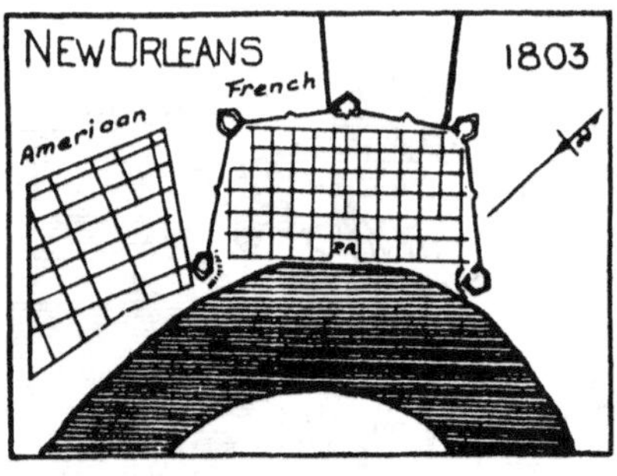

FIG. 106. The plan of New Orleans about 1803. (*After Campbell.*)

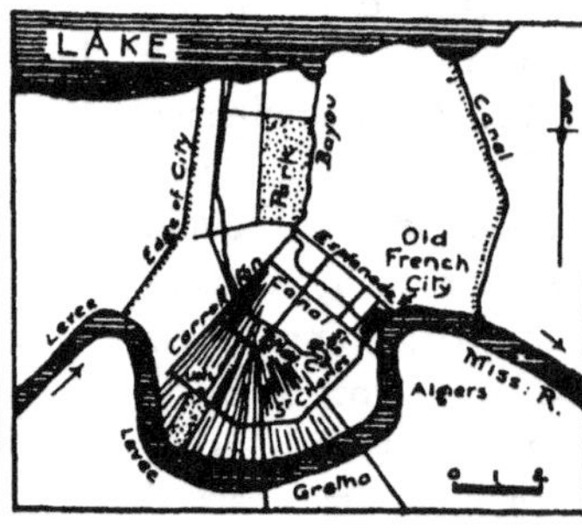

FIG. 107. The main features of New Orleans to-day. St. and P.=Station and Post Office.

being the low site, so that in floodtime there was always danger of the natural levees breaking and allowing the river to flood the lower parts of the city. (The levees are fifteen feet wide and fourteen feet high.) Sewage and drainage were very real difficulties in the early days, and accounted for the rather unhealthy reputation of New Orleans. Bienville founded the settlement in 1718, and the early French city was in the form of a rectangle, surrounded by the ramparts and bastions in the approved military style of the late 17th century. The central portion alongside the river was the Place d'Armes, and the streets within the ramparts were arranged in the conventional grid.

There is still to-day the 'French quarter' to the north-east of Canal Street (Fig. 107), where the houses and gardens are quite unlike those of the usual American city. By 1803 (as the map, Fig. 106 shows) there was an 'American quarter', which had developed immediately to the south-west of the French city. Here again a grid formed the basis of the plan, with the streets at right angles to the river bank. To-day the city extends for four miles up the river from the oldest quarter, and about two miles down stream. It has spread north along the Carondolet Canal as far as Lake Pontchartrain; and important suburbs such as Algiers and Gretna lie on the south bank of the river.

One peculiarity of the western portion of the city is the radiating character of the streets, which seems to converge on the junction of Canal Street and Carrollton Avenue (Fig. 107). This reminds one of the 'fan' pattern of Carlsruhe, where the centre was the Palace. But here there is no building of note at the centre of radiation, and the 'fan' seems to be due to the streets being laid out at right angles to the curving meander of the river. In the remainder of the city the various quarters have adopted the conventional grid, with the usual complete inattention to diagonal streets leading to the centre of the city. To-day this commercial centre lies about one mile west of the old 'French quarter' (p. 352).

Bern, on an incised Meander

Having now considered two cities on *senile* rivers, we may turn to juvenile rivers, i.e., flowing in narrow valleys. One of the most interesting river towns of Europe is Bern, the capital of Switzerland (Fig. 108). It is situated on an incised meander of the River Aar, which here flows to the north in a winding gorge cut 130 feet below a moderately flat promontory on which the city is built. Lofty bridges connect this promontory with the east bank of the river, where many suburbs have developed, almost wholly in the 19th century. The city has gradually spread along the neck of the promontory, as we have seen in classical examples of such a 'promontory site'. Thus in 1191 the town in general included the tip of the promontory, but by 1250 it had spread back to the west about half a mile. By 1350 another extension to the west took place (Fig. 108), so that the whole promontory was now covered with houses.

The town was founded in the 12th century, and a wall was built across the narrow part of the promontory, where the two bridges are shown in Fig. 108. The older part of the city in the east is said to preserve the character of an 18th century town better than most in Switzerland. The chief street still runs west-east along the middle of the promontory, from the railway station to the famous Bears' Den. A number of the old fountains date back to the 16th century. The Federal Palace (H. in map) was built around 1900, and rises from the cliffs of the gorge on the southern side of the old city. The Grosse Schanze was the site of a strong redoubt, built in 1623, but pulled down in 1840, and on this height is the University. To the south of the town the hill called the Gurten rises to a height of 2,800 feet, i.e., about one thousand feet above the town.

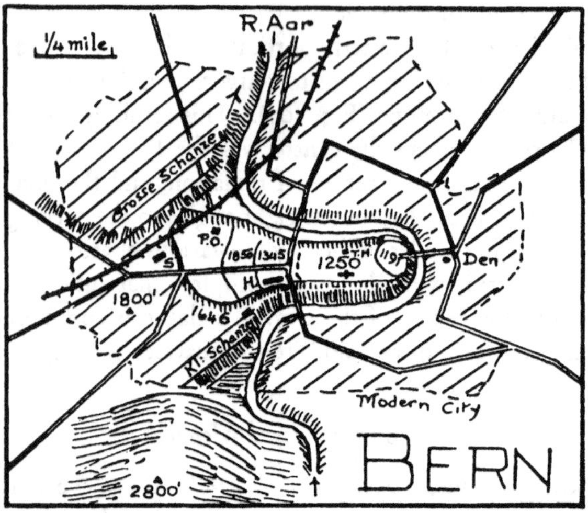

FIG. 108. Site-View of Bern showing a promontory city expanding west from 1191 to 1646. (*S*=Station; *H*=Federal Palace.)

To-day the city spreads far beyond the promontory as the map indicates. On the promontory the streets naturally run parallel to the long axis, but in the suburbs they radiate in a commonsense fashion from the high level bridges leading to the old city. There is, in these extensions of an ancient city, no rigid adherence to the 'checkerboard' plan which spoils so many American cities. The population in 1930 was 120,000.

Constantine and its four Natural Bridges

One of the most remarkable town-sites is that of Constantine in eastern Algeria, which was surveyed by the writer in the summer of 1938. Its general situation is akin to some of the 'promontory sites' already considered, but it is unique in the character of the gorge which separates the town from the adjacent plateau. The general features can clearly be made out from the sketch appearing as Fig. 109. Constantine is a city of more than 80,000 people, of whom five-eighths are European.

Sheer cliffs protect the city on three sides; to the south-west is a narrow isthmus by which the general level of the main valley floor can be reached. The main part of the city is lozenge-shaped in plan, with the great gorge of the Rummel River protecting the east. The Arab fort (*kasba*) occupies the highest part, while the southern part of the city is nearly 300 feet lower. The cliffs below the kasba are more than 500 feet high. The gorge is crossed by three bridges, of which the central and oldest one is above the ruinous Roman bridge.[1]

The most striking feature of the gorge is the presence of no less than four *natural bridges* across it. At one time the river flowed in a gorge 200 feet deep. Later it cut out a much deeper subterranean passage (probably partly by solution), leaving a platform between the two. 'Windows' broken through this platform have produced the four natural bridges. (See inset in Fig. 109.) On some of these platforms are small vegetable gardens. The length of the gorge is about one mile.

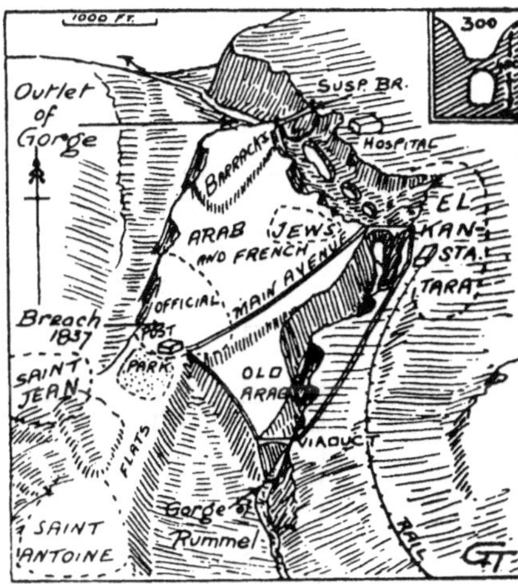

FIG. 109. Block diagram of Constantine showing the four natural bridges across the Rummel Gorge. *Inset* is a section of the gorge.

Constantine was founded by the Romans about A.D. 330, though earlier towns had occupied the site before. Before the siege by the French in 1837 the town was confined to the lozenge mentioned previously. A maze of narrow crooked streets threaded the congested houses, as they still do in the Arab quarter in the south of the old town. But at the higher level the French have driven through several broader streets; and have built the civic centre, 'Place de Nemours' in the vicinity of the 'Breach', where they broke through the wall in 1837. The government offices and the chief hotels surround the Place, and the better shops are close by. The Jewish quarter is in the east of the town. Better-class French and Arab residences, mostly of three or four stories, fill the northern half of Constantine; while the former kasba is now utilised for barracks. By 1884 the city had spread beyond the Breach, where the suburbs of St.

[1]Griffith Taylor, "Sea to Sahara"; *Geog. Review*, April, 1939.

Jean and St. Antoine have grown up. Near the station is another new suburb, El Kantara (i.e., the bridge).

Prince George: a Confluence Town

To the writer the early stages of a town are perhaps on the whole more interesting than the fully developed condition where a large city of the normal western commercial type has evolved. For this reason a brief sketch of the development of Prince George in the centre of British Columbia is included. There can be little doubt that this town will grow considerably in the future, though its development up to

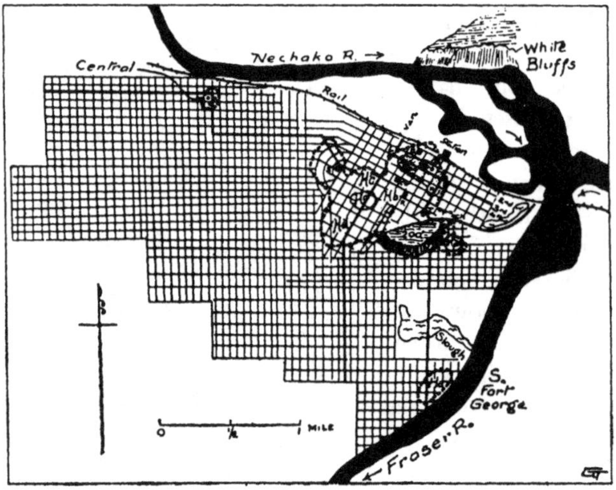

FIG. 110. Prince George in the interior of British Columbia showing a 'confluence town' as planned in 1915. The broken line shows the 'built-up' area in 1941. (*Geog. Review*, 1942.)

1940 was not at all remarkable. A glance at the map of the province will show that it is the natural centre of the northern portion of the 'elevated basin' (the badly named 'inland plateau') which occupies so much of the interior of British Columbia. Here the Fraser River makes its pronounced bend to the south, and along its valley passes the famous 'Cariboo Road' of the early mining days of the region. Across from east to west at Prince George passes the Canadian National Railway, en route from Edmonton to the natural ocean-outlet at Prince Rupert. To the north a railway is being constructed to reach the rich wheat lands of the 'Peace River Block', across the Rockies but still in the Province of British Columbia.

Prince George occupies the low, flat, silty triangle between the Fraser River and its large tributary the Nechako (Fig. 110). It shows us the start of a typical 'confluence town'. The remarkable bluffs of white silt to the north are relics of an earlier flood terrace; the isolated hill 200 feet high just south of the town may have a similar origin.[1] Prince George is a most interesting example of a 'boom town'—or rather of three boom towns that have not come up to expectations. In the decade following 1910 it was hoped that the Pacific and Great Eastern Railway would join the Grand Trunk here at the confluence of the two large rivers. As is not uncommon there was much rivalry concerning the site of the railway station; and about 1914 there were three independent centres, each large enough to challenge the other two to hockey matches! These were *South* Fort George in the south, '*Central*' (Fort George) in the west and modern *Prince George* in the east. To-day the first two have almost vanished; and Prince George in 1941 has dropped to 1,989 citizens, after reaching the figure of 2,479 in 1931.

Only about one-sixth of the streets shown in the 1915 plan were ever utilized; the remainder of the land is in practically virgin bush. The built-up area in 1941 is enclosed by the dashed line in Fig. 110. Third Street is the main shopping centre, and the Dominion Building, where are the main Government offices, is perhaps the main structure. 'Central' Fort George now has only two large empty stores and one small active post office and perhaps a dozen settlers. *South* Fort George has an empty theatre, dating back to the coming of the railway in 1915, one large active store, and about a dozen small farmhouses in the vicinity. (Lyons in eastern France, Coblentz in western Germany, and Namur in Belgium are well-known 'confluence' cities.)

Ashcroft: a Terrace Town in British Columbia

The outstanding topographic features in the province of British Columbia are undoubtedly the extremely numerous and important terraces. Three of them can be made out in the sketch given in Fig. 111. They run for miles along the great glacial valleys, like railway embankments at various levels. Near Lake Okanagan I have seen as many as a dozen of these terraces, one above the other. In the sketch the lowest terrace is about sixty feet above the river. It is the broadest and the best preserved, and upon its surface the little town of Ashcroft—with which

[1]Griffith Taylor, "British Columbia," *Geog. Review*, XXXII, 1942.

we are concerned—was built. A larger terrace runs for miles along the river at a height of about 200 feet, while a third, much dissected, is about 100 feet still higher.

In my recent book on Canada (London, 1947) I have suggested that

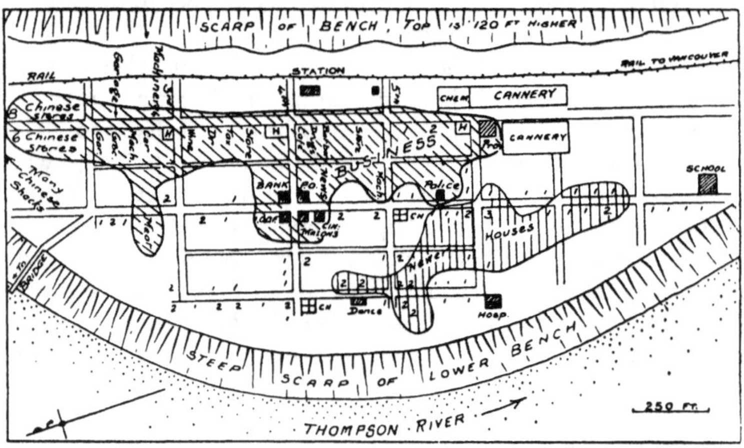

FIG. 111. Site-View and plan of Ashcroft—a small town on the Thompson River in central British Columbia. The terraces are very pronounced.

FIG. 112. Functional plan of Ashcroft—a small 'terrace town' showing a grid plan and a few factories.

these are not 'river terraces' of the type so common in Europe; but are due to the special glacial features of the Ice Age in British Columbia. They represent lake deposits in elongated marginal lakes, at a time when huge glaciers filled the main part of the large valleys. Whether this theory be true or not, there is no doubt that these terraces offer the best sites for agriculture and often for towns in the interior of British Columbia.

The plan of Ashcroft is quite simple. The Cariboo Road ran along the lowest terrace parallel to the river, and here crossed the Thompson by the bridge shown in the sketch. The Canadian Pacific Railway was built in 1885 and is shown in the sketch, while the C.N.R. arrived about 1915, and was constructed down the west bank of the Thompson. There is a good deal of irrigation on the west bank, and the sketch shows Chinese gardens of this type growing tomatoes. About 200 Chinese live at the north end of the little town, where there are a dozen tiny Chinese shops. The 'white' shopping centre is along the main street parallel to the railway; and at the south end is a cannery, which works for about six weeks canning the local tomato crop. Other details as to the public offices are charted in Fig. 112. There is a good deal of ranching in the vicinity, but no normal agriculture, since the rainfall is below ten inches. The total population is about 600 including the above 200 Chinese.

Fall Towns: Rivière du Loup and Grand Falls

Throughout medieval and modern history the advantages of a site near a waterfall have been recognized. In southern Ontario, for instance, one finds that the first act after a tiny village has evolved is to build a mill to grind grain, usually followed by one to saw lumber. These are still in use in towns not far from Toronto, though in the great city itself only the ruins of one or two of the old mills remain (together with traces of the sluices) to show the origin of such a name as Lambton 'Mills' (on the Humber). At a radius of forty miles or so there are a number still in use; and in many cases further afield their gradual conversion to mills run by power transmitted from Niagara can be studied.

'Fall Towns' are legion in the regions of good rainfall throughout the east of North America. I have therefore picked out about half a dozen typical examples of greater or less complexity. Rivière du Loup in eastern Quebec illustrates a small French industrial town based on the falls in a small river. Grand Falls in Newfoundland is a town wholly concerned with paper-pulp obtained through power from the local fall. Battle Creek in Michigan is a larger town, again depending at first on the local falls for its importance; while at Sault St. Marie we have a huge river harnessed to produce power for industry.

Rivière du Loup is a picturesque town on the south bank of the Saint Lawrence, about 120 miles to the north-east of Quebec. It is strung along the sides of a definite gorge, a mile or two inland from the coast of the Gulf (Fig. 113). Here the river descends from a broad inland shelf, about 400 feet above sea level, by several falls of which the largest is sketched in the inset in Fig. 113. This fall has a drop of about sixty feet, and the power house is placed on a ledge just to the west of the falls.

A French Mission was placed here in 1683, and after the Conquest in 1763 the district came into the hands of the Governor, James Murray. Later it was owned by the Frasers, who were descended from one of Wolfe's Highlanders. In 1765 there were fifteen houses in Riviere du

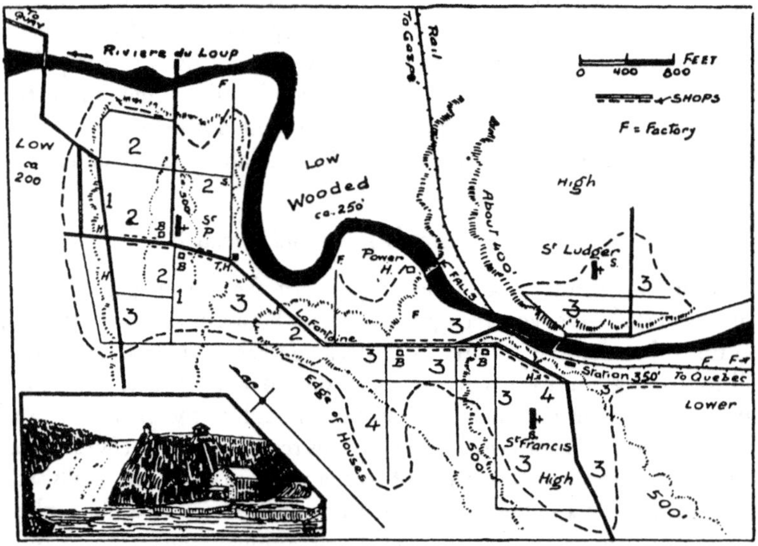

FIG. 113. Rivière du Loup, a town in S.E. Quebec, based on falls. Notice the four 'quarters' strung out along the river. *Inset* a sketch of the 60 feet falls looking south.

Loup with a population of sixty-eight people. Somewhere about 1800 a controversy raged as to whether the first large church should be built near the earliest church or near the trade centre. The latter site won, and the fine church of St. Patrice dominates the parish of the same name in the north of the little town.

To-day the town straggles along the river, from the lower portion near St. Patrice to the higher suburbs of St. Francis and St. Ludger near the railway station. In 1859 the Grand Trunk Railway reached the town; and it has always been a 'division' town, where repairs, etc., are made to the rolling stock. It is also the headquarters of the little Temiscouata

Railway to Edmundston; so that railway men form a considerable portion of the citizens.

In 1881 a company was formed to utilize the power of the main falls. At first this was used only to produce pulp from the adjacent timber; but the power is now employed by other factories and by the towns for various purposes. There are three smaller falls, and they in part supply a furniture factory, a foundry and the railway machine shops. The two latter are situated to the south of the railway station.

A good view of the present town can be obtained from the southern plateau near the church of St. Francis (Fig. 113). Immediately to the east is the fairly mature valley of the upper river along which runs the railway. The station is just below, and is reached by a steep street flanked by some small shops. Across the valley is another flat-topped hill crowned with the tall steeple of the church of St. Ludger. Here another parish has developed, and like St. Francis it contains for the most part small third-class houses (labelled '3' in the map).

The main street of the town runs parallel to the river, well above its west bank. Here are the chief shops and two of the banks. To the west of Lafontaine Street the land descends a little to fields bordered for the most part by fourth-class houses. The large brick Hotel de Ville (TH) crowns the cliffs just at the main bend of the gorge. Here is St. Patrice, and nearby a small English church. Most of the official buildings, such as the Palais de Justice and the hospital, are in this parish, as well as another cluster of small shops. Two large hotels are placed at the foot of the hill, where the main cross road runs along the edge of the coastal plain.

Thus the pattern of the town is rather unusual, since it consists of four units strung out along the river gorge. The railway was placed in the south so as to cross the river above the falls, and this has 'dragged' much of the industry and all the workers' dwellings in that direction. There are three large parish churches, as noted, and the fourth section is close to the power house and several of the small factories. In the writer's classification of towns it is to be assigned to the 'adolescent' class (page 85).

In few parts of the world is water-power of such importance as in North America; and for our second example of a 'fall town' we may consider Grand Falls in Newfoundland, which the writer investigated in the summer of 1945.[1] The chief river in this large island is the Exploits, which reaches the sea about half way along the north coast. The famous airport Botwood is on the estuary, and here the paper from Grand Falls is exported to the rest of the world.

[1] *Newfoundland*, Griffith Taylor, C.I.I.A. booklet; Toronto, 1946.

Grand Falls paper mills were established by the Northcliffe interests of London in 1907-8. The mill produces newsprint and pulp, and the power is mostly obtained from the falls adjacent to the mill. Here

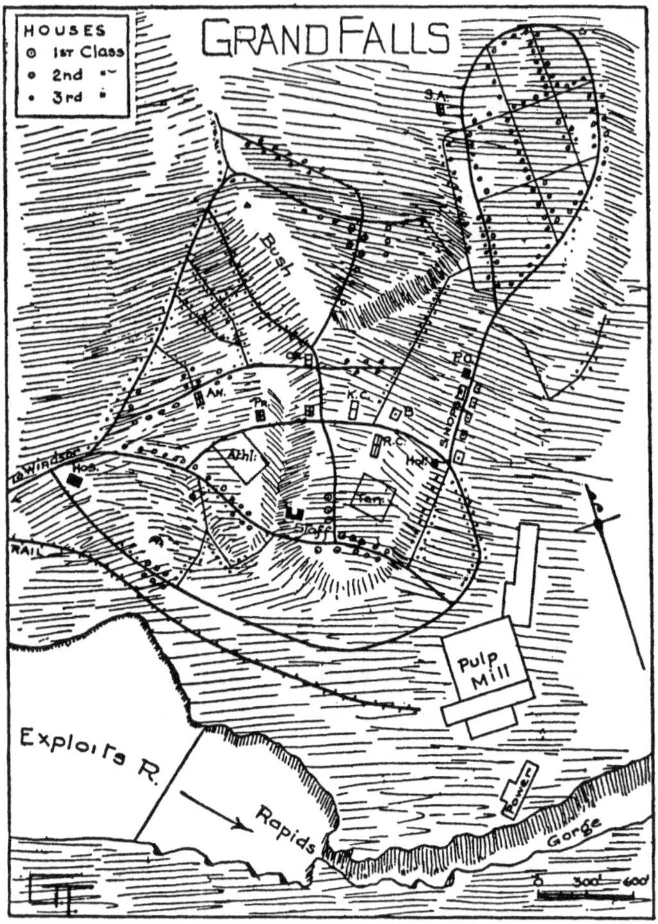

FIG. 114. Sketch survey of the 'Company Town' of Grand Falls on the Exploits River. The topography is suggested by the form-lines. (*From 'Newfoundland'*, 1946.)

70,000 horse power is generated. A dam was placed across the Exploits river just above the gorge at Grand Falls (Fig. 114). On a flat in the bend of the river the paper mill was erected, while the low ridge to the north was chosen as the site of the town.

Grand Falls is therefore a 'Company Town', in which the houses are erected by the Paper Company and rented to employees. There are many such throughout Canada, and one or two more of these most practical illustrations of 'town planning' may be considered later. The plan of Grand Falls is a welcome change from the 'checkerboard' adopted in almost every North American town (Fig. 114). The roads curve around the hill in response to the contours. The civic centre is to the north-east, where the post office is situated near half a dozen well equipped stores. The four imposing churches have been built on one of the main roads leading from the civic centre to the railway station, two miles away to the north-west at Windsor. There is a private line from the mill to Botwood, but this does not carry public passengers.

The Staff House is an imposing building of several storeys in a large garden, and nearby are most of the first-class houses. There is a small hotel near the shops. Another group of first-class houses has been built to the north near the highest land, which has been left as bush up to the present. The position of the hospital and athletic grounds can be gathered from the plan. The third-class houses are close to the mill, or else at a considerable distance away to the north. On a flat, north-east of the mill, is a circular driveway with many second-class houses. In general there is no definite separation into zones of residences, though the shopping centre is isolated. In the writer's system of classification of towns (page 85) this town of 5,000 inhabitants falls into the 'adolescent' class.

It is worthwhile to contrast the site of Grand Falls with that of the rival mill in Newfoundland at Cornerbrook.[1] When the former town and mill were planned the technique of long-distance transmission of power was not so well understood as it was twenty years later, when Cornerbrook was established. In the earlier mill the works were placed close to the falls, which meant that everything needed for plant and workers has to be carried inland from Botwood for some twenty-five miles. Moreover there is the same journey for all the paper and pulp before it can be placed on shipboard. At Cornerbrook, on the west coast of the island, the mill is situated alongside a deep-water port, and all the power is transmitted from Deer Lake power house about thirty miles to the north-east. The latter is found to be the better plan, and the little town of Cornerbrook on the steep southern slopes of the Bay of Islands has a charming situation. Both mills are on the main railway, so that in the depth of winter when the ports freeze, the products can be shipped to open harbours such as Port aux Basques or St. John's.

[1]Site-views and full discussions of Cornerbrook and also of Port aux Basques will be found in the author's recent book on Newfoundland (Toronto, 1946).

Fall Towns: Battle Creek and Sault St. Marie

We owe to H. T. Straw a very interesting study of the town of Battle Creek in the state of Michigan, known to many readers as the home of a special type of cereal preparations, and for the famous sanatorium. Neither of these activities is particularly suited to the region, for there are no medicinal springs, nor is it situated in a particularly rich grain area. Straw shows that the evolution of the town must be studied to understand this interesting location of Battle Creek industries.[1] The city also illustrates the disuse of local water-power in later factories.

The region close to Battle Creek consists of a series of moraines, with axes from north-west to south-east. This series was cut across by a glacial river valley with a rather broad flat bottom, in which to-day the modern streams wander in the typical misfit fashion. Battle Creek city is situated at the junction of Battle Creek stream and the Kalamazoo River; which reaches the city from the south, passing through two small lakes called the Upper and Lower Millponds (Fig. 115). It seems clear from the width of the valley, that the main glacial river consisted of the present-day Battle Creek and the *lower* Kalamazoo. The Upper Kalamazoo drops considerably as it enters the main channel, and it is this drop which in large part determined the water power available at the site of the present town.

Straw in his second paper (published in 1939) deals with the evolution of the settlement. At first the roads were planned in rigid west-east lines, but the swampy character of much of the region caused various deviations. Battle Creek is situated on the so-called 'Old Territorial Road' from Detroit to Kalamazoo (now the 'U.S. 12' route). About 1831 several pioneer groups were interested in the water power in the district, but the first actual settlers were farmers near Lake Goguac (Fig. 115). However the power was utilized within a few years, and a mill race was cut from the Kalamazoo to the lower level on Battle Creek. This early power was used in mills for grinding grain, for wool-carding, and for wagon building. To-day this race is used very little, though it can still be traced through the city.

The railroads entered the city about 1845, and a number of the early industries declined as the result of competition with factories in other cities. By 1870 the town had nearly 6,000 inhabitants. A number of Quakers had settled here, and prior to the Civil War the town was an important station on the famous 'under-ground railway' for freed slaves.

[1] "Battle Creek," Michigan Academy of Science, 1938, and a second paper in 1939.

Another important cultural group in the city is the Seventh Day Adventist, and their leader built a sanatorium at Battle Creek in 1866 on the site shown in Fig. 115. Later Dr. Kellog organized this Institution with its special emphasis on certain diets. In 1874 the College of this sect was founded, and in 1895 C. W. Post started to produce a substitute for coffee called 'Postum'. From 1900 to 1905 some twenty other food companies were attracted to the city, by the prosperity of the Kellog and Post activities, of which only four are now operating.

To-day nearly one-half of the people in the city derive their incomes from the manufacturing and mechanical industries; and of these about thirteen per cent are connected with cereals, and eleven per cent with

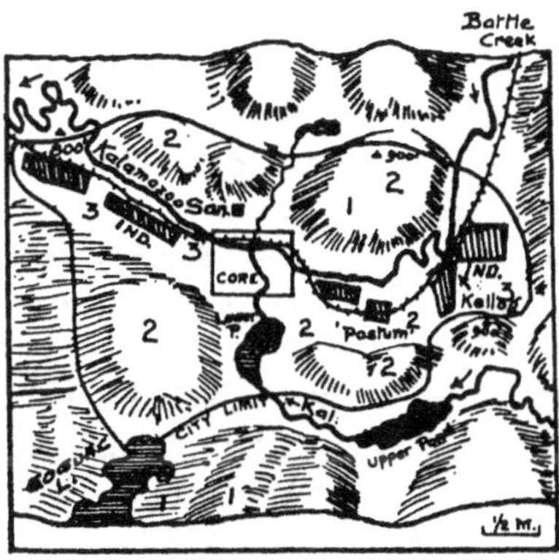

FIG. 115. Battle Creek showing the upper Kalamazoo river dropping into the wider glacial valley. 1, 2, 3, refer to residences. (*After H. T. Straw.*)

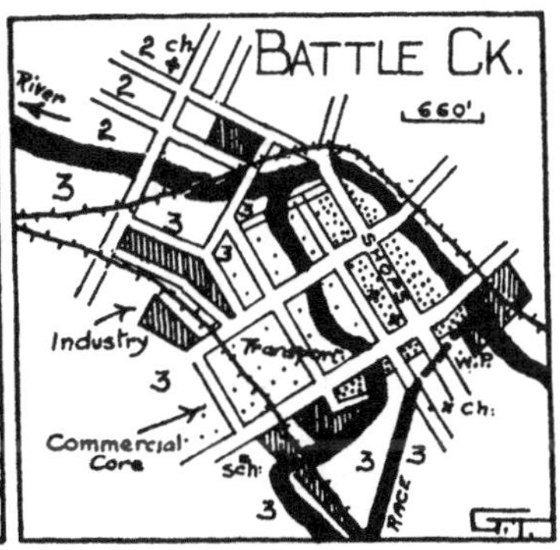

FIG. 116. The 'core' of Battle Creek showing the functional plan, and the almost disused mill-race. (*After H. T. Straw.*)

machinery. The factories are located almost exclusively upon the valley floor near the railways and the main roads. The metal-working factories produce presses, pumps, registering machines, etc., and are mostly in the commercial core; while the huge cereal mills are further afield. The residences are naturally on the higher land, i.e., the rounded moraine hills rising a hundred feet or so above the valleys. The best houses are east of the sanatorium and east of Lake Goguac (1, 1, in Fig. 115). The poorer houses (3, 3,) are close to the factories and railroads. To-day the population is about 44,000.

The town of Battle Creek illustrates clearly the shift from water power in the early days to other kinds of power as industry advances. Mr. H. T. Straw has kindly furnished me with further details on this shift. The use of water power was largely limited to the pioneer industries, and the advent of the railroad brought about its decline. The mill race never had a large volume of water and there were many difficulties attendant on its use. As early as 1870 the current maps show a shift of industry away from locations where water power could be used. Indeed to-day there is only one site (marked W.P. in Fig 116) where water power is used. It is maintained by the Consumers Power Company to retain their rights to the water, which is diverted to, and used in, cooling a steam-operated electrical plant.

The power used by the Battle Creek factories is largely steam power, augmented by electrical power purchased from Consumers Power Company. A surprisingly large amount of gas is also used. Gas heats the ovens in the cereal factories, and the two most famous ('Postum' and Kellog's) are charted in Fig. 115. Moreover during the recent war gas was used to harden small metal items such as valves, which were turned out in the plants normally producing machinery.

Sault St. Marie, commonly called 'Soo', illustrates both riverine and fall towns; but the river is on such a large scale that the site has not much in common with those already considered. Lake Superior has a mean level 602 feet above the sea, while Lake Huron to the east is 581 feet. The difference of twenty feet is due to the falls in the St. Mary River, which indeed gave rise to the French name of the cities which we are considering. At this point the great cuesta of North America—the sickle-shaped ridge which runs from Albany (New York) to Chicago —impinges on the Canadian Shield as it surrounds the well-defined Michigan Syncline (Basin). The falls are at the outlet of Lake Superior, for there is a narrow island-filled channel for some thirty miles to the east of the Soo.

At the Soo the river is about three-quarters of a mile wide, with red Cambrian formations outcropping in places, though the shores are low especially on the Canadian side. Terraces, linked with earlier lake levels are conspicuous, and one such—the Nipissing Bluff—is sketched in Fig. 117. The granites of the Shield are to be seen about three miles to the north of the Soo.

This strait in the system of the Great Lakes with its falls, was naturally an important portage and was early occupied by the Indians, and the first mission arrived about 1670. Before 1788 there were villages on both sides, but the southern side contained the trading stores, since it offered the better portage route. Whitefish were caught largely in Whitefish Bay by the Indians and early settlers. About 1798 the North-west Fur Company built a primitive locked canal to enable small boats to ascend the twenty-foot Falls.

There are two main Navigation canals, the most important being on the Michigan side, which indeed is of double width as the map shows. This was built around 1855, and has been much enlarged and improved, so that in 1908 it would take vessels of twenty-four feet draught. The Canadians built a ship canal on their side of the falls in 1895, but this has never equalled the American canal in the volume of cargoes carried. There are power canals at each side of the Falls, which develop a great deal of electric power; and this is transmitted to factories and plants in the vicinity. In 1934 the Canadian Hydroelectric Company had a capacity of 29,000 h.p. These changes in the neighbourhood of the falls have much altered the topography, and in 1887 a railway bridge (3,607 feet long) was constructed right across the falls to link the two countries.

There are, as stated, two cities of Sault St. Marie, one on each shore. Of these the Canadian is the larger with 23,000 inhabitants in 1930, as opposed to 14,000 in the Michigan City. The two progressed about equally until 1908; but for a few decades the Michigan town has not grown to any extent.[1] The American city has much the greater canal traffic; but the ships in general go right through the locks, only being delayed an hour or so owing to the passage. Hence the Soo towns are based more on the power supplies than on the canal, though these functions are of course inter-dependent to some extent.

The Ontario city is laid out along the old portage road, and the railway runs at the foot of the bluff at the back of the town. The core of the town is shown in solid black in Fig. 117. The town plan consists of three sections each forming a grid, with the main axis parallel to the curving shore. The importance of the town depends mainly on the large group of factories above the falls. Here is one of the largest steel plants in Canada, though the first supply of power was used for a paper mill. A blast furnace was erected in 1902, and much of the ore comes from the Mesabi field in Minnesota. The site was chosen above the falls, since in winter the ice jams in the narrow canals, whereas traffic across Lake Superior is relatively open to the plant. The various Canadian plants amalgamated in 1912 to form the Algoma Steel Corporation, and they occupy 185 acres in the west of the city.

The town buildings are almost all of wood, usually in two or three storeys. There is a six-storey hotel in each town, as well as many smaller establishments. The poorer houses ('3' in Fig. 117) are naturally in the west near the factories on both shores, while the best houses ('1' in Fig. 117) are on the top of the terrace (bluff) which is found behind each

[1]See the memoir by Whitaker, Philadelphian Geog. Soc., 1934.

city. The Canadian 'Soo' is the more thriving, mainly because it is the outlet for a very large district to the north along the Algoma railway. The Michigan 'Soo' has many trade rivals, such as Marquette, so that it has not gone ahead so rapidly. The tourist traffic is quite important, since the bridge is the sole crossing of the great river-lake system to the west of the bridges at Detroit. As far back as 1931 there were 7,000 cars crossing the bridge in a single summer month. The ice blocks the canal, usually after the last days of the year, for several months.

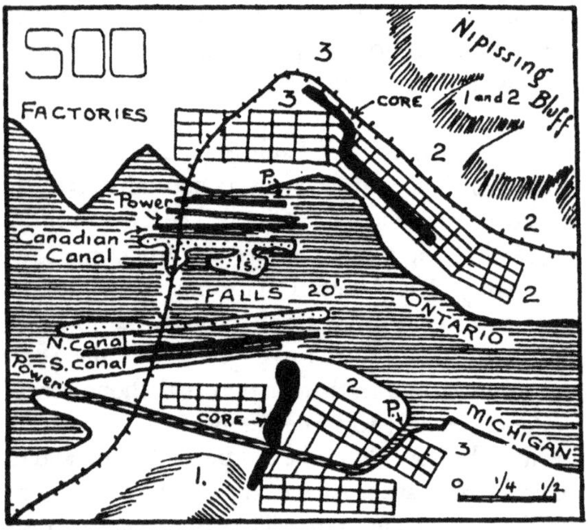

FIG. 117. The twin cities of Sault St. Marie ('Soo') on each side of the Falls at the outlet of Lake Superior. *1, 2, 3,* refer to house-types. *P.P.* are Power Stations. (*After J. R. Whitaker.*)

As far as Canada is concerned the canal at the Soo carries much less traffic than that at Welland, the figures for 1942 being three million tons as, opposed to eleven million. But the American Canal has a trade of about 115 million tons a year. The three great commodities are iron ore, wheat, and coal. Much of the latter is carried to the west in the barges which brought the iron ore from Mesabi. Wheat is only about seven per cent of the tonnage of the iron ore, but its value is generally greater. The total traffic through the Soo canals has been approximately twice as heavy as through the Panama Canal during the last few years.

Many other studies of 'fall towns' could be discussed, such as the detailed study of Lowell by M. T. Parker (1940), but fairly typical towns have been described, and lack of space prohibits further description of this interesting type of town evolution (see also Fig. 92).

CHAPTER XIV

PORTS ON SEAS AND LAKES

Introduction

In the present chapter an attempt is made to discuss most of the types of ports which occur around the world. First of all some description of the topography of coasts—whether stationary or the result of recent emergence or submergence—is given. Then a group of interesting ports in the Pacific is described, due partly to coral reefs, or to relatively recent earthquake and volcanic action. A fairly general study of the ports of Africa follows, which gives us some idea of the less elaborate ports occurring in regions mostly far removed from modern industrial conditions.

In the New World a number of ports were occupied in the early days of settlement, and in some cases have changed very little from those primitive times. Such conditions obtain in many of the ports around the Gulf of St. Lawrence, as at Tadoussac or in Gaspé. Of much larger and more modern ports, St. John's (Newfoundland), Quebec and Vancouver give us a series in increasing importance.

The changes which may be observed in ancient European ports are studied in connection with Bruges and Venice, while the evolution of the new port at Gdynia in Poland is described. Finally the details of the changes in the waterfront—which are characteristic of every large port—are worked out in some detail, using Toronto as an example.

Various seaports have already been described in earlier chapters. Thus the descriptions of Bergen developing on a fiord coast; of Santa Marta behind a high headland; of Tuktoyaktuk on a moraine coast; and of Sydney in a rather complex *ria*;[1] should all be consulted and compared with the ports described in this chapter.

The Topography of Coasts

We have seen that early towns almost necessarily developed near rivers, and it is only in recent years that water supplies from great distances have to some extent enabled large towns to develop away from a

[1]A *ria* is a drowned river valley, usually with steep shores and many branches.

plentiful local water supply. To some extent the advances in technology have also altered the conditions of harbours in regard to coastal towns. In early days sheltered harbours with small depth were satisfactory. To-day a much greater depth is essential, but man can erect gigantic breakwaters, which can change what were once unfavourable situations into useful ports.

The character of the harbour is mainly determined by the natural topography of the coast. Geographers recognize two main coastal types, those which are emerging, and those where drowning is dominant. In general the latter type of coast is the more common, mainly because our civilization has developed in the waning years of an Ice Age. The characteristic feature of our coasts is the *drowning* due to the melting of vast ice caps, which a few thousand years ago covered much of North America and Eurasia as well as far greater areas in the Polar Regions.

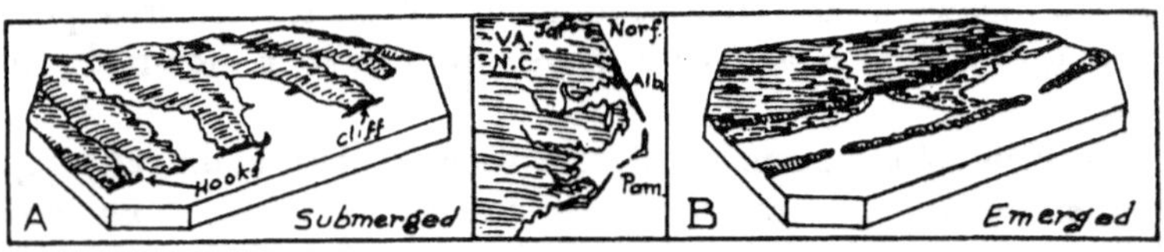

FIG. 118. Coasts of Submergence and Emergence, and Composite coast in centre. (*After V. Finch.*)

In general it is believed that the gigantic supply of thaw water so produced was sufficient to fill the oceans some 200 feet above their level during the maximum period of the Ice Age.

However, there are plenty of examples of *emergent* coasts, for it is precisely the coasts, at the edge of the great continental masses, which are most subject to the ups and downs of crustal buckling. Thus in any one continent we have perhaps more lengths of submerged coastlands, but there are always stretches of the coast which show the features of recent emergence. The sketches given in Fig. 118 (based on Finch) illustrate the differences in the two types of coast.

The character of the coast depends of course in no small degree on the type of landscape which is raised or depressed with regard to sea levels. If the landscape is mature (as in Fig. 118 at A) then drowning Produces a number of fairly deep coves or *rias* with rounded ridges between, as on the southern side of Sydney Harbour (Fig. 94). If the landscape is juvenile, then we may see deep narrow inlets with high rocky ridges between, as on the northern side of Sydney Harbour. If the land to be drowned is a plain, then the resulting sheet of water will be a rounded shallow bay, as in the case of Botany Bay (Fig. 94); or Port Phillip near Melbourne.

Coasts of emergence necessarily do not give rise to good harbours. They are shallow coasts, bordered often by long offshore sandbars, as shown in the sketch in Fig. 118 at B. Yet it is often possible to make an artificial harbour behind such a bar, and some examples may be cited later. In some cases, as in the 'gulf and bar' coast of North Carolina, we find that the rather complicated 'ups and downs' of the coast have produced a sort of mixture of the two types (Fig. 118, centre). This type of coast was quite suitable for early coastal towns, and Jamestown was founded in this region as early as 1607, near the modern port of Norfolk.

Coral Reefs, Volcanoes and Earthquakes as factors in Harbours

Coral reefs can only grow in warm tropical seas, but here they build reefs which often enough lend themselves excellently to the formation of harbours. Since the coral polyps do not grow vigorously on the inner side of fringing reefs—because food is less abundant here—there is bound to be a large development of shallow lagoons behind the growing reef. Furthermore corals cannot grow in fresh water, so that there is often a fairly deep passage through the reef opposite the mouth

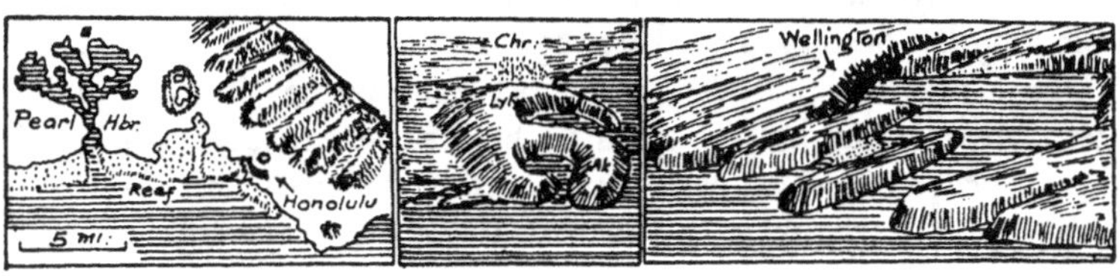

FIG. 119. Site-Views of three interesting Ports visited by the author in the Pacific. Honolulu Harbour and Pearl Harbour due to coral reefs. Lyttleton and Akaroa (near Christchurch) crater harbours; Wellington is a 'fault' harbour.

of a river, just where human settlement is most likely to occur. Finally reef growth is common in areas of rather rapid elevation or subsidence of the crust, which also affect the character of the harbours within the reefs. All these features can be illustrated in the case of the harbours of Oahu, which the writer visited in 1907.

There is perhaps more coral about the island of Oahu than around most of the Hawaiian Islands (C. W. Baldwin). Here a large portion of the coastal plain around Honolulu and Pearl Harbour is composed of uplifted coral reefs (Fig. 119). At one time the island was submerged

700 feet lower than its present level, and during this stage immense thicknesses of coral limestone were laid down by the polyps. Later there was an uplift of about 200 feet, and this elevated much of this relatively soluble limestone above sea level. The remarkable harbour, in which the great disaster to the American navy occurred in 1941, is eroded out of raised coral reefs. Probably ordinary surface erosion has played a part, but a vast amount of rainwater reaches the sea through the porous limestone, and this tends to keep the various arms of the harbour free from growing coral to-day. Honolulu harbour has somewhat the same origin, but is not so striking an example of the effect of a raised reef.

New Zealand offers some of the most varied and remarkable landscapes in the world. These are due to the fact that the islands are situated on one of the most mobile portions of the earth's crust. Earthquakes are commonplace here, while volcanoes are still quite active in the North Island. In the South Island volcanic action has died down in our times, but the old volcanoes stand out on the shoreline near Christchurch and Dunedin, much as they do in the North Island. One of these dead volcanic centres is sketched in Fig. 119 (centre). It lies close to the Canterbury Plains, and to the large town of Christchurch. Here are wide agricultural and pastoral interests, but the plains do not offer good harbours, for reasons described elsewhere. However close at hand are two volcanic craters, each of which has been breached and occupied by the sea. Lyttelton is the nearer to the city, and accordingly a tunnel a mile long has been excavated through the wall of the crater, and now the citizen of Christchurch takes a short train journey into the extinct crater, where he finds large steamers berthed at extensive wharves in the deep water of the drowned crater. Here the writer left civilization on his journey to the Antarctic in 1910; and it was to the adjacent crater of Akaroa that we returned in 1912.

The third example in Fig. 119 illustrates the sort of harbour produced in a region where earthquakes are very common. They take place along great vertical cracks in the earth's crust called 'Faults'. Often the crust breaks into fault-blocks, and these may be depressed to form long valleys called 'graben'. Wellington is situated on the side of such a graben, and the headlands near the mouth of Port Nicholson have all the appearance of other similar fault-blocks. The great fault-scarp runs in a straight line for five miles to the north-east of Wellington. In 1855 the entire coast rose five feet in one night, and created a natural road all round the west side of Port Nicholson. To-day a railway uses this shelf below the steep scarp, and it is difficult to see where a convenient line could have been built if it had not been for this startling elevation in 1855.

The Harbours of Africa

A useful article classifying the harbours of a continent from a topographical and economic point of view was contributed by George F. Deasy to *Economic Geography*, in October, 1942. He considers eighty-eight first class harbours, scattered along the 16,000 miles of African coastline. Each is well protected from wind and sea on all sides; and each has more than fifteen feet in depth, which is about the minimum draught of ocean-going steamers. He points out that among these harbours are a number fully developed, as in Algeria; while others, as in Mozambique and Angola, have hardly any port facilities. Deasy summarizes the condition of these harbours in the following table.

HARBOURS IN AFRICA

Nation	First class	Developed	Undeveloped	Artificial
British	32	19	13	8
Portuguese	22	6	16	0
French	22	18	4	13
Italian	7	3	4	3
Spanish	3	2	1	2
Belgian	2	2	0	0
Total	88	50	38	26

The major features of a number of these harbour towns are described by Deasy, and I have summarized these, as regards a dozen, in the small maps appearing in Fig. 120. They may be classified topographically as follows:

Type	*Examples*
1. Lagoon and Bar	Durban, Tunis and Bizerta (Port Said)
2. Drowned Valleys	Douala, Mombasa, Lorenzo Marquez
3. Rocky Headlands	Dakar, Capetown
4. Tidal Inlet	Lagos
5. Delta and Bar	Alexandria
6. Mostly Artificial	Suez (Port Said)

Three of the first four harbours shown in Fig. 120 have been visited by the writer. All four take the form of *rounded* bays, due to the drowning of relatively level coastlands, and separated from the open ocean by narrow tongues of land, often of the nature of sand bars. Durban is the

FIG. 120. Twelve Harbours in Africa illustrating topographic control. (*Based on G. F. Deasy.*)

only good harbour along the eastern shore of the Union of South Africa. To utilize the shelter of the bay it is necessary to dredge the channel cut through the bar; but vessels drawing thirty-five feet can berth at the Durban docks on the north side of the bay. As is often the practice, two long jetties have been built out into the sea, at the mouth of the entrance, to help (by the scour) to keep the channel open. Durban is an important coaling station, and has a large floating dock.

The coast near Tunis is of great interest, for here was the former great port of the Mediterranean, Carthage. To-day nothing but the outline of the two land-locked ports remains of the huge city of the Carthaginians. In modern times it was replaced by La Goulette at the mouth of the Lagoon of Tunis. This in turn has been almost abandoned, since the six-mile canal was cut across the shallow lagoon to reach the important capital town of Tunis.

Bizerta has been developed by the French until it is one of the best-equipped naval bases in the world. It is a land-locked basin, receiving a number of small rivers, but they pass through large swampy areas before entering the basin, and so do not deposit much silt in the latter. Deep water extends across the lagoon as shown in the sketch map, right to Ferryville; but the chief centre (i.e., Bizerta) is at the entrance to the lagoon.

Port Said at the northern end of the Suez Canal has been built on the sandy stretch of coast separating the lagoon of Menzala from the Mediterranean (Fig. 120). Its site is therefore somewhat like those of the preceding examples. But the port makes no use of Lake Menzala, and is essentially man-made. Deasy points out that the three-fathom line (eighteen feet) lies three miles off shore. Two long jetties have been built into the Mediterranean for over a mile, and dredging keeps the channel between them deep enough for the largest vessels. The early wharves were all on the west side of the canal near the town of Port Said; but lately a number of wharves have been constructed on the east side at Port Fuad. This port, commanding one of the great trade arteries of the world, is naturally well equipped, and is a well-known coaling station.

The harbour near Suez at the south end of the canal is also entirely man-made. Formerly the chief wharves were at the town of Suez, at the head of the Red Sea, but to-day a new port has been built at Port Taufig; and the passenger for Cairo usually motors along the two miles of road which separates it from Suez proper. It is rather a small harbour, for most boats passing through the canal do not leave much cargo here, or indeed wait longer than is necessary to disembark the passengers. However, Suez harbour has a depth of thirty feet, and has all facilities for re-fuelling ships.

Alexandria has had a very interesting history, for it was founded in 332 B.C. on a sandy bar which separates Lake Mariut from the sea (Fig. 120). It is situated about fourteen miles to the west of the Canopic mouth of the Nile, and so received little of the silt brought down by the great river. The little rocky island of Pharos was linked to the sandbar by a long mole, and thus the Greeks obtained a splendid harbour, crowned by the famous lighthouse ('Pharos'). At the east end of the city was the great Library, and two of the 'Cleopatra's needles', which have now been moved to London and New York. Caesar conquered the city in 48 B.C., and it long remained the greatest seaport of the Empire, with a population of three quarters of a million. But later the misrule of the Turks and the discovery of the ocean routes to the Indies ruined the city, and about 1800 it had only 6,000 inhabitants.

To-day the former narrow mole has been broadened to an isthmus of silt half a mile wide; and the West Harbour is protected by strong breakwaters. This harbour is now a mile long and thirty-six feet deep, and was of vital importance as a naval station during the recent war.

We may now consider the three African Harbours which are essentially drowned river valleys of the 'Sydney' type. None of these is so well known as most of those so far considered. Douala is at the head of the estuary of the 'drowned' Cameroon River in French Cameroons. The inset in the chart (Fig. 120) shows the topography fairly clearly. The outer harbour, more than ten miles wide, rapidly narrows near the port, where a fifteen-foot channel extends past Douala. This channel has been dredged to accommodate vessels of twenty-eight feet draught. Konakry, Freetown and Calabar, all on the west coast, have somewhat the same type of drowned river harbour.

Mombasa is placed in a harbour due to the drowning of the lower portions of several small rivers. It is situated on a small island, which is linked by rail to the mainland. It is the port of entry for passengers proceeding to Nairobi and other portions of British East Africa. The old Portuguese port dates back to 1500, but it has silted up; and Kilindini, on the other side of the small island, is where the main wharves are now situated.

Lorenco Marquez has a somewhat similar topography, since it has been developed on the north bank of the drowned estuary of the English River. A dredged channel enables boats of twenty-five feet to reach the wharves. Its chief importance is its connection by rail with Johannesburg and the great mines of the Transvaal. It is very well equipped with warehouses, refrigerating chambers, cranes, etc.

Two of the most interesting ports are Dakar and Capetown. They are somewhat exposed to certain winds, for both are rather open roadsteads though sheltered by headlands from most winds. Dakar has been created on the southern side of Cape Verde Peninsula, so-called because here the navigators came on green forests after the lengthy desert coasts of the Sahara. However, two breakwaters have made this a safe harbour, and it has a depth of twenty-seven feet over a considerable portion. Its chief importance is due to the fact that here Africa is only 1,600 miles from Brazil. During the war it became a great airport, and is now of outstanding strategic value as a corridor from the Old to the New World. (A block diagram of this port constitutes Fig. 174.)

Capetown was originally quite open to the north, though protected on other sides by the spurs of Table Mountain (Fig. 120). However, huge breakwaters now shelter it from north-west and south-east gales. It has a commanding position at the 'corner' of Africa, on the routes to India, the Far East and Australia. It is of course the chief port of entry for the Union of South Africa. Vessels drawing thirty-five feet can enter the harbour, and there are all kinds of fuelling and repairing facilities, including a dry dock.

Lagos is the best of the Nigerian Harbours, and is situated on a deep tidal inlet behind an offshore bar. The channel is dredged to maintain a depth of twenty-four feet; and the deepest water is off Apapa. It is the port of entry for British Nigeria, and a railway here links the coast with the far interior (Fig. 120).

Small French Ports around the Gulf of St. Lawrence

In the summer of 1942 I investigated a number of the small ports which have developed in the east of the province of Quebec along the shores of the Gulf of St. Lawrence. Sketch maps of four of these small ports are reproduced in the present section, from a long article which I contributed to the *Canadian Geographical Journal* for June, 1945. They give a fair idea of the development of small French settlements, dating back a century or more, in an environment which could not be expected to give rise to large towns.

The common factor in all of them is an unfavourable hinterland, consisting in general of the uplifted relics of an 'old' mountain area, on the south, and of an elevated portion of the great Canadian Shield on the north. When the not very abundant timber is cut out, there is no agriculture in most of this hinterland, and the folk depend for the most part on fishing and tourism.

Of the four ports discussed, Madeleine is the smallest, and represents the earliest stage in development. It is purely a 'terrace' village, and is not much affected by the Madeleine River which enters the sea nearby

(Fig. 122). St. Anne des Monts is considerably larger, but has much the same situation. However, here there is a small coastal plain, and there is still a good deal of timber cut in the hinterland, so that it has saw-mills which support a considerable proportion of the inhabitants. Tadoussac is about the same size, but it is primarily based on the tourist trade. Matane is the largest of the four settlements; and is situated in lower country (as Fig. 122 indicates). It is primarily a lumber town with several large mills. It occupies the shores of a deep estuary closed by a bar; and is the centre of a fairly important agricultural area in the low gap which here reaches across the Gaspé Peninsula to the south shore near Campbell-town (Fig. 122).

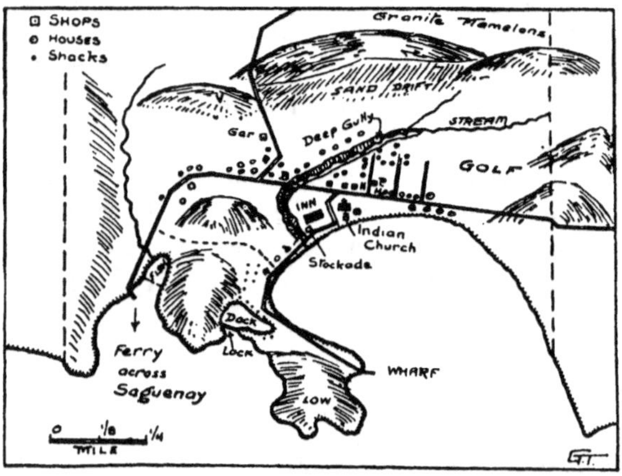

FIG. 121. A site-view of Tadoussac looking north across the mouth of the Saguenay.

Site and Pattern of Tadoussac

At Tadoussac, Cartier dropped anchor as far back as 1535; but Chauvin seems to have made the first settlement about 1599, when he built a guard house so as to collect the furs from the Indians. It is believed that the name of the place refers to the mamelons of granite which are so characteristic a feature of the hinterland (Fig. 121). The composite diagram given in the illustration shows this famous French village, as viewed from the mouth of the deep Saguenay fiord. It is situated just where the Saguenay enters the Saint Lawrence, whose coasts extend to the north-east.

Although Tadoussac is the oldest town in Canada it has grown very little in the 340 years of its existence. Situated in a semi-circle of breast-shaped hills, it clusters along one main road and has not yet needed or developed a complicated street plan. Let us alight at the wharf from the steamer and make a survey of the village. The wharf is on a low rocky point about half a mile from the centre of the town. We soon cross the low isthmus linking the low point to the higher land. Here a small land-locked cove has been converted into a dry dock by building a stone wall or lock across the narrow entrance. A small sailing vessel was in dock at the time of my visit, the gate being shut at low tide.

Beyond the dock we pass a store and soon reach the grounds of the

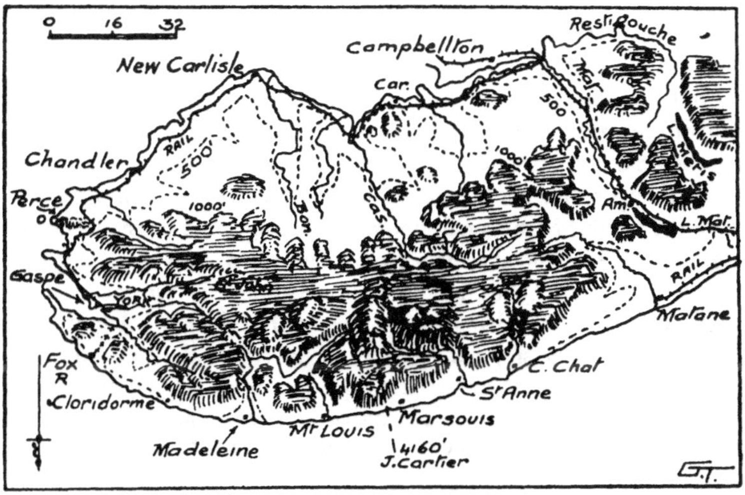

FIG. 122. A block diagram of the peninsula of Gaspé, looking southward from the Gulf of St. Lawrence. The 500 and 1,000 feet contours are indicated.

huge inn which is much the most imposing feature of Tadoussac to-day. About 1870 a company was formed to build a hotel, and in the words of the interesting little book by J. E. Roy, 'thanks to judicious advertising tourists flocked there from all parts'. Lord Dufferin built a summer home here, and of recent years the hotel has come under the control of the Canadian Steamship Co. Their large and comfortable steamers call here frequently on the voyage from Quebec to Bagotville up the Saguenay, and many passengers stop over for a day or two at Tadoussac.

It must be confessed that there is very little to see dating back to the early days of French settlement. There is an interesting stockade and hut near the hotel, but, alas, it is a modern reproduction. Just across the road, however, is one of the oldest churches in the Dominion. It was built of cedar in 1747, and was in use till the new church was erected in 1884. The old wooden church is kept in good repair, and services are occasionally held there.

Just behind the church we reach the main road, built more or less parallel to the coast as usual (Fig. 121). The main stores, the bank (B), and the post office are on this road. Here also are half a dozen pensions (one or two of brick) and some modest hotels which cater to the summer traffic. To the west the road curves down to the next little bay (Anse a l'Eau) where there is a fish hatchery. This was first established in 1875, and just alongside is a small wharf whence starts the ferry to cross the mouth of the Saguenay. Thus contact is made with the main road to Saint Simeon, Malbaie and Quebec.

There is not much traffic beyond Tadoussac. A motor bus goes some sixty miles to the north-east to Laval Bay, passing through tiny villages about fifteen miles apart. At Laval Bay the improved roads of Canada reach their eastern terminus. As the map suggests, this inland road passes close to a deep ravine cut behind the town, effectively preventing expansion to the north. Many small houses, including some pensions, occupy the grassy slope behind the main road. The town ends about a quarter of a mile to the east of the church. Here are some large summer cottages, shaded by trees and overhanging the steep cliff of the bay. A golf course occupies much of the slope at the head of the steep ravine already mentioned. Thus, to-day, there are less than a score of large buildings (including six stores) and perhaps twice that number of small houses in this settlement, founded by Chauvin in 1599, and continuously occupied almost ever since.

In his little book *In and Around Tadoussac* (Levis, 1891), J. E. Roy gives a census of the town about 1885. There were then about 590 people, belonging to ninety-one households. These folks were engaged as follows: 25 farmers, 15 navigators, 23 hunters, 3 merchants, 6 carpenters, 2 blacksmiths, 1 shoemaker, 2 masons, 4 carters, 1 baker, 1 miller, 1 pilot, 1 doctor, 1 magistrate, 1 postmaster and 1 Crown Land Agent.

To-day, I imagine the population is much less. I saw only an acre or two of oats, and there are few farms inland until the next village is reached, some twelve miles away. In spite of its early start and the importance which it possessed in the days of the fur trade, and to a lesser degree in the lumber period, the lack of good agricultural lands has prevented its growth. It is of interest that the entrance to the Saguenay remains free from ice through the winter, though the gulf to the east is not then navigable. It has often been used as a refuge for ships in the gulf; but there is no evidence that the sanguine predictions of some early writers, that Tadoussac will rival Montreal or Sydney as an Atlantic terminal, will ever come true.

Villages on the Coast of Gaspé: Matane

I have drawn a special block diagram of the peninsula of Gaspé to show the structure of this very interesting section of the Dominion. In Fig. 122 we are supposed to be looking to the south, directly at the most picturesque part of the region. The plateau-like character of the northern part is well brought out. The highest points near Mt. Cartier (4,160 feet) are formed of hard igneous rocks which have resisted erosion a little more than have the ancient Palaeozoic rocks which build up most of the Shickshocks. It will be noticed that the divide is much nearer the northern shore than the southern, hence there are steep slopes on the northern shores and much gentler slopes on the southern shores.

In two places, i.e., near Marsouis and Cloridorme (Fig. 122), the plateau almost overhangs the sea. Here, clearly there is no room for agricultural development, and the scenery is correspondingly wilder. Indeed, just east of Marsouis, cliffs of a height of a thousand feet are washed by the sea. In general this height is not reached within a distance of one mile of the sea, and in some places (notably at the mouths of the main rivers) there is a much greater margin of low land. For instance, at Ste. Anne and Madeleine the thousand-foot contour has receded about four miles from the sea. Here, and at Mont Louis and Cap Chat for somewhat similar reasons, there is a greater development of farm lands.

The ancient French method of subdividing the land is still in evidence along this coast. A zone of about a mile and a quarter wide is marked off parallel to the coast. This is divided into extremely narrow strips, about 150 feet wide, running at right angles to the coast, and to the main road of the region. Usually the southern ends of these strips reach up toward the plateau, and are of very little use for anything but wood-lots. All crops must be grown at the northern end of a strip.

The little town of Matane at the end of the local railway illustrates many of the features of a Gaspé settlement. It is situated at the mouth of a river of the same name, which has cut a broad valley in the ancient rocks of the region. The estuary has been drowned by the rise of waters, since the melting of the Ice Cap. But there have been upward joggles of the crust, due to the removal of the weight of the ice, and the gradual approach of the crust to a state of equilibrium. This last phenomenon is indicated by the remarkable series of terraces (or elevated shore-lines) visible all along the Gaspé shore. A well-marked terrace about eighty feet above the sea is shown in Fig. 123. A shore current from east to west has piled up sand and gravel across the drowned valley (or *ria*) and produced a very typical bar, which the French call a *barachois*. Furthermore, Matane has several large sawmills supplied by timber brought down the river from the forests of the interior.

Matane has a much more elaborate pattern of settlement than the little villages next to be discussed. There is the usual main road running parallel to the coast, with a rather close series of wooden houses as we approach the town. There is the usual large church, which in Matane is adorned by an enormous cross on the façade. The river is not crossed at the bar, but inland where the ria is narrower. The presence of the

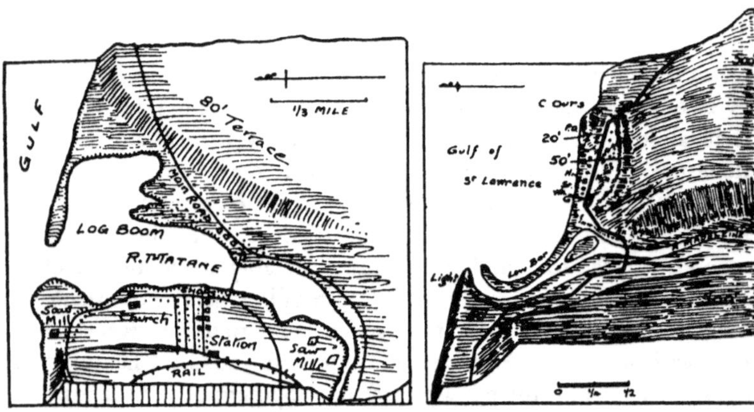

FIG. 123. (*Left*) Harbour of Matane at the west end of Gaspé. (*Right*) Village of Madeleine on the north coast of Gaspé, built mainly on the 20-foot terrace. Initials indicate the three hotels, the store, wharf and garage.

railway station and the sawmills has led to the development of several cross-streets which lead from the main road toward the terrace on the south-west side of the ria. The main shops and official buildings are to be found near the bridge on the west side. The country is relatively low inland, and here an important road to Amqui and Campbellton ascends the river to the southward. There is no other road yet completed across the Gaspé, and indeed this Amqui road bounds the peninsula rather than crosses it (Fig. 122). The harbour is shallow and kept open by dredging, and much of the water within the bar is covered with a boom of logs. The town contains only a few hundred inhabitants, and belongs to the juvenile class.

Ste. Anne and Madeleine

The largest of the villages for a good many miles is Ste. Anne des Monts, whose chief features are given in the map (Fig. 124). Here, as at Matane, the settlement has grown up at the mouth of a large river. Indeed, the R. Ste. Anne is one of the largest of the northern rivers, and has a peculiarly zigzag course for about forty miles. Its estuary has been drowned, and there is the usual bar at the mouth, which, however, is not used by the road. The shore here is relatively flat for about half a

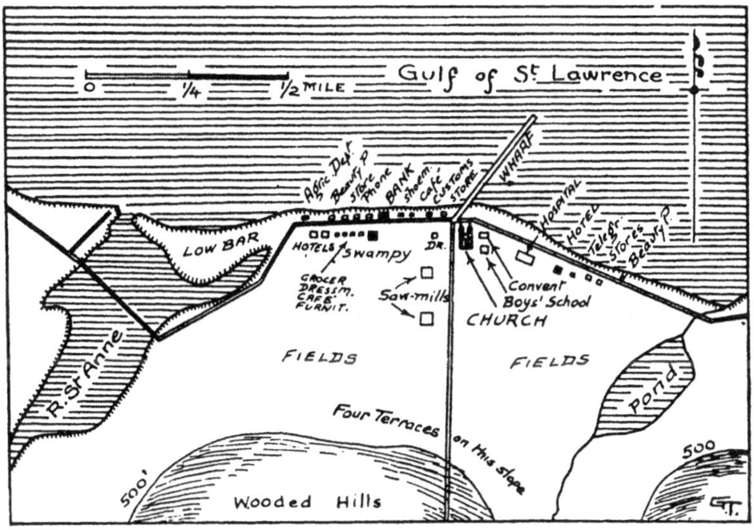

FIG. 124. A sketch map of the vicinity of Ste. Anne des Monts (north Gaspé) showing a broad area of low coastland. All the important buildings in the village are shown, but a dozen of the smaller houses are omitted.

mile back from the sea, and is covered with fields of grass or oats. The main road is close to the beach, and seems in places to run on a ridge or beach with swampy ground in the rear. There is a large church with twin steeples, flanked by a convent, hospital, and two large schools. Just behind the town are two busy sawmills which account for most of the activity in the district. Along the slopes of the wooded hills behind the town four terraces are rather prominent.

In the map I have added the character of the shops, so as to give a clear picture of the anatomy of a Gaspé village. There was a fair-sized steamer at the wharf taking timber aboard at the time of my visit. A leading citizen of Ste. Anne (M. Henri Roy) has very kindly sent me a note dealing with life in the little town. He tells me that the pioneers came from the parishes to the west about 100 years ago. They made a living by fishing, hunting, and some farming. The two former industries are of much less importance now, and farming and saw-milling are the chief interests. It is the second town in the County of Gaspé, and is the natural and official centre for the western portion of this large area. About two-fifths of the timber is exported to Montreal, while the rest goes, for the most part, to England. Some of the pulpwood is sent to the United States, but most goes across the gulf to Trois-Rivières or to Port Alfred (Saguenay).

The next village is Madeleine, whose main features are given in Fig. 123. There is an important lighthouse at the mouth of the big river, but the village is built on a terrace some way from the estuary. Curiously, there is no church at Madeleine, and the habitants must walk two miles to the next village to attend service. There are three well-defined terraces at Madeleine, and most of the houses are built on the middle terrace about twenty feet above the sea. Alongside a low cliff to the east are small houses containing the post office and the telegraph station. It is well to note that small as Madeleine is, it is about as important as any of the settlements which occur, at intervals of about five miles, along this coast between Mont Louis and Fox River. Both Madeleine and Mont Louis are in the infantile stage of town development.

Site and Pattern of the City of Quebec

Few cities in the world have a more commanding site than Quebec. It lies six hundred miles west of the entrance to the great Gulf of St. Lawrence, where the gulf rather suddenly narrows. Here the St. Lawrence can be bridged, though salt water extends eighty miles farther west to Lake St. Peter. At these narrows a most picturesque bluff faces the ocean. It is really the eastern end of a brick-shaped portion of the earth's crust (which geographers call a horst) which has been pushed up during mountain-building movements. This horst is about 300 feet above sea-level, two miles wide in the centre, and about seven-and-a-half miles long. The eastern half of the horst can be made out in the lowest figure of the stage-diagram given in Fig. 125. The St. Lawrence valley has been a marked geographical feature possibly since Ordovician times, i.e., about 350 million years ago. It has been a long-lived depression, between the ancient Shield and the fold mountains, which time and again have surged upward along the site of the Appalachians. Between the

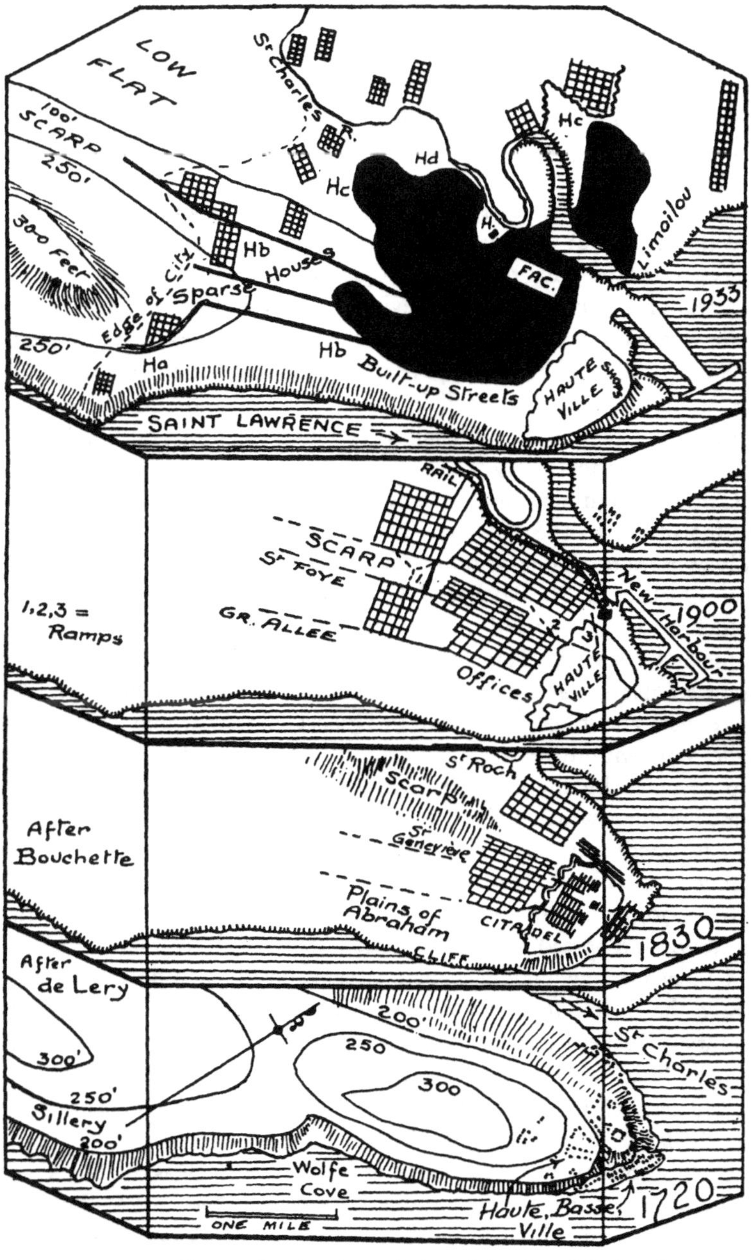

FIG. 125. A stage-diagram showing the evolution of the city of Quebec from 1720, through 1830 and 1900, to 1933. The main features of the crustal block (horst) are shown in the *lowest* diagram. The later areas covered by built-up streets are shown in solid black.

two has been an area of lowland, less than twenty miles wide, in which younger deposits have at times been laid down, notably in the period since the Ice Age. This depression is bounded by cracks in the crust (called faults); and the St. Lawrence lowland is a sort of graben (i.e., a depressed block), just the antithesis of the horst which borders it on the north.

To the north of the Quebec horst is the broad valley of the St. Charles, and the estuary of this small river has of late years been converted into a deep-water port for the City of Quebec. The St. Lawrence is three-quarters of a mile wide at the Levis ferry, but is less than half a mile wide where the great bridge crosses the river five miles west of the city. The steep cliffs of the horst have been notched by little coves, and of these the most interesting is Wolfe's Cove, up which the British soldiers climbed just before the momentous battle of the Plains of Abraham on the 13th September, 1759.

Cartier inaugurated the fur trade at Quebec in 1535, when he visited the Indian village of Stadacona. This seems to have been near the present railway station on the shores of the St. Charles. He spent two winters near Quebec, at St. Roch and Cap Rouge respectively; but no real French settlement started until about 1608, when Champlain built his house and stockade. This 'habitation' was erected under the cliffs of the present Citadel and was the first house in the 'Basse Ville'. In 1663, there were about 800 folk living in Quebec, almost wholly in the Basse Ville, but, within the next half-century, the houses had spread up the gap occupied by the Cote de la Montagne to-day. Some of the fine stone buildings which make Quebec the most interesting historic town in North America date from about this time. Among them are the Ursuline Convent, the Hotel Dieu and the Seminary.

By 1770, as the lowest map in the stage-diagram (Fig. 125) shows, the town had spread up to the plateau 200 feet above the Basse Ville; and indeed a few buildings were situated nearly a mile to the west, where the earlier fortifications had been built around the end of the century. De Lery erected the main French walls about 1720, but they were of little value in the vital struggle of 1759, since the British completely defeated the French about one mile to the west of the fortifications and the latter were quickly yielded.

By this time many streets had been constructed such as the Rue de St. Pierre which skirted the cliffs in the Basse Ville. The Rue St. Louis led to the Plains of Abraham, where it became the Grande Allee, and ultimately reached Sillery and Cap Rouge. The other main street, the Rue St. Jean, passed through the walls of the Haute Ville, and became the Chemin St. Foye. This latter was built along the scarp which bounds the Haute Ville and the horst on the north side. Even to-day only three roads descend this scarp and enable vehicles to reach the low valley of the St. Charles to the north of the older part of the city. These steep 'ramp' roads are labelled 1, 2, 3 in the 1900 map.

About 1771, there were 356 families in the Basse Ville and 610 in the Haute Ville. Timber export became of great importance after the British conquest, and the new suburb of St. Roch, along the south shore of the St. Charles, was largely devoted to this industry. By 1830 this was well established, as a glance at the map of that date in the stage-diagram (Fig. 125) will indicate. The great walls of Quebec mainly date from 1820 or thereabouts, and were built by the British at a cost of about thirty-five million dollars (Blanchard). Shipbuilding was a great feature of the town's industry, especially in the decades preceding 1860. But about this time the development of steel vessels increased rapidly, and the wooden ships of Canada were in much less demand. However, the trade in grain and timber, and the growth of small factories (making shoes, etc.) to a large extent replaced this loss.

As Blanchard points out,[1] by 1871 there were 42,000 folk out of a total of 59,700 in these new suburbs beyond the walls of the Haute Ville. A period of relative stagnation followed, from 1871 to 1901; but thereafter the population increased rather rapidly, and there are over 150,000 folk in the city to-day. The northern suburb of Limoilou started about 1905, and is largely inhabited by industrial workers. The upper map in the stage-diagram (Fig. 125) shows (in black) the area covered by close-set houses. With considerable spaces, however, the city extends far to the west of this area. The more important clusters of streets are shown by small 'gridions', but scattered houses with large gardens occupy much of the intervening region. In general we may say that to-day the merchants' quarter is in the Basse Ville, especially near the huge docks constructed mainly around 1890 and 1930. The largest shops are in the Haute Ville, especially in the Rue St. Jean. There are, however, many smaller shops along the whole course of this street, as it extends west and merges in the Rue St. Foye. The offices and legal sections are found mainly along the other main avenue of Haute Ville, the Rue St. Louis and its extension, the Grande Allee. In this part of the city (west of the walls) are most of the folk of English descent.

Quebec has been termed a city of churches, and the modern Catholic foundations are numerous and imposing. This is especially the case along

[1] *L'Est du Canada Français*, Montreal, 1935, which all students of Canadian Geography should read.

the Rue St. Foye, where half a dozen huge convents, hospitals and allied institutions are to be seen in little over a mile. In the upper map in the stage-diagram (Fig. 125), I have indicated where the larger mansions (Ha) are numerous, surrounded, as usual, with attractive houses of the Hb class. Rows of apartments and semi-detached residençes are a feature of the Grande Allee, especially about half a mile from the walls. The quarters of the poorer folk, almost exclusively French, are to be found in the lowlands near the St. Charles River, and in Limoilou.

It is interesting to compare the later history of the three earliest settlements in Eastern Canada, i.e., Tadoussac, Quebec and Montreal. The first had no hinterland, and never reached any size. The second was the great city of Canada until about 1850, when its rival, Montreal, began rapidly to forge ahead, and now has five times the population of Quebec. Quebec had the great advantage at first of being at the head of navigation. The narrowing of the river (the origin of the Indian word 'Quebec') made the waters to the west unsuitable for sailing vessels. With the advent of railways, and with the deepening of the channel by canals, many of the advantages of Quebec were lost. This was especially the case when steel ships, propelled by steam, replaced the sailing ships of early days. Moreover, the Lake Champlain route from the south was more important than any leading directly to Quebec. To-day, Quebec is emphatically the French centre in Canada, since the proportion of English-speakers is insignificant; and it is industrially much more important than it was a few decades ago. As a centre of tourism it has hardly a rival on the continent. On the classification used in this volume Quebec ranks as a *mature* city.

The port of Quebec is formed of several expanses of water, including the mouth of the St. Charles River, the huge docks off the Basse Ville, and the open St. Lawrence itself. The course of the latter changes somewhat as Quebec is approached from the ocean, changing from nearly east-west to almost north-south at the city itself. Since the river flows several hundred feet below the general level of the horst-block of the city, and below the plateau level of Levis to the east, it can be understood that shelter may be obtained from winds of almost any quarter. The entrance to the little river of St. Charles was obstructed by a rocky bar; and this helped in the construction of the huge docks of to-day.

In 1775 a jetty was built at the Basse Ville for the royal navy; and between 1880 and 1890 the Louise Basin was constructed. The river is 160 feet deep in the middle of its stream off Quebec, and about thirty feet deep at the point. The river is blocked by ice from the end of November to the end of April, though the dates vary somewhat from year to year.

From 1925 to 1931 a series of great quays has been constructed along the foot of the horst between Wolfe's Cove and the Basse Ville. Nowadays gigantic ships, such as the Empress of Britain, can berth here readily. One of the largest dry docks in the world is available at Quebec.

With the great increase in hydro-electric power in this part of Canada, factories are rapidly increasing in numbers in Quebec. It is in fact recovering some of the importance it had in the early days of Canada, though it is never likely to reach the size of Montreal. It should not be forgotten that Quebec is the terminus of the Canadian National Railway, which runs directly from Quebec to Prince Rupert on the Pacific Ocean. To-day Quebec has a population of about 155,000. In 1942 2,714 vessels entered the port, giving it second place on the eastern seaboard of Canada since the figure for Montreal was 3,456.

The Harbour of St. John's—a breached early mature valley

The harbour of St. John's is an interesting illustration of the dominant corrugations in the island of Newfoundland,[1] and reminds one of a miniature Golden Gate. Its evolution is explained in the inset in Fig. 126. The main valley 'B' is bounded on the east by a high rocky ridge, and this has been cut through by river and tide to form an entrance only 200 yards wide, which in early days was blocked by chain. To the north is Quidividi (*Kiddy-Viddy*) Lake (A in the inset), which shows an earlier stage of the harbour; while to the south is Freshwater Bay (C in the inset) where the future condition of St. John's harbour is indicated, when marine erosion has destroyed most of the eastern ridge.

The entrance to the harbour was in the early days guarded by a fort, placed where now is situated the huge Newfoundland Hotel (Fig. 126), and another fort was built half a mile to the south-west. Water Street and the Military Road (linking the two forts) were the earliest lanes in the little town, and ran parallel to the waterfront. An upper path later became Duckworth Street. In 1806 the city lay between Gower Street and the harbour, all the houses being to the north of the present post office. A few very steep streets run at right angles to the waterfront, and the Catholic Cathedral (R.C. in Fig. 126) is 200 feet above the port.

The main features of the functional plan appear in the block diagram (Fig. 126). Warehouses lie next to the wharves on the west shore of the harbour, and the railway enters the city from the south. The main shops, usually of brick and three or four storeys high, occupy the central portion

[1]For a full discussion of St. John's, Cornerbrook, Port aux Basques and other ports, illustrated by diagrams and photographs, see the author's book *Newfoundland*, Toronto, 1946.

of Water Street below the Anglican Cathedral (An.). Small shops are found in this street near the post office and north of the Cathedral. Duckworth Street contains offices, cinemas and a few small shops. All the rest of the city, on the slopes between Government House (G.H.) and the railway station, consists of congested rows of three or four-storey wooden houses (labelled 3) which extend to the edge of the city.

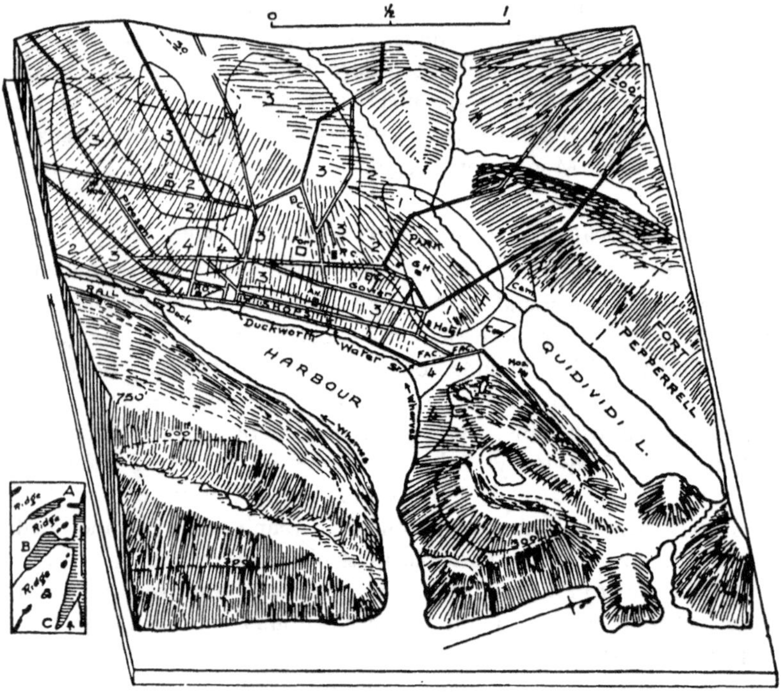

FIG. 126. Block diagram of St. John's—looking west. *Inset* is a sketch map of three adjacent coves.

Newer and better houses (labelled 2), most of them of small size but with small gardens, are being erected on the south-west margin. The best houses (labelled 1), which are rarely as imposing as the mansions of Canadian cities, are found in the vicinity of the Park in the north-west. Here also are Government House and the fine old Colonial Building, which houses many of the government departments. The city houses do not extend beyond the brook entering Quidividi Lake. On the north shore of this pretty expanse of freshwater is Fort Pepperell, where thousands of Americans garrisoned the island during the recent Great War;

but the barracks are for the most part empty now. One remarkable feature of the city is the absence of the usual checkerboard plan. The centre of the town is approached by radial streets on all sides, and it is a pity that this sensible arrangement is so rarely found in North American cities.

Vancouver the chief port of Canada

On the eastern coast of Canada, associated with the great St. Lawrence waterway, are a number of important ports, some of which such as Toronto and Quebec have been considered in detail. But shipping is quite important at many other ports, such as Duluth, Port Arthur (1,200 ships a year), Fort William (900), Marquette, etc., on Lake Superior; Chicago, Gary and Milwaukee on Lake Michigan; Detroit, Midland and Sarnia (1,344) on Huron, Cleveland, Erie and Buffalo on Erie; Hamilton (703), Toronto (2,914), Kingston (1,000), and Oswego on Ontario; Montreal (3,456), Three Rivers (1,924), and Quebec (2,714), on the St. Lawrence; Sydney (1,578), St. John (1,581), and Halifax (2,122) in the Maritimes. By contrast, on the Pacific coast of the Dominion, there is no rival to Vancouver with its 21,600 vessels a year, though Victoria (4,977), Nanaimo (3,909) and Prince Rupert (3,802) rank ahead of many of the eastern ports mentioned above. However, Seattle and Portland in the States no doubt capture some trade which might otherwise reach Vancouver.

Vancouver, as regards population, is the third town in the Dominion, yielding place only to Montreal and Toronto. Its high rank is rather surprising, since there was hardly a house there before 1870, and in seventy years it has grown to a population of 271,000. It has many advantages, such as its terminal position for two great transcontinental railways; its excellent harbour, and its site on a vast delta of good agricultural land, where dwell about sixty per cent of the population of the whole province of British Columbia. The Fraser, however, does not enter that harbour, but reaches the Pacific about six miles south of the city. This means that it is not necessary to dredge the harbour on account of silts brought down by the large river. Unlike many large cities it is situated in the midst of grand mountain scenery, for the northern suburbs lie on the slopes of peaks 5,000 feet high.

Vancouver is built on two morainic ridges running west-east with a shallow valley between them, along which the Canadian National Railway enters the city (Fig. 127). The western end of this valley is drowned to form False Creek, and in recent years a good deal of the shallow bay has been filled in for building purposes. The main wharves are along the northern side of the northern ridge, where the Canadian Pacific Railway follows the shore to its terminal station right in the core of the city. The northern ridge ends in a small peninsula—almost cut off by a lagoon—and this has been dedicated as Stanley Park, the finest city park known to the writer. The fine stretch of deep water to the north of the city is called Burrard Inlet. It is quite narrow, and is further constricted at two places where it has been crossed by bridges. Between these two bridges the ships from all over the Pacific Ocean can berth alongside the rails of the C.P.R.

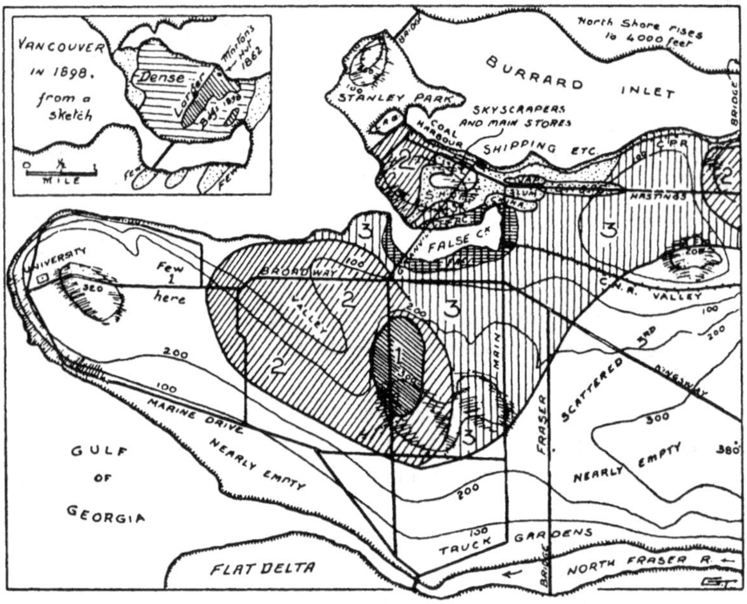

FIG. 127. A simplified functional plan of Vancouver. The large figures *1, 2, 3, 4*, refer to residences. Contours are indicated. *Inset* is a map to the same scale of the city in 1898.

The first town in this region was New Westminster, which was built on the Fraser River about 1857, in the expectation that it would become the capital city of this part of the Dominion. A trail led to Burrard Inlet, and here was a pioneer summer resort in the early sixties near the eastern bridge of to-day (Fig. 127). Morton built the first house in 1862 near the present C.P.R. station; and by 1870 there were six houses, a jail, and an hotel in the vicinity. A large saw mill was erected a mile or two to the east in the late sixties.

In 1866 Victoria became the capital of the vast region including Vancouver Island, as well as the mainland hitherto controlled from New Westminster. It was, however, the coming of the transcontinental railway which decided the evolution of Vancouver. At first it was intended to make Port Moody the terminus on the Burrard Inlet. But the shores were swampy and the water shallow hereabouts, so that it was wisely decided to make the terminus some twelve miles to the west near Coal Harbour, where Morton had settled in a district he called Liverpool. Here the first train from the east reached the Pacific in 1886. A disastrous fire destroyed almost every house in the little town in this same year, but by 1890 the population was 13,000.

The interests of the little town centred largely in lumber, and timber merchants and millowners soon occupied the shores of False Creek—as indeed they do to this day. the population reached 100,000 about 1912, and by this date the functional plan was taking on the pattern it exhibits at present. The main buildings are clustered on Granville Street, which climbs 150 feet up the moraine ridge from the main wharves on Burrard Inlet, and then descends to the bridge over False Creek (Fig. 127). Large houses were built in the vicinity which are still in use, but largely as apartments or second-class residences (2 in map). Third-class houses, factories and lumber wharves crowd all sides of False Creek. The west-east street near Burrard Inlet is called Hastings Street; and near this are definite Chinese and Japanese quarters. The lower lands in the vicinity are filled with third-class residences built of wood, but the higher land some three miles away is occupied by second-class houses. The best quarter to-day is near the summit of the southern moraine ridge, which rises to 300 feet. Here a high-class residential suburb (called Shaughnessy) was laid out mostly during the 1920's. (It is labelled '1' in Fig. 127.) There are also many fine residences on Point Grey in the vicinity of the University.

The inset map in Fig. 127 shows how the city has spread from 1898 to 1940. Truck gardens occupy the low lands on the northern bank of the Fraser River. The main road to the south is the Kingsway, and many houses are being built in this direction which will ultimately link up with New Westminster.

I owe to a recent paper by D. P. Kerr some notes on the development of the port of Vancouver. Until the arrival of the C.P.R. in 1886, the trade of the port consisted almost wholly of lumber from the various saw-mills. At this date twenty-nine million feet of cut timber was exported from the port in twenty-three ships. In 1887 the first trans-Pacific ship arrived in Vancouver with a cargo of tea, and regular steamship service with the east was established in 1891. Lumber remained the chief export even after the Panama Canal was opened in 1914. But by 1921 there was a marked change in the trade, which now became directed to Europe. The first grain elevator was built in 1914, but the World War blocked its use for some years.

There was much difficulty in overcoming the railway rates for grain from the Prairies, which at first strongly favoured export from eastern ports, though Vancouver is nearer many of the wheat areas. Since 1927 much of the grain from Alberta has been shipped *via* Vancouver; though if the distances are at all equal, the rates still favour the eastern route. In 1921 one million bushels of wheat left Vancouver; but by 1932 this had risen to 105 millions. In the period 1921 to 1935 the cargoes from Vancouver *via* the Panama Canal had risen from sixteen per cent to sixty-five per cent of the total. In 1939 Vancouver was connected to forty-seven different countries by regular steamship lines. It is primarily an exporting centre, but the enormous amount of crude oil imported into Canada here helps to balance the outgoing grain and lumber cargoes.

Owing to the extensive supplies of lumber there are many factories in the Vancouver area, including fourteen saw mills and fifteen shingle mills, as well as sixty other wood-working factories. A good deal of coal is brought to the mainland from the Nanaimo coalfield of Vancouver Island. Another great asset of the city is the abundance of hydro-electric power, based of course on the heavy rainfall and the character of the mountain and valley topography just north of Vancouver. It must also be remembered that the average monthly temperature in winter does not fall below freezing point. Hence the port is always free of ice, and the climate attracts Canadians from the chilly east!

Bruges and a varying coastline

One of the most interesting coasts in the world is that of Belgium, which has passed through many changes during historic times. Some of these are suggested in Fig. 128, where we see the position of the coastline at three periods in the last millennium. About the dawn of our era the sea coast was about a mile west of the site of Bruges; but in the next thousand years the land rose until the sea was driven back beyond where it is to-day. During the next 200 years there seems to have been a period of submergence, and the sea once more advanced close to Bruges. The next 200 years was the period of Bruges' greatest prosperity, when the clothmakers of Flanders were perhaps the richest and most powerful citizens of the day. One of the great naval battles of those days was fought just off Sluys in June, 1340. At that time this was an open roadstead capable of holding a large fleet, but it has now been silted up by the River Eede. For a time Bruges was the capital of Flanders, but after 1180 Ghent became of more importance. During the 14th century Bruges was the chief commercial city of northern Europe, but when the Zwyn began to silt its trade diminished, and it was of little note after 1490. Partly for this reason Bruges preserves its early medieval character better than any other Belgian city. A number of buildings

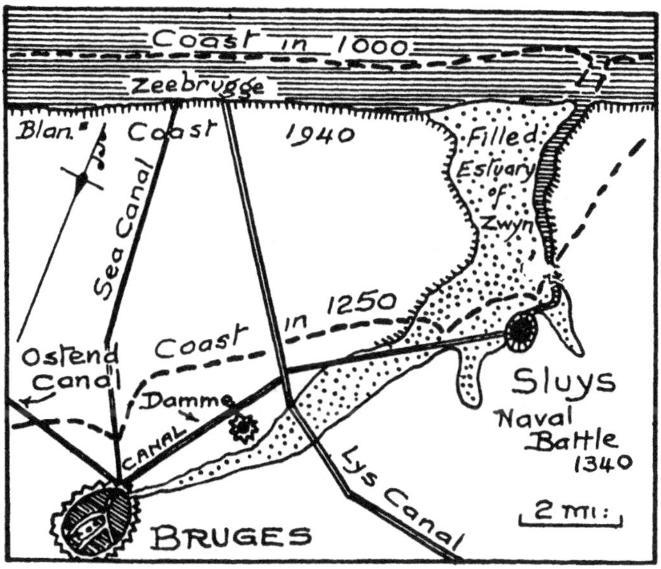

FIG. 128. The coast near Bruges in Belgium, showing alterations since A.D. 1000. Also the silted Zwyn and modern Canals.

have changed little since 1400, and are to be found mainly in the south-west quarter of the city.

The name of Bruges means 'bridges', and refers to the numerous canals in the city and to the bridges across them. Relatively early three canals connected the city with Sluys, Ghent, and Ostend; and around 1900 a ship canal twenty-three feet deep was constructed directly to the coast at Zeebrugge, a distance of nine miles. In spite of the spread of railways and of this new canal Bruges has not increased much in importance in the last century. About 1880 its population was 50,000, and remained much the same by 1940. As the plan of Bruges in Fig. 128 suggests, the early oval outline of the city still bounds most of the residences, and the encircling moats and the position of the ancient ramparts are clearly visible.

Venice: an island port expanding to the mainland

Probably the most remarkable site for a large port in the whole of Europe is that of Venice. At the head of the Adriatic Sea the downfold in the crust—due to late Tertiary crustal folding—is being gradually filled by the Po and other rivers descending from the Alps. Since the sea was first called the Adriatic—from the former port of Adria—the land has advanced twenty miles to the east. Bordering this delta-like area the Adriatic Sea has thrown up a series of sand-bars called 'lidi', and many small islands dot the lagoons between the sandbars and the

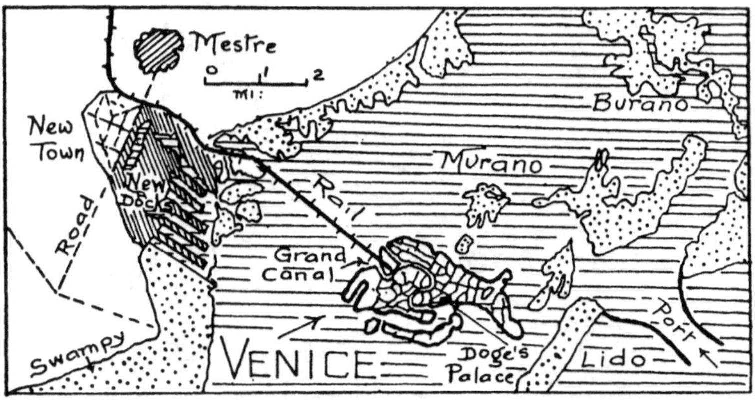

FIG. 129. Map of Venice, the Lido and the new Port (Marghera) on the mainland. (*After Von Seydlitz.*)

delta-swamps (Fig. 129). Venice is said to be built on a cluster of 117 of these islets. The houses are constructed over piles, and there are 150 canals of varying sizes permeating the city, which are crossed by 378 bridges. In general it may be stated that there are lanes but no streets in the city, and only one or two piazzas; all traffic is carried on by boats on the canals.

The settlement dates back to very early times, when the Veneti fled from the mainland which was being attacked by northern invaders. About A.D. 800 the settlement of 'Rivoalto' became the head of a union of these refuge settlements under the Dux (later Doge), and the town became known as Venice. It soon became allied with the Byzantine

Empire, and rose to great prominence in the Crusades. Indeed Constantinople (though Christian) was taken in 1204, in part to help the Venetian merchants in their rivalry against the other trading folk in the Mediterranean. In the 14th century Venice successfully attacked Genoa, and in the 15th century reached the zenith of her power. 'She was the focus of the world's commerce, numbered 200,000 inhabitants, and was universally respected and admired.' At this time she had a commercial fleet of 3,300 vessels and a navy of forty-five galleys. Her decline began about the time the Turks captured Constantinople (1453), and this was accentuated by the shift of trade to the New World somewhat later. By 1718 she had lost almost all her wide possessions.

In the middle ages vessels lay alongside the wharves at the east end of the Grand Canal (Fig. 129), but as trade fell off the entrance at the Lido silted up, and external trade vanished. After the Austrians controlled Venice some revival of trade took place. When the railway reached Venice a new opening to the sea was made at Malamocco, and a new dock was made at the western end of the Grand Canal near the railway station. The opening of the Suez Canal in 1869 brought a renewal of the eastern trade to Venice, since this was the Mediterranean port nearest to the centre of Europe. The former Lido opening was dredged and widened, and long moles were erected as shown in Fig. 129. By these improvements the trade of Venice rose from sixteen to twenty-five millions sterling between 1886 and 1905. The population declined steadily from 158,000 in 1548 to 132,000 in 1881, then it rose slowly to 147,000 in 1918, but by 1928 was 207,000.

Before the first Great War the chief interest in Venice centred in the tourist trade and in the production of small objects like lace and jewellery; but about this time it was felt that an important port could be developed in association with the ancient city. It was decided not to interfere with the plan of Venice itself, but to build a new port on the adjacent mainland. Between 1923 and 1929 the Port of Marghera was created across the lagoon, some three miles to the east. It was linked by rail to the little town of Mestre (Fig. 129), and the proposed plan will cover an area three times that of the port of Genoa. Three industrial zones were arranged in the plan, with a special exit to the sea for each. There was also built a special dock for inflammable oils. Nearby a garden city was laid out for 50,000 people; but the onset of the Second World War has blocked some of these worthy projects. In 1938 no less than 6,271 vessels entered the port, a number only exceeded by those for Trieste and Naples.

Gdynia—a new Baltic port to serve the Polish Corridor and Poland

After the first Great War a corridor to the sea was given to the Poles, cutting right across the German territory of East Prussia, and flanking the west side of the important port of Dantzig. The latter city was almost wholly occupied by German citizens, and though the Poles were given considerable rights to use Dantzig, they preferred to develop a completely national port at Gdynia. This was a little fishing village about fourteen miles to the north-west of Dantzig, and the site is protected from the north by the long sandy peninsula of Hel (Fig. 130).

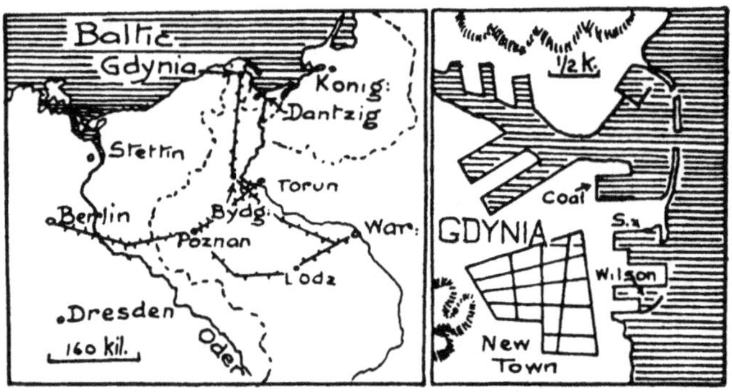

FIG. 130. (*Left*) The site of Gdynia in relation to the Polish corridor, Stettin and Dantzig. (*Right*) Plan of the docks at Gdynia in 1938.

In 1920 Poland was engaged in warfare with the Soviets, and was unable to use the port of Dantzig. They made some temporary wharves at Gdynia for the purposes of this struggle. The depth of water near the shore is over thirty feet, and from that time on the Polish Government carried on with the construction of the port, though the main contracts with a French-Polish Company were not signed till 1924. Long moles and breakwaters had however been completed by 1924.

In 1939 Poland had a population of thirty-four millions, and her sole port was Gdynia (called Gotenhafen by the Germans). This town had grown from a tiny village to a total of 114,000 by 1938. At the same time the trade of its rival Dantzig had tripled since 1914. The following table shows the growth of the trade at Gdynia.

1924	10,000 tons	Ratio, exports to imports—			9 : 1
1928	1,957,000 ,,	,,	,,	,,	8 : 1
1932	5,194,000 ,,	,,	,,	,,	10 : 1
1936	7,743,000 ,,	,,	,,	,,	4 : 1
1937	9,006,000 ,,	,,	,,	,,	4 : 1

The port is divided into external basins and internal basins, as the chart in Fig. 130 shows clearly. The three external basins were built first, and are named Coal, South and Wilson Basins. They are used primarily as passenger landings, for coal and oil, and for the local fishermen. Then to the north-west the new basins have been excavated, and one of these is a free port, others are used for the export of timber. It is expected that there will be a much greater trade in industrial products in the near future. There are very modern refrigerating plants on the wharves, as well as up-to-date cranes of the largest size. Warehouses for grains, cotton, and all sorts of food such as herrings, fruits and rice, are available.[1] Naval construction was being developed to a considerable degree before the invasion of Poland by the Nazis.

Coal is the chief export and with the timber export constitutes most of the trade of Gdynia. All kinds of goods are imported, as will be gathered from the brief references to the warehouses. Not only is the port used by the Poles, who have built special railways thereto to by-pass Dantzig; but the adjacent states of Rumania, Czecho-Slovakia, Hungary, and to a lesser degree Austria and Yugoslavia were making increasing use of this new port. The plan of the new town is indicated in the chart, Fig. 130, but no very accurate map was available to the writer, so that it is only approximate.

To the geographer interested in the evolution of towns the future of Gdynia is not at all clear. Since the last war the Poles have been awarded Stettin, the former port of Berlin (Fig. 130). This is a city of 250,000, and a port of much greater importance than Gdynia in the past. It is specially noted for the great Vulcan ship-building yards, and for the trade in grain, timber, chemicals, and oil. No doubt the vast extent of new territory occupied by Poland in the hinterland of Stettin will continue to be served by Stettin, while Gdynia will still be much better placed for the more eastern traffic. Dantzig also is under the control of Poland, so that this nation has three large ports in the place of none as in 1920. It will be of great interest in a decade or so to see how these three rival ports on the Baltic have progressed.

Toronto—a Sandy Hook converted into a harbour

In earlier chapters little attention has been paid to the origin of the present large harbour at Toronto. It is, as usual, partly natural and partly man-made; and it will be of interest to separate these aspects of the present-day harbour. The early administrator, Colonel Simcoe, wished

[1] C. de Kovnacky, "Le Port de Gdynia," *Jnl. des Économistes*, Vol. 108, Paris, 1938.

for a site removed some distance from the American territory, which in 1788 had only recently broken away from the British Empire. Several sites were considered, but the sandy hook on the northern shore of Lake Ontario offered considerable advantages. It was the southern terminus of a well-used trail to the north. The bay within the hook was large enough and deep enough for the small boats of the day, while the hinterland showed promise of good farming land.

If we glance at a map of the lake we find no other sandy hook along either the north or south shore. The western end of the lake is cut off by a marked sand bar, which has led to the development of the port of Hamilton, though much later than in the case of Toronto. However, in Lake Erie there are three sandy hooks somewhat resembling that at Toronto, and necessarily with somewhat similar origin. In effect the hook is due mainly to along-shore currents tearing away cliffs of soft material. There is the closest connection between the sandy hook (now called Toronto 'Island') and the striking sandy cliffs of Scarborough Bluffs, some ten miles to the east. But one other condition is necessary to account for the large sandy hook; and that is the great change in the level of the lake, due to the sinking of the water level when the ice dam near Prescott broke down some 15,000 years ago. When the waters rested at the present level they found a great mass of morainic debris piled up at the Bluffs, which was quickly torn down and driven to the west by the along-shore current. This debris was deposited to form the sandy hook when the speed of the current diminished.

The appearance of the harbour shortly after the first settlement is illustrated in the map in Fig. 131. The harbour was square in outline and about one-and-a-half miles across. The older portion of the hook, i.e., that in the east, was occupied by the marshy lands of Ashbridge Bay, which was separated from the lake by a ridge of higher and firmer sand. The hook clearly showed the stages in its evolution by the numerous curving ridges of which it was composed together with intervening ponds and marshes. A wide zone of very shallow water surrounded the end of the hook, so that the entrance at the west end was only half a mile wide. The northern shore was shallow and filled with rushes, while a clay cliff a dozen feet high separated the water from the plain along most of the harbour.

The first settlement was laid out at the north-east end of the harbour, with King Street as the first main street (Fig. 131), and George as the chief north-south street. About this time the deepest part of the harbour was one mile south of the centre of the settlement, where the water was over thirty feet deep. Most of it was, however, about eighteen feet deep

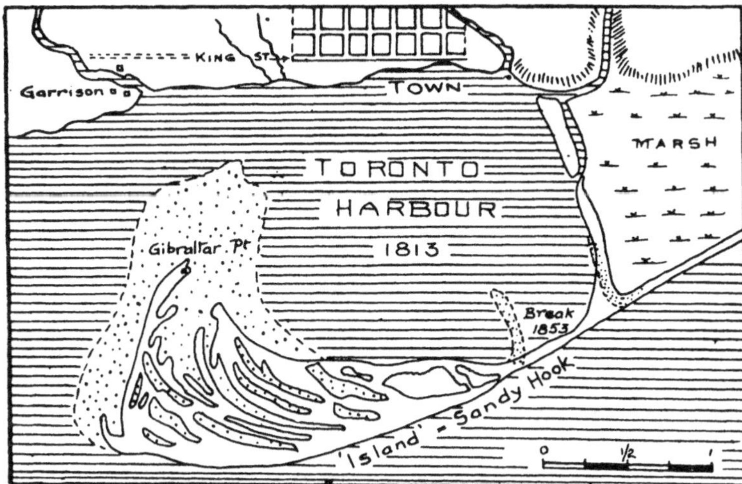

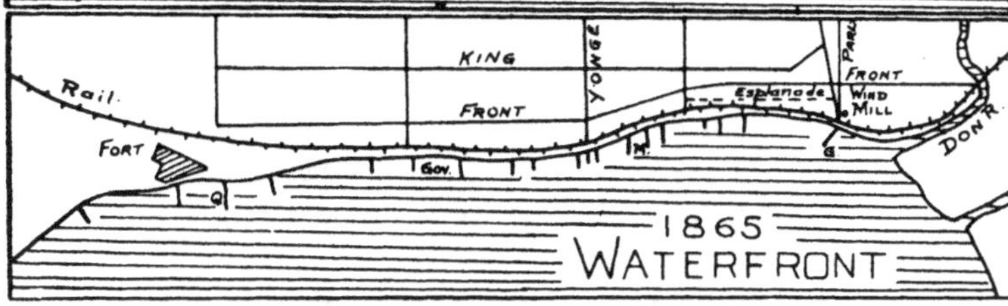

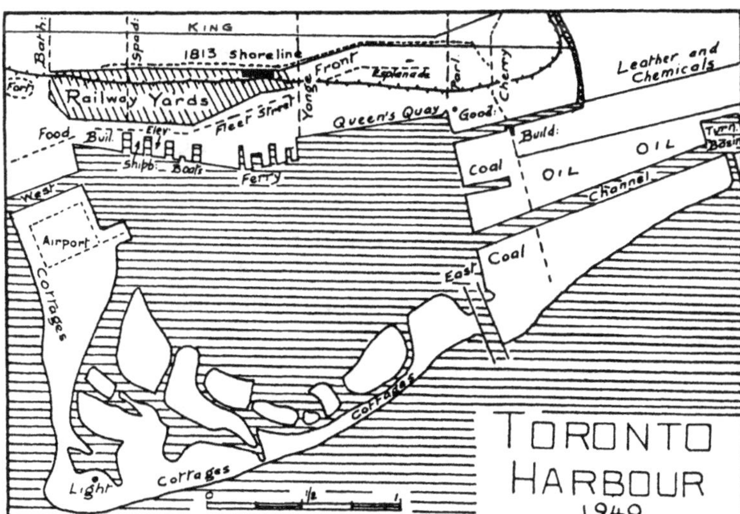

FIG. 131. Toronto about 1813 showing early streets, and the shape of the Sandy Hook (Island).
FIG. 132. The second Waterfront at Toronto, widened to take the railways about 1865.
FIG. 133. Toronto Harbour in 1940 showing the new land added south of the 1813 shore-line. Note also new land filled in for air port and fuel reserves.

though considerably less in the entrance. A military garrison with block houses was established just north of the entrance, and this portion of the plain was made a military reserve. In 1793 Aitken surveyed the harbour, as well as that at Queenstown (now Kingston) at the east end of the lake. He pointed out that the latter was soon frozen up in winter, whereas from Toronto it was possible to reach the open water with naval vessels much more readily.

The first wharf was built just east of George Street in 1796, and a little canal was made near the garrison to enable stores to be landed there. In 1818 several other wharves are indicated on early maps, and some of these are reproduced in Fig. 132 where the Government Wharf, Maitland Wharf and Gooderham's Wharf in the east are charted. In 1833 a Commission was appointed to improve the wharves, and a little later Queen's Wharf was built, to remain a prominent feature of the harbour until it was buried by reclamation work in 1917. In 1834 a line marking the front of future wharves was drawn from the Windmill in the east to the Old Fort, and this for many years was known as the 'Windmill Line'. In 1841 there were thirteen wharves along the harbour front, and five more were added by 1847. During the fifties the Esplanade was built along the waterfront (see footnote, p. 195).

The coming of the railways in the fifties and sixties completely altered the pattern of the waterfront, since several lines entered the city along the shore, necessitating many alterations to the existing wharves. The limits imposed by the old 'Windmill Line' were extended, so that wharf owners could make their wharves longer, and fill in the portions to the north. In 1903 there were forty wharves between Queen's Wharf and Parliament Street (Fig. 133).

During all these early years the 'sandy hook' had not changed much in its plan, though further increments were being added at the western tip. A substantial stone lighthouse was erected at the end of the hook in 1808, and still survives, though it is now half a mile east of the end of the hook. The sandy beach was soon used as a pleasure resort by the citizens, and after the war of 1812 a block house was erected on the south side of the entrance. In 1853 after a heavy storm an entrance was breached in the main sandy ridge, and this has since been improved to constitute the Eastern Entrance (Fig. 132). Between 1849 and 1854 a steam dredge was purchased to keep the western entrance clear for shipping. After 1882 the eastern channel was deepened, the swamp (known as Ashbridge's Bay) to the east of the harbour was sealed off by a dyke, a new outlet for the Don River was opened in the bay, and the western outlet was altered in position and deepened.

In 1911 a much stronger Commission with ample powers was appointed to take charge of Harbour improvements. The whole of the swamp was now to be reclaimed and utilized for factories and wharves, etc. This involved an area of over 1,000 acres, in which a channel and Turning Basin was to be excavated. The 'windmill line' was pushed about 1,000 feet south again, giving a much larger area for industrial expansion to the south of the railway lines, which blocked the access to the lake everywhere along the shore. Around 1925 a street eighty-six feet wide was constructed to the south of the railway lines, now known as Fleet Street, and south of this is another street called Queen's Quay. Both are being lined by warehouses and factories.

Reference to the map in Fig. 133 will show how far the city buildings have encroached upon the former harbour in the last 150 years. The 1813 shoreline is indicated, and the Front Street of to-day marks approximately the water's edge in the days of the first settlement. The Esplanade (Fig. 133) shows a further advance, linked with the coming of the railway, while Fleet Street and Queen's Quay are creations of the last thirty years. The varied character of the factories on the 'made ground' can be seen from the labels on the map. From west to east we pass buildings devoted to food and builders' products, shipbuilding, wheat elevators and Ferry wharves; while to the east of Yonge Street on the former swamps of Ashbridge's Bay, are the coal reserves and the tanks filled by the oil boats from U.S.A.

A few words as to the shipping using the port of Toronto will not be out of place. A yacht called the *Toronto* was built in 1799, and many boats up to 100 tons in weight were built before 1820. The first steamer on Lake Ontario was the *Frontenac*, launched in 1819, and she plied from Prescott to Kingston, York (Toronto), and Niagara. In 1826 there were eight steamboats on the lake. Toronto was visited by 2,194 vessels in 1942, being surpassed in this respect in the east only by Montreal and Quebec.

CHAPTER XV

MOUNTAIN TOWNS AND VILLAGES

Introduction

OF all topographic controls affecting settlement those depending on mountain conditions are the most complex. It would be impossible to discuss all the variations which have been in part determined by elevation, by varying slopes, by the character of the rocks of which the mountains are composed. Some of these factors have already been discussed; e.g., the effect of a dissected sandstone plateau on the spread of the settlement near Sydney, Australia. So also the effect of the cuestas of the Weald is illustrated in Fig. 95. I have, therefore, decided to choose certain classical examples of mountain settlement, which will, I hope, help my readers to realize how manifold are the variations of this type of environment.

The first section attempts to show how settlements vary in the best investigated mountain area of all, i.e., in the European Alps. My own experience is mainly based on a survey of the Adige Corridor through the very heart of the Alps where they are widest; and this traverse from Trento to the Reschen Pass shows many types of settlement as we rise from 600 feet to 6,800 feet on the main divide. The great importance of fan structure in determining the site of settlement in a glacially-eroded trough is made clear. Conditions affecting the limits of trees and pasture are developed also.

Roderick Peattie has given much attention to this problem in many memoirs, and I have borrowed frôm him a map dealing with the valley of the Doron on the French slopes of the Alps. This shows that vegetation depends on slope as well as on elevation, and that grain may be preferable to vines where conditions suit both. One of the largest fans known to the writer is that on the north flank of the Pyrenees. The types of settlement on this giant fan of Gers have been investigated by H. M. Kendall, and his study shows how the depth of the valleys and the slopes of soil affect the pattern of settlement, when there is more of a soil covering than is usual on mountain slopes. The most typical horst in northern Europe is probably the Harz, rising as a plateau 1,000 feet above the north German plain, and capped by the much higher Brocken. Here the shift from a mining economy, based on the ancient rocks and their minerals exposed in this horst, to a tourist economy after the minerals are largely exhausted is discussed at some length.

The foregoing Old World examples deal with regions where the population pressure is considerable; but if we turn to the New World we find somewhat similar conditions where settlement is only in a pioneer stage. Thus a study of the way in which the topography of an 'eroded dome' can affect early settlement is given by O. E. Guthe in his research on the Black Hills of Dakota. Here the topography very definitely decides the distribution of mining, lumbering, ranching and tourist interests, which are all evident to-day even though the settlement of this area is only seventy years old.

In the Kicking Horse Pass region we find the Canadian Rockies exhibiting some of their finest scenery, as well as one of the most interesting sections along the great Canadian Pacific Railway. Here the various interests can be readily understood from a detailed block diagram based on the author's familiarity with this region. Lastly, in far away Australia, we find the best illustration of a quite considerable population settled on a sterile sandstone plateau. This result is due mainly because the climate is considerably cooler owing to the height of 3,500 feet above the great city of Sydney. Here on the Blue Plateau is a continuous series of tourist towns, where the main attractions are the unique 'bottleneck valleys' and their falls and ferns, which lie a thousand feet or more below the towns on the plateau. This type of settlement is illustrated by a large block diagram which depicts clearly the scenery in the vicinity of Australia's most popular tourist resorts.

Topography and Settlement in the Italian Alps

In July, 1938, I was engaged in a survey of various towns and villages to be seen along a traverse of the Adige Corridor for about 100 miles from Trento in the south to the Reschen Pass in the north. During this traverse the bed of the river Adige (*pron.* 'Adger') rose from 600 feet at Trento to 5,000 feet at the Pass, while small villages and farms were to be seen about 1,600 feet higher. I contributed a lengthy paper on this survey to the *Geographical Review* in April, 1940, where a number of photographs illustrating the region may be consulted.

The district investigated is particularly interesting from a cultural point of view, since here Italian and German cultures mingle, for there are nearly a quarter of a million German-speakers south of the Reschen

Pass, in territory which was given to Italy after the first World War; and which apparently is not to be returned to the control of Austria after the second World War. The character of the villages alters considerably as we pass northwards, the actual cultural boundary being near the little towns of Ora and Salorno. The rather crowded and untidy buildings of the Italians here give place to the isolated, single ridge-roof farms of the Germans. At the time of my survey the Italians were trying hard to obliterate as much as possible of the German culture —so that all the towns and street names were changed, as well as the trade names over the shops; and only in the cemeteries the German family names showed that the inhabitants were almost wholly German.

Among the many communities studied it will serve in this section to confine our attention to a few typical settlements, whose characters are summarized in the following table.

Community	*Population*	*Elevation*	*Position and Features*
Trento	60,000	600 ft.	South end, in broad trough
Merano	20,000	1,045 „	Tourist town partly on a fan
Glurns	700	2,975 „	Walled town in Vintschgau
Laives	300	900 „	Large fan village below gorge
Messo-corona	200	650 „	Small fan village below notch.
Reschen	150	5,000 „	Village in Pass
Roia	20	6,700 „	Tiny hamlet above Reschen
Buffalora	10	6,800 „	Alpine farm at Ofen Pass

Settlement in the mountain corridor of the Trentino naturally depends to a large extent on elevation. For in climbing up to 9,000 feet—as when one crosses the Stelvio Pass—a great many different types of environment are traversed. These are shown fairly clearly in Fig. 135. We see in most of these deep glacier-cut valleys the phenomenon of *adret* and *ubach* (i.e., sun and shade) exercising a special influence on human settlement. This will be specifically illustrated in the description of Davos in Switzerland (Fig. 163). Near the Stelvio Pass the alpine cabins are found on the average up to about 2,160 metres on the sunny slopes facing south-west, but only as high as 1,755 metres on the slopes facing north-east.

The lowest level of the permanent ice near the Stelvio Pass is about 2,800 metres (9,000 feet), and in favoured spots some grass will grow almost up to this level (Fig. 135). The tree-line is at 2,340 metres on the adret slope, but drops to 2,170 or less on the ubach side. Throughout the region the steeper rocky slopes are bare or forested at lower levels; but wherever possible small patches of meadow hay are grown and

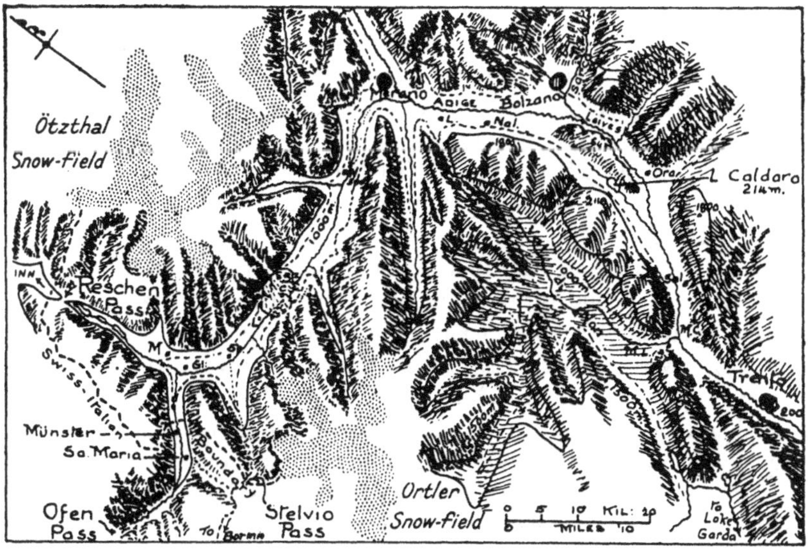

FIG. 134. Sketch map of the Adige Corridor. The 1,500 and 1,000-metre contours are shown, and snow fields are indicated by stipple; the large Alliz (A) fan is also represented. The Reschen, Ofen, and Stelvio passes and the more important population centres are marked (by name or initial letter).

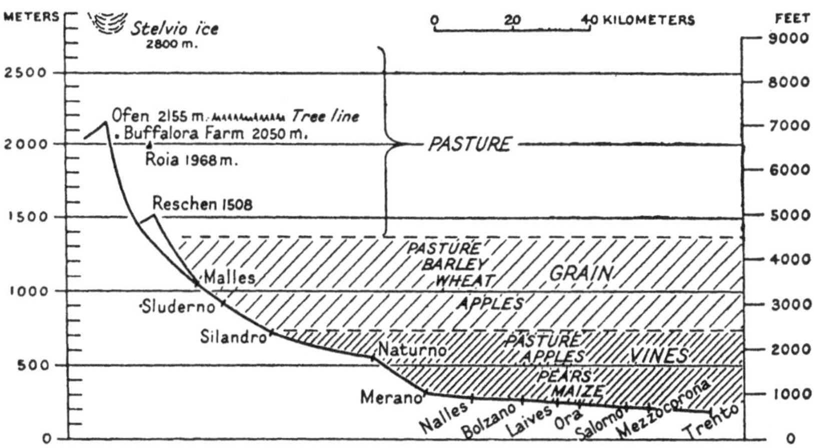

FIG. 135. Vertical section from Reschen to Trento showing progressive changes of crops with elevation.

harvested. The fans, at the foot of the steep tributary gorges, are of course the best sites for settlement in the higher portions of the Trentino.

Grain in small patches is found at altitudes as high as Reschen (1,500 metres); but pasture and orchards seem more important than grain above Merano, and below this town vines are ubiquitous. Vines reach their limit on the adret slopes of Silandro about 750 metres, and I also noticed vines at about this height on the sloping sides of the main trough near Bolzano.

Small huts are found occasionally at the highest levels at which any pasture is to be found, for the cattle of the lower farms are grazed at these alpine levels in the summer. The highest permanent farm examined by the writer was just near the Ofen Pass, and is known as the Buffalora Farm (Fig. 135). Here is to be seen a long rough stone building with a simple ridge-roof in which the farmer dwells, and alongside were several equally long low barns built of logs. Often enough some timber-cutting is done on these upper slopes and a steel cable runs down the mountain-side to the village beneath; and down this timber and faggots slide to the rough saw-mill below. Such a means of transport is illustrated in the sketch of Laives (Fig. 138).

In the early days of settlement much of the lower flat floor of the glacial trough was occupied by marshes in which the river ran sluggishly in 'braided' streams. This condition is visible in the Vintschgau above the great Alliz fan to this day (Fig. 134) Under such conditions the better-drained debris of the fans was that first occupied by primitive farms. In many places in the debatable cultural zone near Salorno, the Germans occupy the older farms on the lower slopes, while the later Italian 'invaders' have drained the floor of the valley and now occupy the lowest levels.

Trento, a fortress town in the Adige Corridor

Trento offers many points of interest to the student of town evolution. It has a unique position in the broad flat trough of the lower Adige, as can readily be observed in the simplified plan in Fig. 136. Hereabouts the valley plain is in general about a mile wide; but a precipitous crag—nearly 400 feet high—the Doss Trento—blocks nearly half the valley. This crag has caused the Adige River to swing sharply to the east, so that it almost cuts across the valley. In early days a castle was built at this constriction of the valley, and a settlement grew up around the castle, which received the name of the local tribe of Tridentini, i.e., Trient or Trento. Hereabouts two important side valleys enter the main Adige

Valley; on the west is the great valley which contains Lake Garda, which is quite distinct from the Adige Valley; and on the east the Val Sugana which leads to Venice. From 1027 to 1803 Trento was under the control of the Catholic Church, and it then came into the hands of Austria. Hence the Italians had no political claim to it, though it was handed to them with fair reason after the first World War.

One of the most interesting streets in the old town is an arcade of shops which formerly bordered the Adige River along the curved line B, C, in Fig. 136. But about 1850 the river's course was straightened, thus wiping out the meander caused by the crag, though the ancient arcade still remains in active use. In the middle ages Trento was a walled town with the castle at the north-east corner, and relics of the walls and towers are still to be seen. To-day, however, the town has spread beyond its ancient walls, and the main zones are indicated in the small map in Fig. 136. On the outer edge is a zone of villas scattered among gardens and merging into vineyards, and this is especially characteristic of the south-east quarter of the newer belt. Within this is a zone of apartments on the east, and a professional quarter with the provincial offices in the south-west.

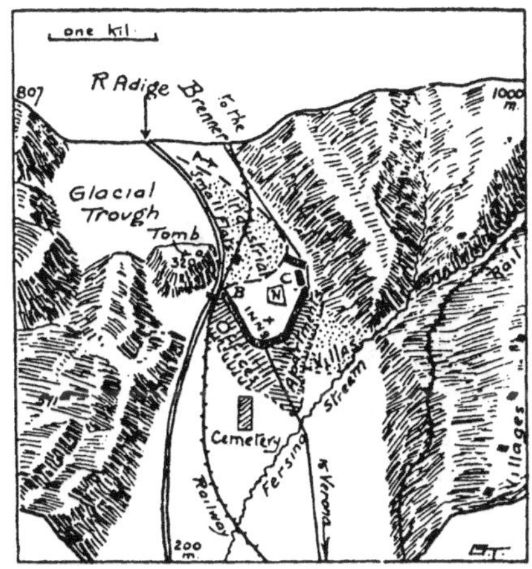

FIG. 136. A simplified plan of Trento, a fortified town in the south of the Adige Corridor.

The newly-made land in the north is occupied by factories and small flats, though a small storm-water channel occupies the position of the former Adige. The ancient portion of the city still preserves the old houses of the 18th century; with numerous churches, such as Santa Maria, where the Council of Trent sat from 1545 to 1563. A common feature of the old Trentino cities is the replacement of some of the narrow old-world streets and congested buildings by modern shops and hotels of larger size and convenience, usually in one specific quarter where the tourists congregate. This area is marked 'N' (for Neotechnic) in the small map.

Trento is the administrative centre of a large district, and there are many blocks of government buildings. Cement works, timber yards, and oil depots cluster at the north end of the town; and huge barracks occupy part of the new ground east of the station. The other industries are not important; for the most part they depend on vines and silk. A medieval map of about 1600 may be consulted in my original paper.

Fan Villages at Messo-corona and Laives

The main features of a fan village can be learnt from my sketch of Messo-corona (Fig. 137). This settlement is eighteen kilometres north of Trento, and is situated where the great valley of the Noce enters from the west. It is a typical example of a 'notch-fan village'; and above it the cliffs tower nearly 670 metres on the west side of the Adige glacial

FIG. 137. Sketch of the notch-fan village of Mezzocorona, looking north-west.

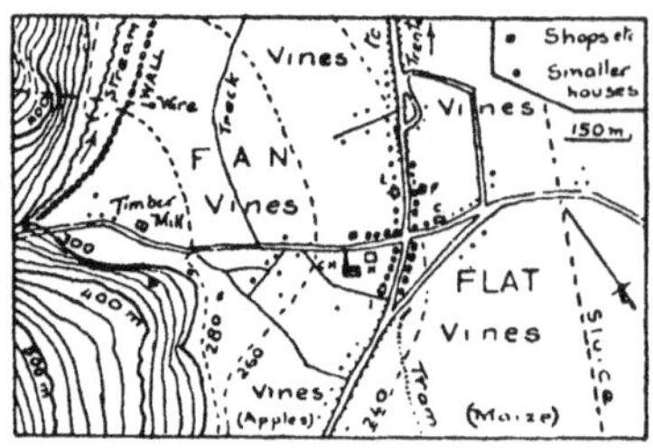

FIG. 138. Sketch survey of Laives, a small notch-fan village near Bolzano (*see Fig.* 134).

trough. A deep notch in the cliffs has led debris down to the fan below, and produced a fairly deep area of soil, raised well above the originally marshy floor of the Adige trough. As the sketch shows, the higher rocky portions of the notch-fan are left in woods. Then below this is a belt of pasture land. A main road crosses the fan at the middle level (as in Laives, Fig. 138) and here the village has grown up. There is a fair-sized church, and the Italian peasants live in a number of three-storey stone houses, with several families to the house. The houses are jumbled together in the village as a photograph in my original paper clearly shows. Trees surround the village, but the lowest slopes of the fan are given over exclusively to vineyards.

At Laives (formerly Leifers), about eight kilometres south of Bolzano, I made a rapid survey of a larger fan village. In this case the fan had developed at the outlet of a rocky gorge through which a tributary stream reaches the main Adige River (Fig. 138). In general the largest fans are on the adret side of the glacial trough, for here melting and erosion, especially towards the close of the ice age, were much more intense than on the ubach side. The torrent from the gorge is directed close to the southern cliff by a heavy stone wall, in places twenty feet high. This leaves the main area of the fan available for buildings, gardens, and vineyards. The fan at Laives is about sixty metres above the floor of the trough and a kilometre wide.

The main road at Laives traverses the lower part of the fan above the swampy floor of the main valley (Fig. 138). Here a village, consisting of a church, a hall, a Fascist centre, two inns and perhaps half-a-dozen large stone houses, and sixty or seventy smaller houses, had evolved during the centuries. At the foot of the fan is a large irrigation channel, and the fan is covered with vineyards and a few orchards. Timber is worked in a saw-mill at the apex of the fan, and two long wire cables from the high cliffs above enable the villagers to send lumber from the Alpine slopes directly to the fan. Above Laives the Adige is not navigable even for small boats.

Meran, a tourist town

The road to the Brenner Pass leaves the Adige corridor at Bolzano (Botzen), and turns to the east up the Isarco valley. Consequently Bolzano is an important junction town, with a population of forty thousand, augmented by 100,000 tourists in the summer. Its plan is described in my paper '*Trento to the Reschen Pass*', and no further reference to it will be made in this section. About eighteen miles higher up the Adige valley is Merano (Meran), which has grown up at a height of 1,045 feet above sea level, where the little river Passer enters the Adige from the east.

Meran is at the head of the wide, flat floor of the Adige, just below the rather steep step up to Naturno and the Alliz Fan. Meran is not built on a fan, but is tucked between the Passer River and the steep slopes to the north. Just to the south, however, is the rather large fan of the Naif stream; and modern Meran has spread across the Passer over much of this fan (Fig. 139). A new road leads north from the town over the Jaufer Pass, and so to the Brenner Pass.

The plan of the town differs in the two sections. In the old town (black in Fig. 139) the main street is parallel to, and 200 yards from, the Passer river, which is bordered by a pretty promenade. The newer part of the town proper is in the west near the station. A large piazza fronts the station with main streets converging to it. Here are the main hotels and several of the largest pensions—for which Meran is noted. Since Meran became so popular for tourists, both in winter and in summer, the great fan of the Naif has been built over. The hotels and pensions are in the north of this fan in Ober-Mais, while private houses border the 'contour-running' streets in Unter-Mais. The Naif stream now runs close to the cliffs, leaving the body of the fan available for houses.

The sketch plan (Fig. 139) shows the functional zones in Meran. The old town lies at the base of the Kuchelberg. At its north-west corner is the 'burg' or castle of the Counts of Tyrol, dating back to 1450. The casino and municipal buildings front on the river.

Next is a zone of hotels, the largest being near the station. Clustering

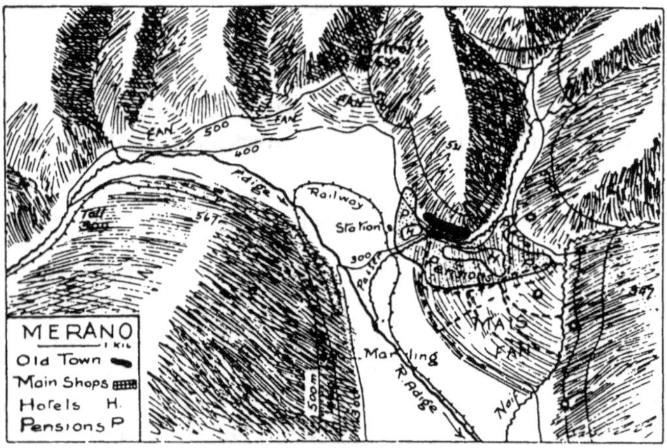

FIG. 139. The site of Merano, between the Passirio and the vine-clad hills. Note the modern pensions and the sunny fan. The rapid rise of the valley to Tel (Toll) is indicated by the railway loop. The private chateaux are shown by circles. Streets by broken lines.

around them, but spreading also southwards are the innumerable pensions (boarding houses), of which there are nearly a hundred of considerable size. On the slopes of the hills, especially above the Mais Fan, are a number of châteaux (German, *Schloss*). These belong to members of the Austrian nobility. The most famous is the oldest, i.e., Schloss Tirol about a mile to the north-west of the town. This dates in part from the 12th century, and was the castle of the ruler of Tyrol, and indeed the name of the whole Austrian province is derived from this Schloss. The town has a population of about 20,000, which is much augmented in the tourist season. It will be noted that the town has little relation to the river Adige, and one of the main advantages is its dry and sunny climate, as well as the splendid scenery in the neighbourhood.

Glorenza and the Reschen Pass

Settlement in the great Alpine valleys, such as that of the upper Adige, depends to a considerable degree on the size of the fan deltas. It will have been noted that the lower villages are placed on the fans; but as the main trough becomes somewhat narrower as we travel north, and the cliffs even steeper, the tributary fans become of greater and greater importance. Two of the largest fans in the Alps occur at Alliz and above Malles; and in each case the whole trough valley is dammed by the fan from the 'adret' (north or east) side. At Alliz the Adige is only thirteen metres wide, and flows in a deep notch in the fan on the southern side. Here the road and railway climb 1,000 feet, and above this fan is the flat, rather empty, sterile plain of the Vintschgau. This plain, about 3,000 feet above sea level, contains a number of villages, of which one, Glorenza (in German, Glurns), is sketched in Fig. 140..

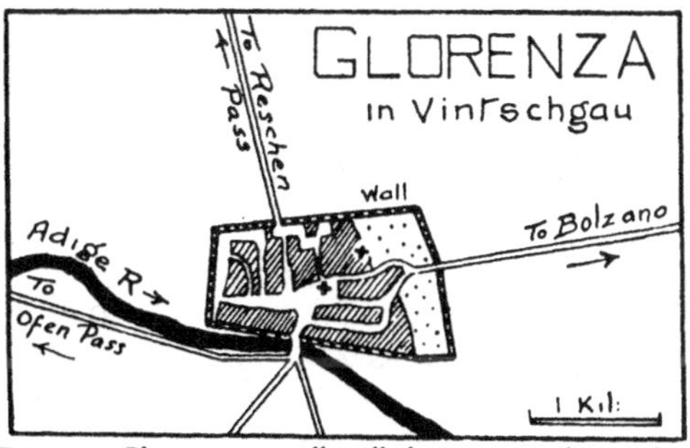

FIG. 140. Glorenza, a small walled town 17 miles south of the Reschen Pass.

Glorenza has only about 700 inhabitants, but nevertheless it is a walled town, which apparently has not changed much since medieval times. The walls and the gate-towers are still in evidence; and I should have liked to have made a detailed survey of the town. Unfortunately I had already just spent a day in a Fascist guard-room at Nalles, near Bolzano, and was warned that sketching, photographing, or anything like a geographic survey, was looked on with grave suspicion. Hence my sketch map is based entirely on the large scale Carta d'Italia (which in 1938 could be purchased in any of the towns!) The little plaza in the centre of the town with the ancient church on the east side is indicated. Here also is apparent the irregular pattern of the narrow streets, while what is rather unusual—there is a considerable area of more open ground in the east within the walls. One may suppose that the sterile character of the Vintschgau—where much of the flat floor only grows alders—has not led to any increase of population. Moreover, the little railway runs on the other side of the trough to Malles, and this town is the general junction for the Ofen Pass and the Reschen Pass. The road to the Stelvio Pass leaves the Vintschgau at Sponding, some six miles to the east of Glorenza.

Most of the details in the environment of one of the lower Alpine passes can be gathered by a study of Fig. 141. Here is a block diagram—drawn with the horizontal and vertical scales about equal—looking north to the actual summit of the Reschen Pass at 1,508 metres. Here the Italian Adige rises in a little grassy meadow, and the site is a splendid illustration of what is meant by a 'valley divide'. The Alps tower each

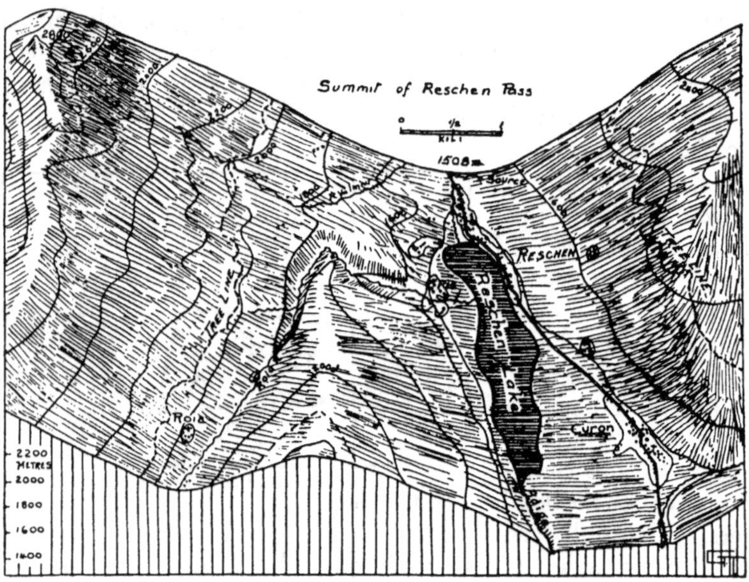

FIG. 141. Block diagram of the Reschen Pass (to natural scale) showing the sources of the Adige in a valley divide. The hamlets are enclosed in an oval line. Isolated dots are cabins. Note the tree line. (Looking north.)

side 4,000 feet above the flat floor of the valley, which obviously has been cut out by a former glacier. Hence the divide is in no sense due to the action of two opposite-flowing rivers, though a tributary of the River Inn rises in the same flat meadow as the Adige, and flows across the Austrian border a kilometre to the north of the summit.

The level character of the pass may be gathered from the lake, which is only thirty metres below the summit, and is 2.5 kilometres long. The two villages shown are Reschen and Curon (Graun). Both are 'strassendorfen', strung out along the main route to the Reschen Pass. Curon has about 600 inhabitants, and Reschen about half that number.

Six other tiny hamlets are shown on the diagram—each enclosed in a circle. Of these Roia is the most interesting (Fig. 141). It contains a small chapel and half-a-dozen houses at a height of 2,000 metres; and is one of the highest hamlets in the Adige Corridor. Scattered about the same level on the western slopes of the glacial trough are many isolated cabins, all of which appear in the diagram. These are much lower on the eastern side of the trough, which is on the whole steeper and more rocky. The tree line near Reschen is about 2,000 metres on both sides of the valley.

In Reschen itself there are about thirty or forty houses, which straggle along the main road for about a kilometre. There are three or four shops, which are dominated by the steepled church, and near the lake are two large and several smaller hotels for tourists. Some of the larger houses have several ornate balconies across the front. External ovens 'hung' on the walls, are characteristic of the Alpine houses.

Villages in the French Alps: Doron valley

Roderick Peattie has made a detailed study of one of the main valleys in the Tarentaise region just north of the Mont Cenis Tunnel.[1] He shows that a number of other factors besides insolation determine the upper levels of agriculture and pasture. The valley of the Doron has an east-west trend, and so has a well defined *adret* and *ubach*. The adret side has vines, grains, and hay (Fig. 142), while on the ubach side hay predominates though there is a little grain. Vegetables may be grown on both slopes. Above the agricultural zone lie the isolated fields of the *mayen* or *vor-alp*, the site of the so-called summer villages. On the ubach side mayen grass predominates and fields are scarce. Sunlight is too slight for best field-culture, and yet grass flourishes well here. The cattle rest on the mayen for a fortnight in spring and again in autumn, and are also found in the mayen stables for a period in mid-winter to eat the hay stored there.

The vines in the Doron valley are confined to the adret of the post-glacial gorge, since here the slope is too steep for other cultures. On the ubach side the gorge is devoted largely to forest. Thus the vines grow on land not suited to other uses, where the insolation is large. Where the slopes are gentler, as above the gorge, they are used for other crops, though vines would grow here. There is another peculiarity in the Doron pointed out by Peattie, for there are twelve villages on the ubach and only seven on the adret. This is quite contrary to the general rule, and is due to the fact that the ubach side is better suited for a road,

[1]Height limits of mountain economies, *Geographical Review*, July 1931.

and this fact determines the sites of the villages. 'Local topography must always be dealt with in mountain topographies.'

Peattie gives some very interesting data regarding other controls in mountain settlement. For instance, he states that the greater the mass of the mountain the longer the growing season, hence plateaux show higher limits of agriculture than do mountain peaks. Thus, rye cultivation is notably higher on the plateau of the Cerdagne (at the east end of the Pyrenees), than on the flanks of the nearby Conflent valley. So also forest limits are higher on the massive Alps than on the Fore-Alps of

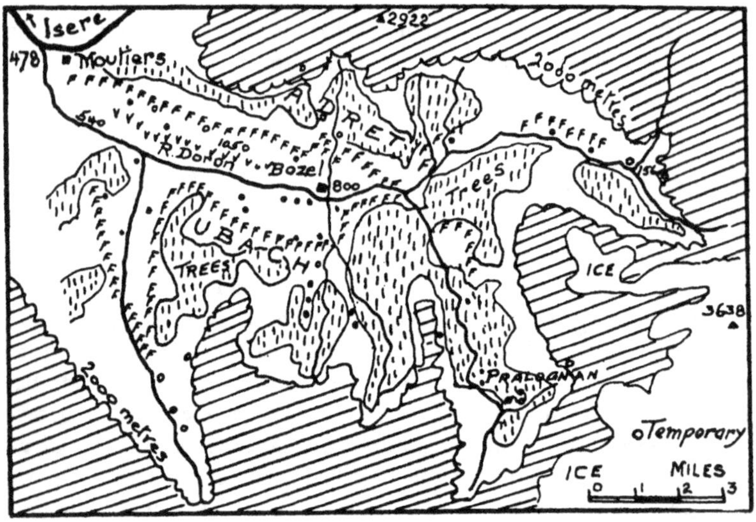

FIG. 142. Mountain villages, trees, fields (*F*) and vines (*V*) in the Doron Valley in the French Alps near Mt. Cenis. (*After R. Peattie.*)

France. In general, greater mass means gentler slope, though if there be heavy snow this may produce too much ground-water for cereals, which in such cases are found somewhat lower on this account.

Snow lowers significantly the period of pasturage, so that in the valley of the Inn at 2,500 metres there are only eight snow-free weeks (Peattie[1]). Permanent snow here begins at 3,000 metres and of course restricts all pasturing. Indeed historical advances of snow have actually ruined considerable areas. In the French Alps are many prairies that now are meagre pastures, but were of first importance before the advances of the ice at the close of the 16th century. In the same paper (1931) Peattie gives a

[1] *Mountain Geography*, Harvard Univ. Press, Cambridge, 1936.

map of the Sierra Nevada in Spain, showing that the villages reach up to the level of about 1,200 metres on the shady northern side, but only to about 900 metres on the sunny southern side. The boundary seems to be determined by the limits of the sterile granite mass of the Sierra Nevada, much more than by the elevation.

Towns and Villages on the great Fan of Gers (South-west France)

It is obvious that mountain topography is not in general very suitable for important settlement, and we have seen that in the Italian Alps the towns are found mainly on the floor of the glacial trough, or on the small fans which develop at the mouths of the tributary valleys. However, in some areas these fans are of large size, and then even under conditions of mountain environment fair expanses of deep soil may be met with. Here widespread agriculture and farming are possible. One of the largest fan topographies known to the writer has been investigated in the region north of the Pyrenees by H. M. Kendall.[1]

The Central Pyrenees form somewhat of a shallow arc with the concave side to the north, and from this arc a remarkable series of rivers flow north to join the Garonne. From west to east, all rising in or near the tableland of Lannemezan, are the following: Gave de Pau, Cabas, Adour, Baise, Gers, and the Garonne itself, which lower down passes through Toulouse (Fig. 143). These rivers flow to the north over a giant fan, which is perhaps eighty miles long, and 100 miles wide. It is composed essentially of fluvio-glacial material; and has naturally been considerably dissected, since the fan is formed of relatively weak debris and the rivers have fairly steep slopes. The southern portion of the fan is the least dissected, since headward erosion has not cut away the original surface to so great a degree (Fig. 144 at A). This is the Plateau of Lannemezan, which lies to the south-east of Tarbes, and is about fifteen miles from west to east. As the rivers flow north their valleys often exhibit a steep eastern bank and a gentle slope on the west. This may be an expression of the Ferrel effect (due to the rotation of the earth) which has been discussed on page 221.

This great region of fluvio-glacial soils was originally covered with a beech forest, but it is now fairly well occupied by farms throughout. Kendall's research dealt chiefly with the villages in the department of Gers, including the river of that name. Auch is the main town in this district; and the total population of Gers is about 200,000, of which

[1] "The Central Pyrenean Piedmont of France"; Michigan Papers in Geography; Ann Arbor, 1935.

120,000 is rural while 70,000 is agglomerated into villages (Fig. 143). The houses fit the landscape, being built of limestone in the west, and of red brick in the east. Most of them are roofed with red tiles. The crops consist of wheat fairly generally throughout, while oats are widely grown in the east of Gers, but are replaced to some extent by vines in the western portion of the department. Not much maize is grown in

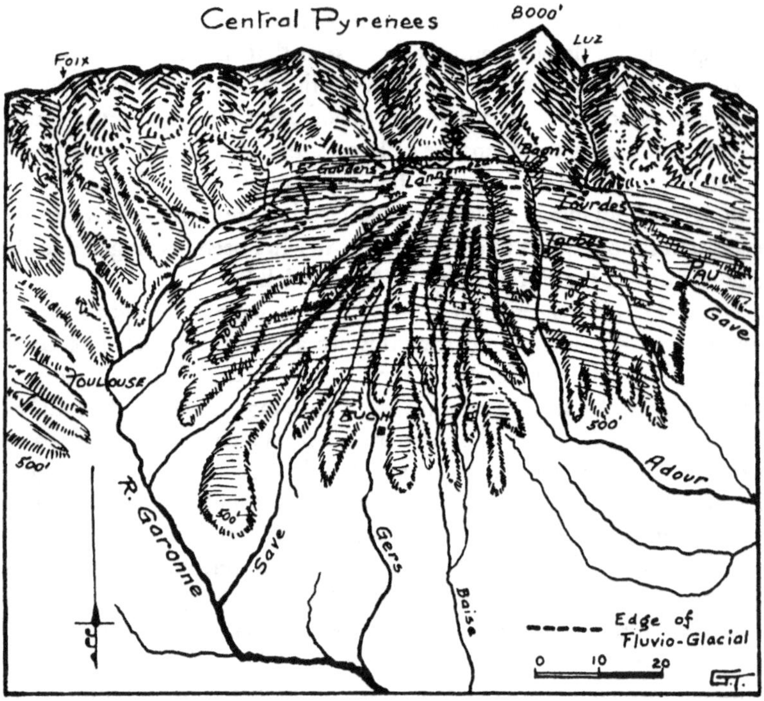

FIG. 143. Sketch (looking south) of the northern slopes of the Pyrenees to show the huge complex fan of fluvio-glacial material north of the Plateau of Lannemezan.

this elevated area. Cattle are reared in large numbers in the north, and hogs in the higher portions, but there are no sheep or horses.

At first a hill in the forest was crowned by a château or a church; and by degrees fields were cleared, the further fields being the larger. Gardens and vines developed around these scattered villages. Good roads seem to have developed only lately, and they have begun to determine the pattern of the villages to some extent. On the Plateau of Lannemezan the villages are chiefly on the flat tongues which project to the north, i.e., between the heads of the rivers. This type of site also obtains lower down on the fan, and not many farm houses are built on the flood plains of the rivers.

In the early days the need for defence determined the site of the farms, and they were sited on the higher interfluves; but in the later peaceful years there is a tendency for houses to spread down into the valleys. However, in the north and east many of the villages cluster round the church or château on the mesa-like hills which are common in the dissected fan (Fig. 144 at B). In the broader valleys of the north of the fan there is a tendency for cross-road villages to arise (Fig. 144 at C). The river is often restricted by artificial banks, and the farms have occupied the finer soils which are often found near the mature rivers.

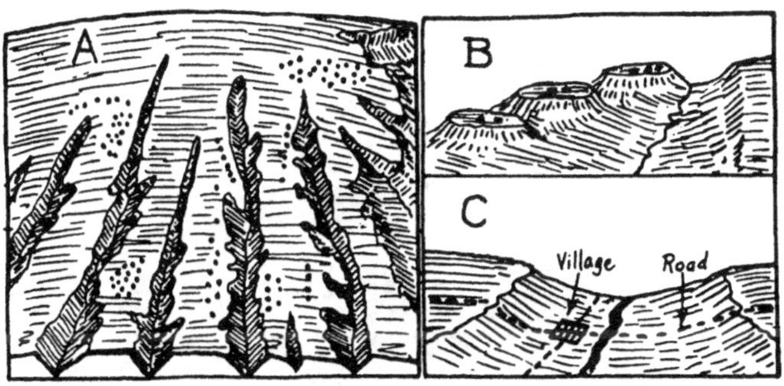

FIG. 144. Villages on the Fan of Gers. *A* on southern plateau (dots are houses *B* on northern mesas; *C* at northern margin. (*Based on Kendall.*)

The earliest rough roads naturally followed down the interfluves and joined the villages to Auch and to other towns in the north (Fig. 143). Auch is of pre-Roman origin and has always been the chief town of Gers. In ancient days it was the capital of Aquitaine, and to-day it has a population of 14,000. In Roman times roads were constructed from Auch to Toulouse (which lies forty miles to the east), and also to the north-east and to the south. The last road led to a pass across the Pyrenees between Mounts Perdu and Maladetta.

Settlement in the Horst of the Harz, in North Germany

One of the best known examples of a horst is found in the north German plains, where the Harz Mountain rises 2,500 feet above the plains in the form of an oval highland, about fifty-six miles long and eighteen miles wide. Its isolation, steep scarps, relative height, and frequent mists give it an importance much greater than that enjoyed by mountains of far superior height. For a long time the Brocken (3,415 feet) was looked upon as the highest mountain in all Germany (Fig. 145).

The rocks have a general north-west to south-east strike; and the whole horst is a relic-stump, belonging to the ancient Armorican fold-mountains which rose above the seas about 150 million years ago. They were worn down to stumps, but during the last period of mountain building (perhaps ten million years ago) this stump was elevated en masse. The rocks in the west consist of Carboniferous and Devonian strata; a granite

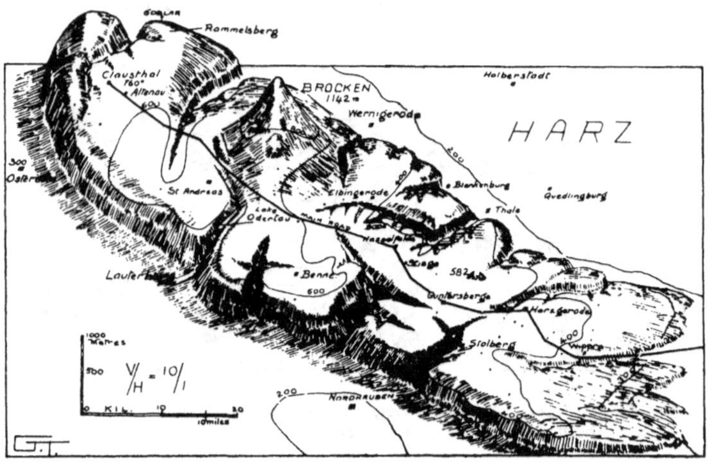

FIG. 145. Settlement on the Horst of the Harz in north Germany. It is 56 miles long and rises 2,500 feet above the plains. The ancient rocks are rich in mines.

boss builds the Brocken; while Devonian and Silurian overfolds form the lower relic-mass in the east. The horst block is somewhat tilted, so that the north-west end is considerably higher than the south-east end. The highest scarps are in the north-east above the famous old mining town of Goslar (Fig. 145). The horst surface is clothed in forest, but the exposed condition of the area near the Brocken, which lies directly in the path of the frequent north winds, permits of little vegetation but lichens and mosses in this part of the horst.

This relic-mass has been famous for ages on account of its rich deposits of various minerals. The argentiferous ore of the Rammelsberg is said to have been discovered towards the close of the 10th century, and already enriched the neighbouring towns during the eleventh.

The Rammelsberg rises 1,200 feet immediately south of Goslar, and has for centuries yielded gold, silver, lead, zinc, and copper. The whole mountain is honeycombed with shafts and galleries. It was the miners of the Harz who became the instructors of those of the Saxon Ore Mountains to the south (Reclus). These mines have given birth to large villages and towns at an elevation which, under other circumstances, would have but few inhabitants. In addition to mining which is now of far less significance than in the early days, the number of tourists is very large and brings a good deal of cash into the villages on the Harz.

The Harz is divided into three regions; the north-west and south-west slopes, with the towns of Goslar, Seesen and Harzburg, are called the *Vorharz*; the lower eastern half of the horst is called *Unterharz*; while the higher western portion, the *Oberharz*, is furrowed by numerous dark-wooded ravines. Several railways climb the scarps to reach the summit of the horst, such as that in the east which links Harzegerode (Ha.) to Hasselfelde (Has.) Iron foundries occur on this railway near the scarp, and chalybeate springs with tourist hotels are near the former little town. Hasselfelde has about 3,000 inhabitants, and was formerly a mining centre.

The valley of the Bode (Fig. 145) offers the most picturesque scenes, and is reached by a railway from Blankenburg at the base of the northern scarp. Here are cliffs rising 650 feet above the stream, and many myths are associated with the queer-shaped cliffs in the region. Elbingerode is another iron mining town with about 3,000 inhabitants. The Brocken is bare of trees, and many granite tors in the vicinity are featured in old German legends. Claustal, with the adjacent suburb of Zellerfeld, was formerly a town of 13,000 miners. It used to produce about 100,000 lbs. of silver, 8,500 tons of lead and 220 tons of copper annually; but in recent years the output is quite small. To-day the peasants have transferred much of their attention to wood-working, charcoal burning, canary-raising, stock and primitive agriculture. In the map (Fig. 145) all the small towns and most of the villages which have developed on the horst are indicated, the former by names, the latter by dots. Thus the Harz illustrates a plateau settlement where mineral resources have largely been worked out and replaced by tourism.

Mountain Settlement in the Black Hills of Dakota

Few regions in the world show the response of settlement to mountain environment better than the Black Hills in the north centre of the United States. Here is one of the eastern outliers of the Rocky Mountain uplifts, which may fairly be described as a 'blister' in the earth's crust, of which the uppermost skin has been eroded since it was elevated above the general level. The whole structure can be understood by a glance at the block diagram in Fig. 146. In more precise geological language we are here concerned with an 'eroded dome'—in which, as usual, the older formations are exposed at the centre of the dome, while the flanks are formed of the youngest rocks. Since there is a considerable difference in the resistance of these formations to erosion, we find a very marked series of zones in the topography, and these are indicated in the block diagram.

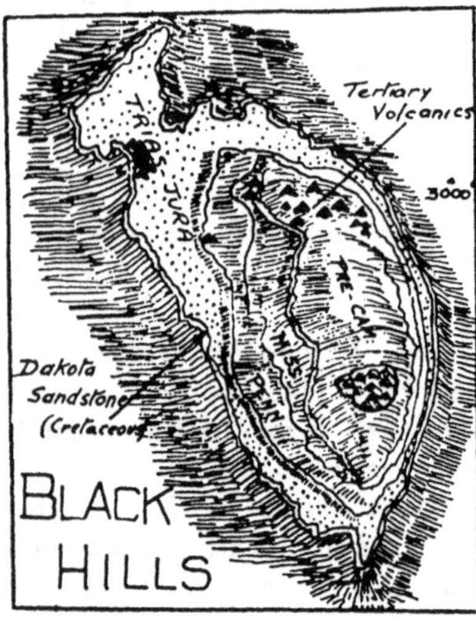

FIG. 146. Mantle map of the Black Hills of Dakota to show the strata in the 'eroded dome'.

FIG. 147. Map of the Black Hills showing the towns and products. *M*=mines. *L*=lumber.

One of the hardest of the Mesozoic formations is the Dakota Sandstone of Cretaceous age. It stands up all along the eastern flanks of the Rockies, and is perhaps the most striking feature in the topography of the Black Hills. The almost continuous rim formed of this hard sandstone projects like a wall all round the Black Hills (Fig. 146). Within this is a belt of softer shaly rocks of Triassic age, and they have been eroded to form a fairly continuous oval trench all round much of the Black Hills. Within this again are the harder limestone rocks of the Carboniferous age (labelled Penn.-Miss.) in the diagram. The central part of the eroded dome is built up of Pre-cambrian formations about a core of ancient granites. The latter are the most resistant of all the formations, and accordingly form the highest mountains clustered around Harney Peak (7,240 feet).

A sort of secondary dome of small size is attached to the main dome at the north-west corner. Here the erosion has not exposed most of the older underlying strata; but the depression formed within the harder Dakota Sandstones has a somewhat southwest-northeast axis, which is drained by the upper waters of the Belle Fourche River (Fig. 147). This flows to the north-east, and then curves to the south-east flowing through the town of Belle Fourche on its way to join the main river, the Cheyenne. The latter exhibits a similar curved course on the southern side of the eroded dome. Many short rapid rivers enter the Cheyenne from the eastern side of the main dome, and towns such as Sturgis and Rapid City have grown up on their banks.

It would be surprising if this diversity of strata and of topography was not reflected in the settlement of the region. We owe to O. E. Guthe a description of the region as it appeared in 1930–32.[1] The present writer's visit was made some years later in 1934, during which he traversed the whole region, but did not have time to make detailed studies. The Interior Basin is covered with a forest of Yellow Pine (*P. ponderosa*), and this is also the common tree in the Limestone Plateau of Carboniferous age. The Inner Lowland is usually occupied by grassland, while the outer slopes of the Rim carry cottonwoods and oaks, especially near the radially-flowing streams.

About 1875 the discovery of rich gold deposits near Custer in the central rocks of the dome led to a rush of miners, so that in 1876 there were perhaps 10,000 people in this district. In 1876 alluvial gold was found at Deadwood about forty miles to the north, and Custer was soon almost deserted. Ranching soon developed as a natural corollary to the mining industry, especially in the Inner Lowland. Various towns such as Rapid City, Spearfish and Sturgis were founded about 1877. Railways reached Deadwood in 1891 and this gave rise to various new settlements such as Newcastle. In 1907 other railways from Chicago reached Rapid City, and from this time this town has been the most important in the Black Hills area.

Of the present population of 42,000 in the whole region, 29,000 live in the little towns, engaged in mining or in commerce, while the remainder are rather widely scattered. For instance the Limestone Plateau has only a density per square mile of about 1.5 persons, as shown in Fig. 147. The Inner Lowland—a narrow zone of better soil—carries a density

[1] "The Black Hills of Dakota," Michigan Papers in Geography, Ann Arbor, 1935.

of 31.1. There are four main occupations in the hills, mining, ranching, lumbering, and recreation. The mineral deposits are found in the south near the ancient granites near Custer and Keystone (Ke). In the north near Deadwood the ores are associated with the Tertiary intrusions shown in black in Fig. 146. The Homestake mine in this district was the largest producer of gold in the United States.

Ranches, which depend a good deal on irrigated alfalfa, etc., are found mainly on the major streams which flow out of the hills. Smaller ranches are scattered in the grassy Inner Lowland, and in clearings in the forests of the Limestone Plateau. Lumbering is most abundant in the Interior Basin, and much timber is cut for use in the mines. The chief districts are marked 'L' in Fig. 147. Recreation brings thousands of visitors to the hills, and in 1930 ninety-five per cent came in automobiles, usually entering at Rapid City. Custer State Park (near Harney Peak) is the largest in the United States. Many game reserves, and the giant statues of the Rushmore memorial near Custer, are some of the numerous attractions of the Black Hills. Rapid City with 10,400 citizens is the largest town; Lead comes second with 5,700; while Hot Springs, Deadwood and Belle Fourche have between two and three thousand.

Settlement in the Rocky Mountains, near Kicking Horse Pass

It should be of interest to study the block diagram given in Fig. 148 for it shows the present state of settlement near the main railway which crosses the Rocky Mountains in Canada by way of the Kicking Horse Pass at 5,339 feet. This pass is just about the same level as the Reschen Pass, and both are near latitude 50° N. The pressure of population is very different in the two areas to-day, so that there is no agricultural or pastoral settlement yet in the Canadian Rockies. Yet the topographies are somewhat alike, with the great glacial trough of the Bow River akin in origin to that of the Adige, and there are glaciers hanging down the steep slopes of the cirques and side valleys in Canada like those near the Rescnen Pass (Fig. 134).

Some day there may be a denser settlement near the Kicking Horse Pass, but at present the residents are concerned mainly with the maintenance of the railway as at Field, or with the very important tourist trade as at Lake Louise. Apart from a few foresters and the folk at the chalets, etc., at Emerald Lake, Yoho Valley, Lake O'Hara, there are few other residents in the region charted in Fig. 148; but not far to the east are coal mines with an important settlement near Canmore—as well as the colossal tourist hotel, etc., at Banff.

Kicking Horse Pass owes its broad low character to bygone glaciers which carved out a considerable gap in the divide here. Just to the south-east of the pass are several cirque-headed valleys, one of which contains Lake Louise and the Victoria Glacier, while the next valley has the large Horseshoe Glacier clinging to the back wall of the cirque. The Canadian Pacific Railway Company has built a very large hotel at the lower end of Lake Louise, and near it in the woods are several more modest establishments where tourists can stay during the summer season. A good road runs through the pass, enabling motorists to reach the Columbia Valley some fifty miles to the west.

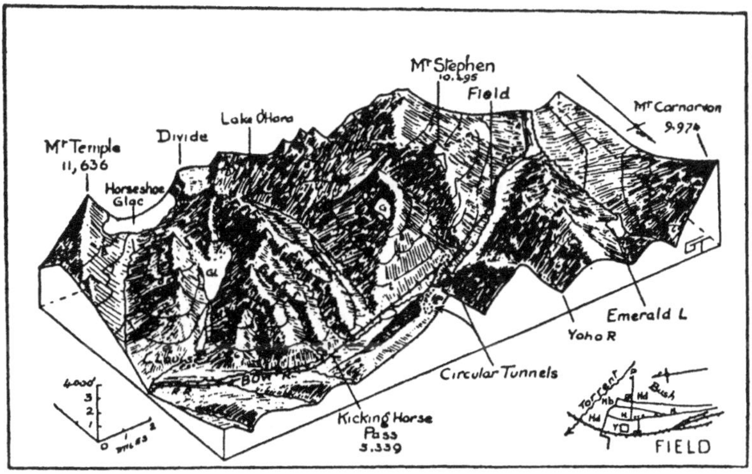

FIG. 148. Block diagram of Kicking Horse Pass, looking to the south-west.

One of the most interesting settlements in the Rockies is Field, which is 4,075 feet above sea level and about ten miles to the west of Kicking Horse Pass. The traverse of the broad pass by the railway presented no special difficulty, but five miles to the west the harder Lower Cambrian rocks are marked by a constriction in the valley, and a drop of 900 feet in four miles. To meet this rapid descent two spiral tunnels, each with a drop of about sixty feet, were built. They can readily be made out in Fig. 148. The lower tunnel is on the nearer northern side of the gorge, then the railway crosses to the other side of the river. The first moderately level portion of the glacial trough on the western side is found at Field.

Each side of this little settlement pyramidal mountains rise to nine or ten thousand feet from the broad glacial trough. A steep torrent flows into the main valley from the slopes of Mt. Stephen (10,945 feet), and the main streets of Field follow the curve of the rather steep torrential fan (inset to Fig. 148). The river follows a braided course over the flat floor of the glacial trough.

Field is a railway town; since it is the 'division station' where locomotives are changed, watered, and fuelled, etc. Here also a considerable number of engineers live, and there are railway repair shops. The inset sketch plan of Field shows the large Y.M.C.A. Hall (Y.) run mainly for the railway men, but also catering for tourists. Shops are shown by dots, and the hotels (H.), church, police offices and other government buildings are also charted, mostly high up on the alluvial fan. There are about thirty one-storey houses (Hd.) along the upper road, and about a score of two-storey houses (Hb.) mostly at the north end of the village; twenty houses are scattered at the south end. The railway yards are also at the south end of the village, which extends for about half a mile along the side of the main valley. It is interesting to note the importance of the fan deltas in this Canadian glacial trough, just as we have seen is the case in the European Alps.

CHAPTER XVI

MINING TOWNS

Introduction

PERHAPS of all towns those which are built in response to mineral deposits are created with the least regard for environmental advantages. Gold is where you find it: usually in lands where the earth's crust is formed of old rocks, and in general in hilly areas where the outcrops are more readily discovered. Regions of close settlement, where soils are deep or where forests are extensive, are precisely those where minerals are not readily visible to the prospector. Often enough mining towns are extremely badly situated from the point of view of food, water, and timber supplies. Transport is often quite difficult, and the mines are often long distances from any other settlement, so that loneliness is a frequent feature in such towns. For all of these reasons the study of mining towns is of special interest, if only because they contravene some of the general rules of urban settlement.

There are an infinite number of mining towns scattered over the world, and the writer has naturally chosen those with which he is most familiar. It is surprising what few descriptions in print are available of such towns; and in general the published accounts deal with relations to other parts of the land, rather than with site, pattern, and evolution, which it is the main purpose of this book to elucidate. The first place described in the chapter is Silver Peak, a small mine in the desert of Nevada near the Californian border. This has endured for eighty years, in spite of the numerous disabilities of such an environment. But it has a water supply, whereas the much larger mine of Blair nearby had no such supply and soon was abandoned.

The second study describes the famous nickel mines in central Ontario at Sudbury. Here much the largest sources of the world's nickel occur, in an environment rather far from other centres, but not so unattractive as that of Silver Peak. The two largest lead-zinc mines familiar to the writer are next discussed. One of these is in British Columbia at Kimberley in a rather arid 'ranche' environment, while the second is the famous field of Broken Hill in Australia. Here a town of 36,000 flourished in the desert around 1914, though it contains much less than that population to-day.

The second type of mining field is associated with coal and petroleum. These occur often in younger formations, and more usually in agricultural lands than is the case with metalliferous ores. I describe first the largest coal field in Canada, near Sydney (Cape Breton); and the Sheffield coalfield in England. Then follows a short account of the rise of cities in the coal and 'heavy industry' region of the Ruhr in western Germany. The last section discusses some of the characteristics which develop in the early years of an oilfield. The example chosen is Norman Wells in the far north-west of Canada, which is a most interesting settlement since it was started as a war measure, largely by the American army, and was developed without much need to consider finance.

Silver Peak and Blair, two small mining villages in Nevada

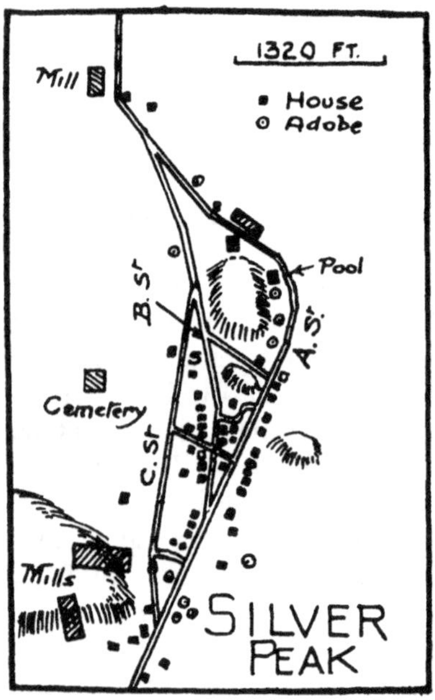

FIG. 149. Silver Peak, a mining village in Clayton Bolson in S.W. Nevada. (*After J. C. Weaver.*)

Not many descriptions of far western mining towns have appeared in geographical journals, hence the paper by J. C. Weaver is of considerable value.[1] Of these two mining fields, Silver Peak was founded in 1863, and still survives after eighty years. Blair in similar country only lasted from 1907 to 1915, and is now a 'ghost town'. Both are about 105 miles east of Yosemite Valley. Weaver draws a useful distinction between the factors which produce a permanent town in the desert, and those which led to the abandonment of the settlement in a few years.

Silver Peak was built at the springs which form a small pool below the butte shown in the map (Fig. 149). Rich silver veins had been found in Silver Peak Mountain some two miles to the west of the town; and the sole water supply in the district decided where the miners should live, and the mill buildings be erected. The town is built along three main streets which converge at either end. These are connected by a

[1] "Silver Peak and Blair," *Econ. Geog.*, Jan. 1939, with map and 7 photos.

number of short cross streets. None of the streets is paved, but each consists of the pebbly surface of the fan below the western hills.

There are seventy-three buildings in the settlement, of which forty-eight are shacks for the miners, twelve are old adobe houses now fallen into ruin, seven are mill buildings, five are shops, and one (S.) is a small schoolhouse. The vicinity is semi-desert or worse, and the village is unshaded by trees, and sited in a region where agriculture is impossible; yet it has maintained a continuous existence since 1863. The population has ranged between 100 and 140 persons.

Let us now turn to the other mine at Blair, which is situated at the head of an alluvial fan three miles north of Silver Peak. This mine was opened up by a powerful mining Company in 1907. They ran a branch railway into the mine from the Tonopah and Goldfield line for seventeen miles, which carried building material for the new town. Within three months a 120-stamp mill, office buildings, stores, a hotel and houses for nearly 500 people were set up in the desert, and Blair was a going concern. Water was piped from the springs at Silver Peak, and electricity came from a substation near the same place. This modern town had become, almost overnight, a far larger settlement than Silver Peak had ever been.

However, in 1915, when the mine ceased operations, the town disappeared as quickly as it had grown up. Every movable item of the smallest value was placed on the railway and carried out of the region. With that done the rails and ties of the railway were taken up also; and to-day only three buildings, made of poured concrete, remain. Hence Blair is not a particularly good example of a 'ghost town', since ghosts are reputed to like creaking wooden buildings better than concrete ruins. To-day the streets are almost obliterated by transverse gullies, and alluvial deposits and scanty vegetation will soon wipe out any signs of human occupation. Thus Blair, depending purely on outside corporate interests, vanishes, while smaller Silver Peak was founded by more or less permanent settlers near permanent water, and has survived.

'Nickel' Towns on the Canadian Shield

Probably the most interesting mining field in Canada is that in the vicinity of Sudbury in Ontario, about 260 miles to the west of Ottawa. In recent years the largest amounts of nickel and of platinum in the world have been obtained from the mines in this area. In 1942 these amounted to 285 million pounds of nickel and 124,000 fine ounces of platinum. Here the rocks of the Canadian Shield to the north of Georgian

Bay (in Lake Huron) occur in the form of a remarkable canoe-shaped basin (see inset in Fig. 150). This basin is thirty-six miles long, and its outer rocks consist of a coarse basic igneous rock called *norite*. This is indicated by the black oval in the inset. Within this norite are softer and younger slates—which have eroded into a hollow, since they are relatively weak. In this hollow silts was laid down in a glacial lake,

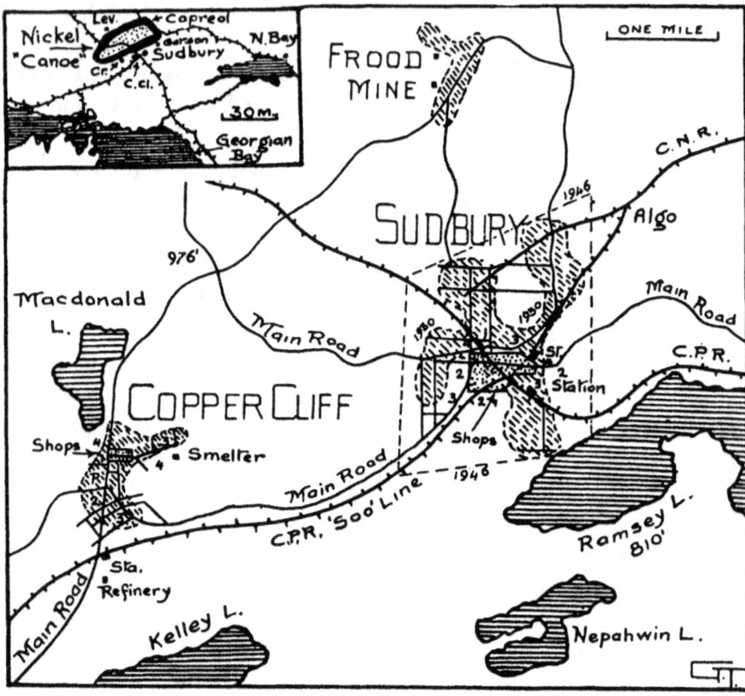

FIG. 150. The nickel region of Sudbury, Ontario, where a group of towns with 50,000 people developed on the sterile Shield. *Inset* is a map showing the nickel-bearing rock (black) and the chief mines. (Distribution of houses 1–4, *after D. Ellis*).

and now there is a good deal of farming within the 'canoe'. The nickel occurs in the norite, and so mines are distributed all round the margin of the 'canoe' (see inset). But by far the chief town of the area is Sudbury.

The origin of Sudbury is rather unusual, for it is several miles from the nearest nickel mine at Frood. In 1883 it was the temporary terminus of the Canadian Pacific Railway, and the usual end-of-rail log town sprang up. Some promising copper deposits were exposed in the railway cuttings, and there was some lumber obtained in the vicinity. This started the town; though it nearly vanished in 1885, when the main railway operations moved west. In 1887 the branch railway to the 'Soo' (far to the west) was built from Sudbury, and by 1893 there was a railway-lumber settlement of 2,000.

About 1888 the first roast-beds were started at Copper Cliff in order to produce a copper-nickel matte. The Creighton Mine was not worked until 1900. The Mond Nickel Company was formed in 1900, and drew most of its ore from the Victoria Mine, and later from the Garson and Levack Mines. Both Mond and the International Nickel Company have mines near Sudbury in the Frood area (E. Moore). The position of these famous nickel mines is shown in the inset map (Fig. 150).

In 1945 the population of Sudbury was 36,000, and in the immediate vicinity of the town are other centres, which make a total of 61,000 folk, who use Sudbury as their centre of supplies.

As the main map shows (Fig. 150) the vicinity of Sudbury contains a number of the small lakes which are so characteristic of the undulating surface of the Precambrian Shield. The level of the town is about 900 feet above the sea, and there are many rounded granite hills in the district. The roads and railways wind among these hills, and the houses follow the roads. Hence the outline of the built-in portion of Sudbury has a curious swastika pattern.[1] No mine is found close to Sudbury, but the chief supplies for the smelters at Copper Cliff come from the Frood Mines about three miles to the north-east. Sudbury is the shopping and social centre, and here are to be found eight churches (four being Roman Catholic), seventeen hotels, several hospitals, four theatres, and 7,685 homes. The houses are of brick or wood, about half of each. Two-thirds of the citizens speak English, and about one third French. The shopping centre is indicated by the dotted area on the map. (Most of these social facts have been kindly supplied by Miss Dorothy Ellis, B.A.)

Copper Cliff is a settlement controlled by the International Copper Company, i.e., it is what the Canadian calls a 'Company Town'. It was developed in connection with the mine to the south of the town, and to-day the population is about 3,500. It is connected with Sudbury—four miles away—by an electric railway, as well as by the Soo Railway and a good road. The mine is now closed, but the important smelting and refining operations ensure a permanent population at Copper Cliff. There is a fine social club house (called 'Inco Club') built of brick in several storeys. Nearby is the Company Hospital and a Community

[1]This pattern changed after 1933: and Mayor Beaton of Sudbury tells me that to-day (1946) the open spaces between the branching valleys are now occupied by houses, so that the outline of the city is more nearly a lozenge.

Hall, both built in 1936. Copper Cliff has a public School and a new High School. There are six general stores, a bank, police station, library, firehall, tennis courts, rink, and stadium. The five churches serve the Anglican, Catholic, United, Finnish and Italian citizens.

Early in the history of the town, when roast-heaps were used, all vegetation was severely affected by the sulphur fumes. To prevent this nuisance the Company constructed the tallest chimneys in the British Empire. One of these is 512 feet high, with an inside diameter of forty-five feet at the top. The central section of the town has paved streets, but the remainder have only board walks. Electric light and power are available throughout the town, derived from falls in the vicinity.

Frood is the most important mine in the Sudbury district, but miners are not allowed to live near the mine, except for a dozen homes where dwell the workmen who may be called on in an emergency. The sole establishment is a small bank, which is opened one afternoon a week (D. Ellis).

The Kimberley Mine in south-east British Columbia

One of the largest mines in Canada is to be found at Kimberley in British Columbia, at the south-east corner of the Selkirk Ranges. During Jurassic times vast masses of granite and allied igneous rocks welled up from deep magmas, and formed the Selkirk Ranges. Later erosion exposed these deep-seated granites, and we find important mines in many parts of the Selkirks, and the similar Coast Ranges, in consequence. The famous Sullivan mine at Kimberley is in a small valley just off the 'Rocky Mountain Trench'. There is very little farming hereabouts, since the rainfall is only about ten inches. The sparse cover of pine was cut for lumber long ago; and ranching is now the chief occupation in the district, except for small irrigated areas where conditions are most suitable.

The Sullivan lode was staked out as far back as 1892, but little prospecting was done until the branch railway to Kimberley was constructed in 1899; and production did not reach a commercial scale until 1920, when a satisfactory method of concentration of the complex ore was evolved. The concentrates are smelted at the huge smelting and chemical works at Trail, managed by the same company. Some of the sulphide is converted into sulphuric acid at Trail and used in the preparation of fertilizers.

A block diagram showing the surroundings of the mine, and the three settlements associated with it, appears as Fig. 151. Two adjacent hills are separated by the juvenile valley of Mark Creek, and each rises about

3,000 feet above the town. In the southern hill the North Star mine was worked for a time; but later the Sullivan lode was found on the Sullivan Hill to the north. The mine is worked by a tunnel, which runs into the hill at the 3,900 feet level for a distance of over a mile, from the bottom of Mark Creek valley. Large crushers were erected at the mouth of this tunnel, but since 1922 much of the ore has been taken to concentrators about three miles down the valley. The concentrates, as stated, are smelted at Trail, which lies about eighty miles to the south-west of the mine.

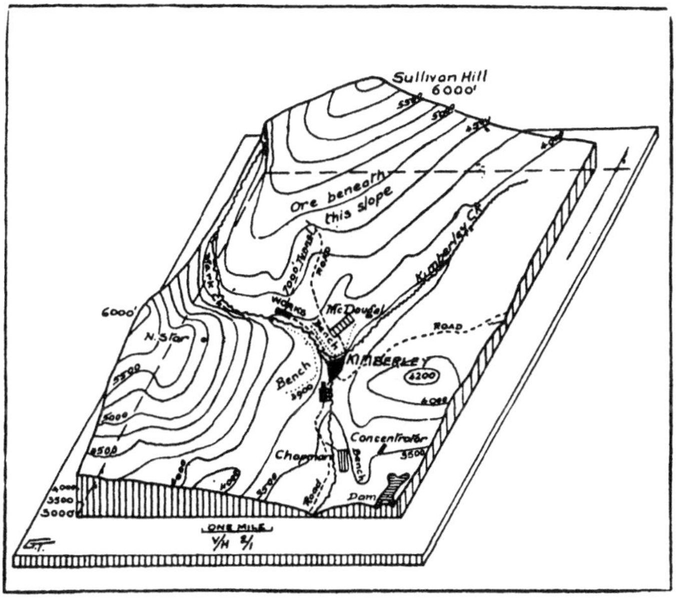

FIG. 151. The three settlements near the Sullivan Mine at Kimberley (B.C.). The black patch is enlarged in Fig. 152.

The valleys of Mark Creek and Kimberley Creek are bordered by the usual terraces which are so marked a feature of the valleys in British Columbia. Two of these benches flank the valleys just at the town of Kimberley at about 3,900 feet above sea level. On the northern bench the little 'Company town' of McDougal has been built, for the benefit of the staff and employees of the big mine. Then about 50 to 100 feet below in the main valley is the town of Kimberley itself (Fig. 152); while two miles down the Kimberley valley is the third settlement called Chapman. This is another 'company town' erected near the concentrators. From here the road runs about thirty miles south to Cranbrook on the main C.P.R. along the border.

The mining town of Kimberley has a population of 4,600, of whom 1,300 are miners; and its functional plan is given in Fig. 152. The deep valley runs nearly true north-south, and the creek hugs the western terrace; so that the floor of the valley slopes definitely from east down to west in the town itself. I was quite at a loss to account for the curious checker-board pattern, which seems calculated to introduce as many zigzags as possible into the traffic from Cranbrook up the valley to the centre of the town and the mine. However, I believe the streets were laid out

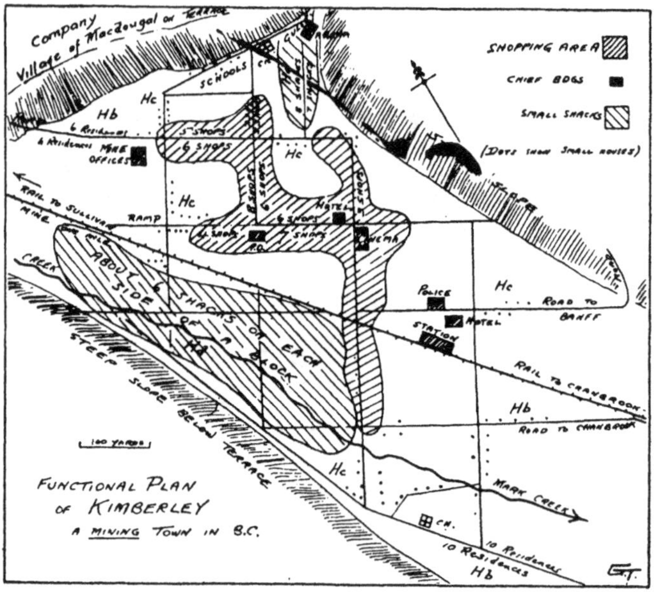

FIG. 152. Functional Plan of the town of Kimberley. (*Both from Geog. Rev.*, 1942).

parallel to the magnetic meridian; though this seems much less convenient than a main road up the valley with cross streets leading direct to the best routes up the terraces at the sides.

All the houses are built of wood with the exception of the post office, which is brick. The shops occupy zigzag roads each side of the post office, and in August, 1941, a census of them gave the following result: food shops, fourteen; garages and tool shops, fifteen; dry goods, etc., eight; clothing, etc., six. A fine group of school buildings occupies a block in the north. The best residences (Hb) are on the north and south edges of the town. About sixty shacks (Hd) are clustered in the area between the railway and the western terrace, and a group of fifteen is situated near the Arena at the mouth of Kimberley valley. Two large hotels and a Cinema Hall complete the picture.

Broken Hill, a large silver-lead mine in Australia

In 1923 the Pan-Pacific Congress met in Sydney, Australia, and the longest journey arranged for the delegates was to the famous field of Broken Hill, some 700 miles to the west of Sydney. The writer was one of the members of this party, and we crossed the whole of New South Wales by rail and motor; for in those days the railway was not yet built from Sydney to Broken Hill. However, the mines had long been established, and were indeed more important then than they are now, though Broken Hill is still a very noted producer.

A glance at the small map in Fig. 153 will show what an interesting environment surrounds Broken Hill. It is well within the ten inch rain-line, and so may fairly be called desert country, yet this locality, by the end of 1925, had produced silver-lead-zinc of the total value of 637 million dollars. In spite of the long distance from sources of food, fuel and equipment a town had developed which ranked third in the state, with 35,000 citizens at its peak in 1913.

Silver ores were first discovered in 1876 in the area of Thackeringa about twenty miles to the south-west of Broken Hill; and by 1882 many small mines were being discovered nearby at Purnamoota, Silverton, etc. (Fig. 153). The general topography of the area can be learnt from the lower diagram in Fig. 153. Here we see a broad low block of the crust, about thirty miles wide, of the type known as a horst, and rising to 800 feet above the plains, bounded by well-marked faults in places. All the rocks seem to be of Archean age, and they were folded in a north-east to south-west direction at the close of Paleozoic times. The harder layers of these folds stand out above the rest of the country a hundred feet or so, and the small creeks of the horst have cut narrow gorges through these ridges (Fig. 153).

Long before 1883 the country had been leased to cattle ranchers, and there was a head station (ranch) at Mt. Gipps, ten miles to the north-east of a long low ridge of brownish quartz-bearing rock of the type called 'gossan' in Australia. Seven of the ranchers agreed, in 1883, to lease this irregular 'broken hill' in the hope that it contained tin. Towards the end of 1884 rich ore was discovered in 'Rasp's Shaft', which happened to be sunk on one of the poorer portions of the great lead-zinc lode, and in 1885 the Proprietary Mine began operations on a large scale.

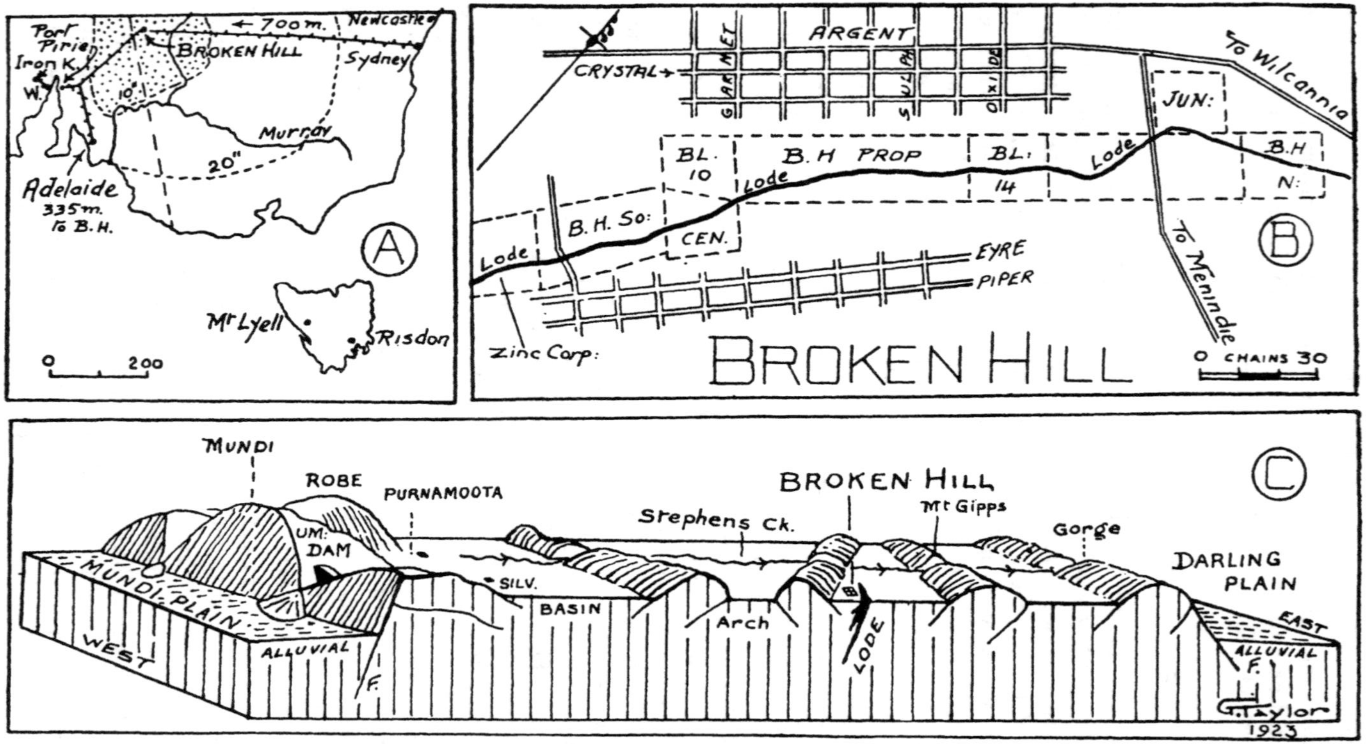

FIG. 153. *A.* Position of Broken Hill in arid country (dotted) in Western New South Wales. *B.* Plan of the ore-body at Broken Hill and of the main streets. *C.* Block diagram of the region, about 30 miles across, looking north-east.

The lode is approximately three-and-a-half miles long, and varies in width from a few feet to as much as 200 feet. The chief mining companies are indicated on the upper map (Fig. 153); and of these the B.H. Proprietary is much the most important; then come the Central and Junction, followed by B.H. South and North B.H. mines. To-day most of the lode has been worked out, and the chief operations are going on at the ends of the great lode. In places the lode has a depth of over 1,500 feet; and it has been compared to a great silver blade driven on edge into the crust, and over three miles long.[1]

Before 1883 the district was thickly covered with the usual scrubby vegetation of the semi-desert, such as acacias, eremophilas, and much saltbush (*Atriplex*) and blue bush. Fuel was in great demand at the mines, and the trees were cleared from miles around Broken Hill. A town soon arose near the 'broken hill' of Mt. Gipps ranch; and the old drovers' trail from Wilcannia to Silverton became Argent Street in the town plan (Fig. 153).

The streets were laid out along an axis parallel to the direction of the lode, and on both sides of the latter. But most of the mining was done on the north-west side of the lode, since the latter dips steeply to the west (Fig. 153), and so the main streets and shops are on this side of the town. The upper map shows only the *main* streets on the north-west side, for they extend somewhat beyond the area charted.

Water supply was a very acute problem at the start of the mines, since the rainfall is only nine inches and there are no permanent streams. However, two reservoirs at Stephen's Creek and Umber-ùm-berka—the latter built in 1915—now ensure a sufficient supply by storing the infrequent rains. In 1887 a railway was completed to the coast at Port Pirie, and in later years most of the smelting of the ores was carried out at this port. Various railways were built to collect limestone, etc., for the smelters, and the large deposit of iron ore at Iron Knob (Fig. 153 at A) was first developed for this purpose. Later much ore was taken—after preliminary treatment at Port Pirie—to the electrical refineries at Risdon, Tasmania. This latter site is one of the few where there is adequate hydro-electric power in Australia. Since 1915 Broken Hill money has been used to promote the huge steelworks at Newcastle (north of Sydney), based on ore from Iron Knob, etc.

Zierer describes the recent appearance of Broken Hill as follows:

[1]The best popular account of the mining field is to be found in the Guide Book, prepared by the leaders of the excursion and others, for the Pan-Pacific Congress, Sydney, 1923. C. M. Zierer has published a good recent account in the *Annals, Amer. Geog.*, June, 1940.

The hill provides an imposing skyline for the squat town development as viewed from many angles. Seldom is so much mining activity to be seen in so restricted an area. The general bleakness of the area, the driving sand storms, and the dependence on facilities only available in the town are factors encouraging compact settlement. Streets are excessively wide in most parts of the town, and are unpaved, rough, and dusty, and lacking side-walks. Ornamental shade trees consist of dwarf eucalyptus and Peruvian pepper trees'. Residential plots are enclosed with sheet iron fences to protect them from sand; and corrugated sheet iron is the standard building material. Brown schists have been used to build many of the older houses. Shade is a much sought feature in Broken Hill, since the average temperature in January and February is 78° F. Verandas cover the sidewalks, and in second-storey buildings verandas also often extend over the sidewalks. No sewage system had been installed in 1939; and the diminution of the ore reserves explains why there are so few recent improvements in a town of this size.

Starting in 1883 the town grew to its zenith in 1913, when the population was 35,000 with 8,700 men employed in or about the mines. The figures slowly declined to a total of about 27,000 from 1925 to 1935. The great lode will last for several decades, and the construction of the railway from Sydney direct to Broken Hill has opened up many pastoral areas, and affords a direct route from east to west of the continent which is not likely to be abandoned even when the largest town on the route in the semi-desert has disappeared.

Coal Mining and Steel Town of Sydney, Cape Breton

Canada extends to the east as far as 60° W. longitude, and it is rather interesting that the best coalfield in the Dominion is precisely at this extreme eastern tip of Canada in Cape Breton Island. Here is the fine harbour of Sydney, named after the same statesman as the even more picturesque harbour of Sydney, Australia. In the vicinity of Sydney (in Canada) the surface rocks have been corrugated by foldings of the crust to form a series of ridges and hollows. As usual, the younger upper layers are preserved in the synclines, while the tops of the ridges have been worn away, and often here the ancient lower rocks are exposed. All the formations dip gradually to the north-east, and so the youngest beds are found actually on the coast. These include a number of coal seams, with a total thickness of forty-six feet. Owing to the dip of the seams they pass under the ocean, and many of the workings are two miles to the east of the coast in consequence. The erosion of the numerous coves, including the fine harbour of Sydney, does not seem to have been determined either by the synclines or by the intervening anticlines.

Sydney is the chief town in Cape Breton Island, and became of considerable importance as the terminus of the Intercolonial Railway. Its fine harbour is open almost all winter, though Louisburg (twenty miles to the south-east) is better situated in this respect. Sydney had only a small population (2,427) in 1891, although the neighbouring coal mines had been worked for many decades, and had enjoyed great prosperity about 1854. However a great change began towards the end of the century. In 1899 the Dominion Iron and Steel Company was organized, which bought large resources of iron ore on Bell Island in Newfoundland. Later the Steel Corporation also acquired many of the coal mines in Nova Scotia, at Springhill as well as close to Sydney.

In 1913 an output of over seven million tons of coal was reached (Forsey, *National Problems*, 1926). In recent years there have been about 4,000 men regularly at work at the various blast furnaces, blooming mills, nail factories and associated quarries. The plant covers about 460 acres in the east of the town.

The present distribution of buildings is charted in Fig. 154. The town has now spread far beyond the few streets which led down to the harbour before 1890. The district which is closely ruled with diagonal lines in Fig. 154 is the nucleus of the city. The chief banks and offices are clustered close to the junction of Charlotte and Dorchester Streets. The former is the main shopping street, and includes three cinemas as well as many imposing stores. There are, however, few buildings more than four storeys high. The city has wisely reserved the end of the promontory as a public park, just as in the case of Halifax and Vancouver. Warehouses are found along the water front, and motor-supply stores on George Street. Many small factories are adjacent to the station.

Much of the small shallow bay called Muggah Creek has been filled, and all the east side of this bay is part of the huge Dominion Steel Works. The wharves of the latter are at the north end, and here vast piles of iron ore from Bell Island (Wabana) are temporarily stored. The enormous masses of slag are carried from the blast furnaces in cars, and tipped in the valleys leading to Muggah Creek. Railways from the coal mines, some miles to the east, bring coal direct to the smelters.

To the east of the city are rounded hills clothed with small trees. These hills rise over 200 feet, as shown in Fig. 154, and several small brooks flow from them to the harbour. It is in the south-east along these pretty wooded valleys, that the later development of larger residences has taken place. These are most numerous near Whitney Street, and at

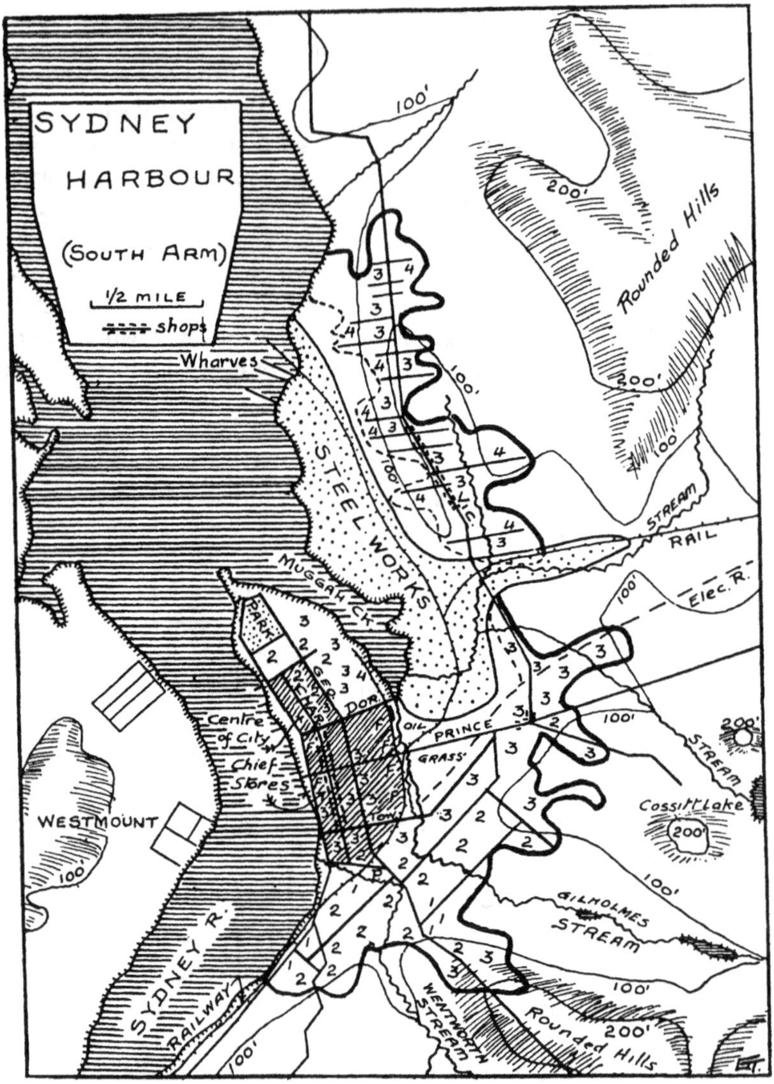

FIG. 154. A functional survey of Sydney (Nova Scotia), showing the early town by close diagonal ruling. The heavy line shows the present limit of houses. Figures indicate class of residences. The 100 and 200 feet contours are given

the extreme south of the city away from the steel works. One huge mansion (Moxham House) is a prominent feature among the trees in this suburb. Other mansions surround the pretty little park marked 'P' in Fig. 154.

On the higher ground to the north-east of the Steel Works a rather unattractive suburb of third and fourth-class houses has grown up. Here live many of the workmen at the Steel Works. Victoria Street exhibits some very monotonous rows of semi-detached houses of a type which are luckily more common in the Old World than in Canada. Here are numerous small shops, banks, and offices, which are quite separated from the main portion of the city by the huge area belonging to the Steel Corporation. The most rapid growth of Sydney occurred from 1895 to 1915, when the population increased from about 9,000 to 20,000. In 1941 the figure was about 28,000.

In conclusion we see that neither the good harbour, the presence of coal, or the terminus of a railway, was such an important factor as the rise of the steel works in adding to the growth of the city. Of course the steel works were placed at Sydney for the geographical reasons just stated. The city has developed asymmetrically owing to the 'pull' of the steel industry. There are, in a sense, two nuclei; one in the heart of the old town, and the other at the blast furnaces. The zones of settlement are fairly regular and obvious in the main city. Shops, offices, and factories and old third-class houses in the city; and zones of better-class houses away from the industrial centre. The various suburbs are linked by a convenient system of street cars.

The development of the Steel City of Sheffield, England

The writer spent some eight years of his youth in the city of Sheffield, one of the most noted centres of certain branches of the steel industry in the world. Since Sheffield illustrates the gradual evolution of a very important type of industrial city, dating back perhaps to Roman times, an account of its growth will complement that of other steel works of only recent origin.

Some unkind writers have described Sheffield as the dirtiest city in the most attractive site of all England's industrial centres! It is situated on the Don River, where it receives four small tributaries from the western hills (Fig. 155). Of these the Sheaf coming in from the south is the chief, and it has given its name to the town. The manor of 'Escafeld' goes back to Domesday Book and further, and to-day the centre of the city lies on the hills to the south of the valley of the Don, while the steel works are down in the hollow close to the main Don valley.

The structure of the district is very interesting, and can be made out from the diagram in Fig. 155. The city lies on the eastern flank of the Pennine Anticline, so that all the strata are dipping to the east. The oldest rocks are the Carboniferous Limestone shown at the left of Fig. 155. These are capped by the Millstone Grit which borders the east side of the Derwent and covers the Limestone in the centre of the anticline near the Peak (2,088) feet. The coarse sandstones called 'Grit' extend along

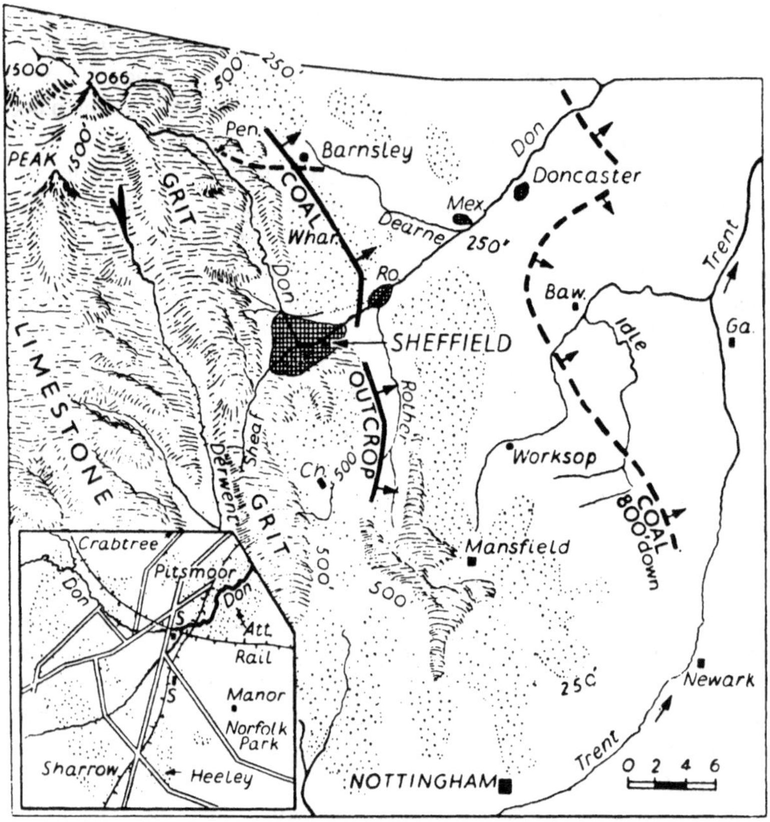

FIG. 155. The site of the steel-producing city of Sheffield on the eastern flank of the Pennine Anticline. The boundary between the magnesian limestone and the millstone grit runs along the Derwent River. *Inset* is a map of the city, showing radiating streets, rather than a 'grid'.

the slopes of the Pennine Range almost to the Upper Don valley, where the productive coal measures in turn cap the Grit. The coal seams outcrop as north-south bands of coal; and the main Barnsley Seam is charted on the diagram, and extends from Barnsley to Chesterfield. It dips underground to the east; and its position where it is 800 feet below the surface (along a line from Doncaster to Newark) is shown on the map. There is, of course, much more coal to the east of this line, but it is still deeper below the surface along the main Trent Valley. This is about the eastern limit of the coal workings at present, which have been developed in this concealed coalfield since 1859. To-day perhaps one-third of the Yorkshire coal is gained from this new coalfield near the Trent.

The Barnsley Seam is eleven feet thick near Barnsley, and out-crops as shown, though the seam is thinner to the south-east. Coal mining in this field began many centuries ago;[1] and to-day the most productive area is along this seam where the upper part of the Middle Coal Measures is exposed. 'There the valuable seams are relatively easy to reach, and there the hideous colliery villages and gaunt waste heaps and general devastation of the country side can be seen at their worst'. (R. N. Rudmose Brown).

Several geographical factors favoured the rise of the steel industry in the middle Don Valley; and turned the unimportant hamlet of Sheffield into one of the great manufacturing towns of Britain with a population of 518,000. In the early days the ore was dug in quarries, and smelted with charcoal obtained from the dense forests of the times. The furnaces were on elevated sites in order to get the strong draught from the winds. Increased air pressure was obtained by water power, and this led to the furnaces moving down hill to the streams. Along the five converging streams at Sheffield as many as 200 waterwheels were in use at one time. Later these turned grindstones, and for this purpose some are at work to-day.

Although steel goods were manufactured more than 400 years ago the local iron ore contained too much phosphorus for steel-making. Better iron was obtained from Sweden and Spain. The hard sandstones were excellent for grinding, and the inherited crafts together with the abundant coal counterbalanced the lack of the best iron ore. Even to-day there are many private grinders producing cutlery, who work in rented premises just as in the early days of the steel industry.

Coke replaced charcoal, and clay and ganister for linings were found in abundance in the Pennines. The plants for making heavy steel goods all moved down into the lowest part of the middle Don Valley where the transport was easiest. The western suburbs of the industrial city extend up onto the higher moors, and are swept by the strong west winds. They are in great contrast to the grimy centre and east of the city.

It is interesting that Sheffield lies on no natural route, for the upper

[1] "The Don Valley and Coalfield," R. N. R. Brown, *Great Britain*, Cambridge, 1938.

Don valley leads, both north and south, merely to empty moors. It is away from the main corridor of communication, which runs down the Rother valley from Chesterfield to Rotherham. Hence for forty years Sheffield was bypassed by the main Midland Railway, and was only served by a branch from the latter. Only in 1870 did a main line traverse Sheffield after piercing the main Sheaf-Rother watershed. Towards the east there is some indication of a conurbation, where Rotherham is being absorbed by its larger neighbour. The tributary area of Sheffield contains about a million people of whom the city itself provides half.

The inset map in Fig. 155 shows the main features of the city, whose core is near the two stations and is left blank. The residential suburbs are dotted, though many houses extend beyond the built-in area. The great steel works such as Browns and Firths are near the centre of the town along the Don. Here are made armour-plate and all varieties of cutlery, as well as silver-plate ware and machine tools.

The Mining Towns of the Ruhr in Germany

The most important mining fields in the world are those of Pittsburg in the United States and the Ruhr in western Germany. In 1937 the Ruhr produced 127.8 million metric tons of coal; three times as much as all France, and more than the entire Soviet Union. Indeed the coal mined in the Ruhr in 1937 contained more power than was produced by all the hydro-electric installations in the world in that year. Expressed in kilowatt-hours the Ruhr gave 200 billion units, while the world's hydro-electric power totalled 172 billion units.[1]

The exposed coal measures in the Ruhr cover an area as much as ten miles wide and are nearly forty miles long. The outcrop extends in an east-west direction. The northward dip of the coal is about three or four feet per 100, so that the most northerly coal pits have to sink through 2,500 feet of over-burden to reach the coal. In the Essen region the upper surface of the coal measures is 165 feet above sea level, but near Hamm in the eastern part it is 2,600 feet deep. There are 130 well-recognized seams spread through a depth of coal measures amounting to 9,000 feet thick. The high volatile bituminous coals occur in the upper seams, the rich coking coals are in the middle layers, while 'meagre' semi-anthracite coals are found below these (see map, p. 381).

Harris gives some interesting data on the history of the coal towns. Far back in the 13th century coal for local use was mined in the exposed

[1]See the informative article with many maps by Chauncy Harris in the *Geographic Review* for April, 1946.

southern part of the field, and in the middle 17th century coal was carried on horseback half-a-dozen miles to the industrial towns in the southern hills. However, the early iron-smelting in the Ruhr used charcoal and water-power in the 18th century. About 1800 there were 127 coal mines in the Ruhr, with a total annual production of 172,000 tons. Later, with the introduction of the steam engine, it was possible to win the coal from much greater depths (Fig. 156), so that the coking coal—which is mostly rather far below the surface—could now be obtained. Coke ovens were soon built in large numbers, and the foundation of the iron-smelting industry on a large scale soon followed.

By 1860 large blast furnaces using coke from Ruhr coal and ore from Siegerland to the south were in operation; and tariffs protected their

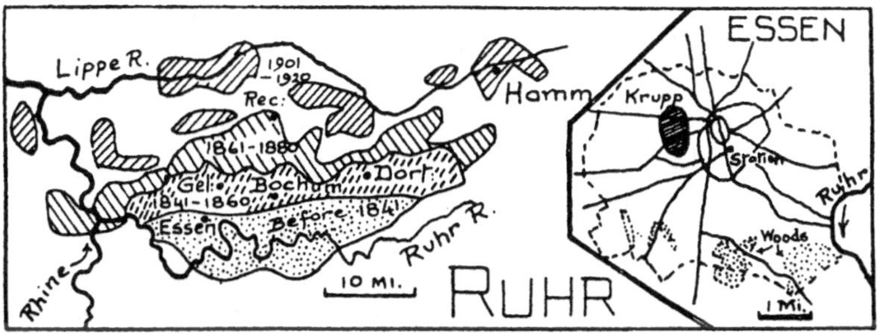

FIG. 156. The spread of the coal towns in the Ruhr from 1841 to 1920. (*After Harris and Spethmann.*) *Inset* is a plan of the radial highways in Essen.

steel from competition with British and Belgian steel which had previously dominated the market in Germany. In 1861 Krupp introduced the British Bessemer process at Essen, and he developed shipyards at Kiel, and brought phosphorus-free ore from Bilbao in Northern Spain in 1875. Later in 1878 the Thomas basic process was introduced from England, and this enabled Essen to use the phosphorus-bearing ores from Lorraine, which had just been conquered from France.

The zenith of the coal and steel production and the limits of active coal mining were reached by 1913.[1] To-day the central zone of the coal from Duisburg to Dortmund is concerned with heavy manufacturing as well as with the production of coal. Thus in 1933 Essen had 34.7 thousand workers in the coal mining and 19.2 thousand in steel pro-

[1]A very interesting description of the vicissitudes of the mining industry in the Newcastle Coalfield of England will be found in the article by A. E. Smailes "Population changes in Colliery Districts" in the *Geographical Journal*, March, 1938. A dozen maps and illustrations are included.

duction. The northern belt (Gelsenkirchen and Recklinghausen) has little manufacturing, and the miners are largely migrants from Poland. The southern belt of towns is beyond the coal belt, and has long been concerned with textiles (as at Krefeld and Wuppertal) and metal-working (as in Solingen and Remscheid). In 1939 there were almost seven million people living in the Ruhr area. The way in which this cluster of industrial towns has grown into a conurbation is described on page 379 (See also Fig. 180.)

We owe to Nelson P. Lewis a description of the great steel centre of Essen as it was before the last Great War. He writes in the volume *The Planning of the Modern City* (New York, 1923):

'Perhaps no city, to which modern planning has been consistently applied, affords a better example of studied irregularity in its street system than Essen, located in the largest coal and industrial district of Germany. The old portion of the town lies north of the railroad, which traverses the city from west to east, and from the old centre there is an excellent system of radiating traffic streets, while several ring streets encircle it. In the newer part of the city lying south of the railroad the irregularity of the minor street system is as conspicuous as in the older town, yet these streets have been carefully planned by Dr. R. Schmidt, the chief engineering official of the city. . . . The aim of the planner was to furnish throughout the city constantly new and varied vistas, not only in respect to the houses and their relation to the streets, but also to the width and arrangement of the individual streets.'

An Oil Field in North-west Canada

At Norman Wells (65° N. latitude) the Mackenzie valley is about twenty-five miles wide. The drainage is poorly developed and there is much muskeg. In this region the permanent frost is found to extend about 100 feet beneath the surface, and in the muskeg area the surface thaws to a depth of only a foot beneath the moss covering. In 1914 Dr. T. O. Bosworth staked out claims near some petroleum seepages on what is now called Bosworth Creek. In 1920 Discovery Well was drilled, and commercial oil was encountered at 783 feet. It is now (1944) 1,025 feet deep, and is rated at 75 to 125 barrels per day. No. 2 Well is 150 feet away and is 1,602 feet deep. It produces rather more oil. There are about two dozen wells now in the vicinity of Bosworth Creek, and a few more on the adjacent islands in the river. The proved field seems to occupy an area of about four miles by three as suggested by the oval in the inset map in Fig. 157. The oil occurs in Devonian limestone

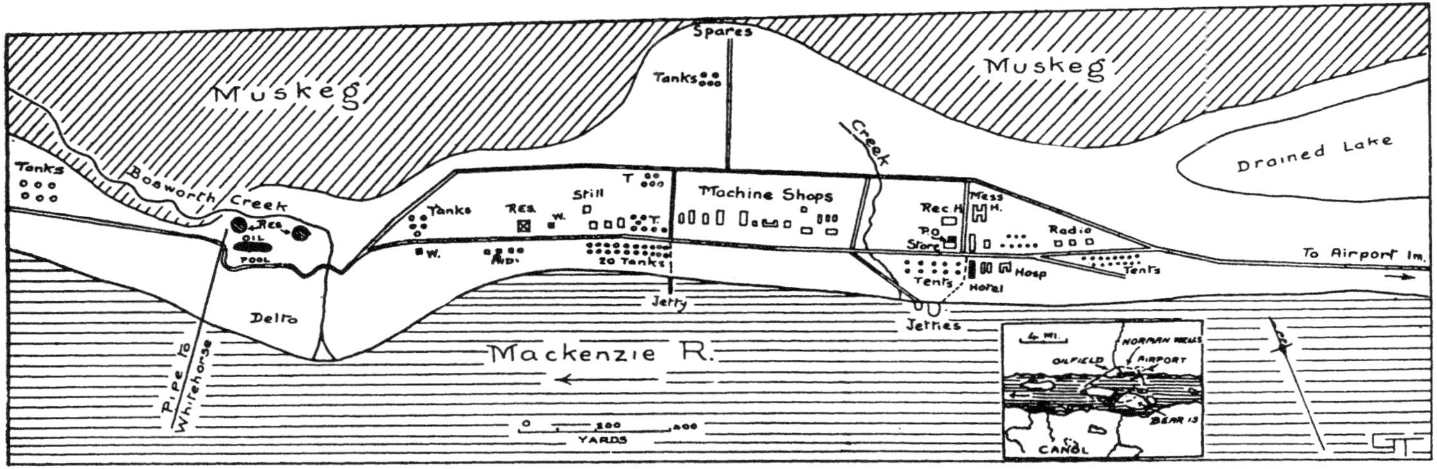

FIG. 157. A sketch survey of Norman Wells, looking north at 65° 20′N. Note the four divisions in the settlement. In the *inset* map Carol and the oval oilfield are indicated. (W. is a well.) *Muskeg* is swampy country, rich in moss, with small conifers.

dipping about 5° S. The present refinery began work in 1940; and in 1942 the United States War Department started the military enterprise now known as the 'Canol Project'. I did not survey Canol, which is mainly concerned with building the pipe-line and road for about 600 miles to the west to the large refineries at Whitehorse. There are about six hundred employees at Canol, which has been built some four miles south of the river (see inset).

Like all the other river settlements, Norman Wells is strung out along the river bank, in this case for about two miles. There is a rather broad sloping beach below a cut-bank about forty feet high, but the silts from Bosworth Creek at the north-west end have built out a perceptible delta. There is a large airport flanked by a dozen or more buildings about a mile to the south-east of the settlement. Planes linking the oil wells to Edmonton or to Whitehorse fly into this airport practically every day (in 1944).

There are four well-defined sections to the town. At the south-east end is the residential quarter surrounding the hotel. This is an imposing wooden building of two-and-a-half storeys, which is controlled by an American army officer. At its north end is a sort of plaza containing the post office, commissary, main mess building, and many of the women's tent-houses. Duckboard side-walks link the buildings, for the dirt roads become deep in mud when it rains. A hospital, recreation hall and the radio station are to be found in this quarter also. A fine clump of poplars has been preserved in front of the hotel, and some small gardens contain a few flowers. There is no church, though service is held in the hall occasionally. No children are allowed in the settlement, and the seventy women are all employees of the various oil units. The total population is about five hundred.

The next quarter further to the north-west is where the engineers live, and where their machine shops and stores have been erected. A further area in the rear of the settlement is used to store many other of the engineer's supplies. Very noticeable around the place were the huge bull-dozers and carry-alls, which were more impressive than I have seen elsewhere. A boiler mounted in a caboose on runners was used to supply live steam, wherever it was necessary to sink posts through the frozen ground. It takes about half an hour to thaw six feet. All surface pipes for conveying water about the area are encased in wooden channels filled with sawdust.

The third quarter centres about the wells and the large still. In this section is the original Mackinnon residence which has been there since 1920, and is close to the discovery well. A few Indians live close by, and they are maintaining a small garden here. There are great numbers of oil tanks, mostly painted with aluminium. One cluster of twenty just by the river is a very prominent sight from boat or plane. The large still is about twenty-five feet high, and breaks the oil into various grades. Other cylinders separate the gas from the liquid, and some of the gas is burning continuously in large flares down on the beach. Here there is a portable jetty made of steel pipes, used in loading oil on to barges, etc. It can be pulled ashore when the river freezes, and so it is not damaged by ice pressure. The main wharf near the hotel is simply silt piled above bundles of poles and branches. This type of wharf is often much damaged by the ice each spring and autumn.

The wells are not imposing after the drilling rig is removed. One sees a black hole about five feet square, containing a sort of stand-pipe with a large valve and a pressure gauge. A one-and-a-half inch horizontal pipe leads the oil to a neighbouring tank. At the north-west end of the settlement the road crosses the rather deep gully of Bosworth Creek; and in a bend in the latter are the two largest tanks on the field, from which the oil flows to Whitehorse 600 miles away. Below these tanks is a pool of oil, about 150 feet long, held up by a rough bank of muskeg; and nearby the four-inch pipe dips into the river, which is over four miles wide here, and so carries the oil on its first stage to Canol. We made a visit to Bear Island, but found rather poor soil with a growth of medium spruce. Poplar and birch in this region are said to indicate the better soils.

It is difficult to place Norman Wells in one of the normal classes of a pioneer settlement, since it is essentially only an industrial camp. No doubt if the oil supplies continue for many years a permanent town will develop, but in the absence of the usual houses, shops, schools, and churches, as well as the Hudson's Bay buildings, it is quite out of line with the other pioneer towns at present. (Canol and Norman Wells were nearly deserted at the close of 1947.)

CHAPTER XVII

RELIGIOUS CENTRES AND RESORT TOWNS

Introduction

ALTHOUGH at first glance there does not seem much connection between towns based on shrines, on health, and on scenery; yet in general we find that all these places attract tourists; even though there is some attribute which separates them from the normal settlement, which is usually established to make a living. In the present chapter a number of localities are described because they were founded in response to a religious urge of one sort or another. Some of these occur in good farming country, as in the case of the Lutherans at Amana (Iowa) and the Mennonites in southern Manitoba. Others have been founded primarily to get away from the rest of the world, as in the case of the Mormons in the desert at Salt Lake. Two localities which are described in some detail originated as shrines. One of these settlements is the Vatican City at Rome, a community of less than 700 people, which vitally affects the religious life of 334 million Catholics. The other site described is a modern example of a vast growth in a town due to the Catholic Church making it a place of pilgrimage. There are many such, but our example is Lourdes in the French Pyrenees.

In the second half of the chapter various scenic resorts are charted and described. The first is Davos in the south-east corner of Switzerland, which acquired fame as a mountain resort for lung patients. But it has superb facilities for winter sports, and is surrounded by very interesting Alpine scenery. Hence to-day its visitors are more likely to be healthy folk than those afflicted with pulmonary diseases. Davos offers a striking example of the importance of sunshine and shade in human settlement. Winter sports are the main attraction in our second example, which is in Quebec. This illustrates how tourism has benefited the farmer who attempts to wrest a living on the more elevated portions of the Shield behind Montreal. A million dollar industry has grown up in a small district here, formerly occupied largely by subsistence farmers.

A profitable comparison may be made between tourism in the Laurentians behind Montreal and in the Rocky Mountains. The region from Banff north to Jasper offers some of the grandest Alpine scenery in the world, and it is crossed by two transcontinental railways. Largely as a result of the initiative and wealth of the two great railway systems of Canada there has developed in this region, far from any large centres of population, an extremely popular 'tourist land'. It is not based on winter sports, though ski-ing is spreading in the Rockies, but motoring on the excellent roads, tennis and golf at the comfortable hotels right on the main railway lines, carnivals arranged at regular intervals, all these combined with the chance to tramp over glaciers and snow fields, have made the Banff-Jasper area one of the most frequented in North America.

Quite different in environment, but equally familiar to the writer is the best known tourist resort in Australia. Here elevation on the Blue Plateau is perhaps the chief asset, for in the hot summer the one-and-a-half million citizens of Sydney (as well as residents from all over the rest of eastern Australia) love to spend some weeks at a height of 3,500 feet, some sixty miles west of the city. On the Blue Plateau there is no snow—except a few flurries in the depth of winter; nor are there any mountains. But cut out of the plateau are giant gorges, which can bear comparison with the Grand Canyon itself. Here in the valleys are numberless walks among tree-ferns and other semi-tropical plants, which gain by contrast with the sterile sandstone plateau surrounding the boarding houses near the railway.

PART I. RELIGIOUS CENTRES

The City State of the Vatican

The smallest city in the world (and certainly the smallest State) is the Vatican, situated in the north-west corner of the city of Rome (Fig. 56). This City State was brought into existence as late as June, 1929, by the Lateran Treaty between the Holy See and Italy. At the beginning of the Italian Revolution the States of the Church extended from Rome to Ravenna, but after 1870 almost all this large territory was annexed to the Kingdom of Italy. The exact boundary of the small area left to the Holy See, and the relation of this territory to Italy, were left ill-defined until 1929.

The Vatican City State comprises only 108.7 acres,[1] and this area is all taken up by buildings, streets, and gardens. Its boundaries are well defined by the strong walls built during the Middle Ages, except on the east where they cross the large Piazza of St. Peter (Fig. 158). The Italian police are required to keep order in this piazza, though the special

[1] "The Vatican City State," U. Toschi, *Geographical Review*, Oct. 1931.

Vatican police guard all the rest of the territory to the west of the broken line in Fig. 158. In the evening the entrance to the city is closed, and nobody can stay in the territory but Vatican citizens. There is no such thing as Vatican nationality, and the 639 citizens belong to the following nations: Italian, 495; Swiss, 118; French, 8; German, 8; and 10 members of five other nations. When a citizen leaves the Vatican service—for instance a member of the Swiss guard—he reverts to his former citizenship.

The main features of the Vatican City are shown in the sketch (Fig. 158). The Piazza of St. Peter is about 300 yards across, and is surrounded

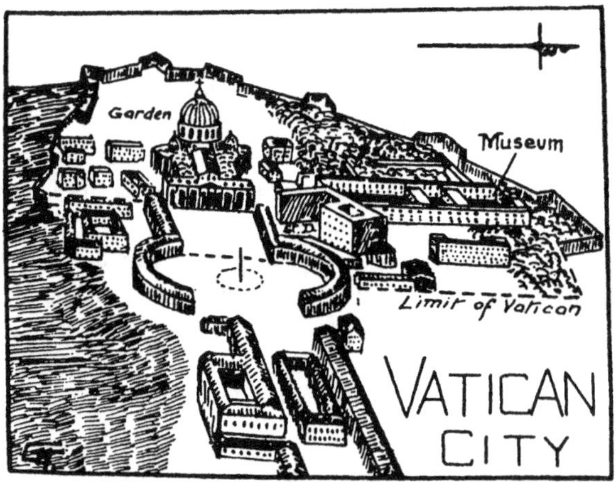

FIG. 158. The city state of the Vatican, with only 639 citizens. '*St. D.*' is the courtyard of St. Damasus. (*Sketched from a photo.*)

by a grand colonnade built by Bernini in 1667. The Obelisk in the centre was brought from Egypt in A.D. 39. It weighs 3,270 tons, and was removed to its present site in 1568. The Basilica of St. Peter is built where formerly stood a church due to Constantine the Great. The modern Basilica dates from 1626, and was designed mainly by Bramante and Michelangelo. It is the largest church in the world, being 187 metres long and more than 100 metres wide.

The Vatican proper is to the north of the Basilica, and is the largest palace in the world. Late in the 14th century Gregory XI moved his residence from St. John Lateran in Rome to the Vatican. About 1450 it was fortified, and it has been greatly extended since. For instance Bramante designed the splendid court of St. Damasus at the south-east corner of the main building. The Palace contains twenty-two courtyards and 11,000 rooms, besides various chapels and halls. The Vatican Museum contains eight separate galleries arranged around the two main courtyards of the Vatican. The apartments of the Pope are in the Courtyard of St. Damasus.

Much more than half of the area of the Vatican State is devoted to gardens in the west and trees in the north. On the east side the close-set houses of Rome encroach on the limits of the Vatican, though most of these houses are omitted in the sketch (Fig. 158). The broad avenue leading to the Piazza from the city of Rome in the south-east is almost completely blocked by a group of tall buildings. To the south, west and north the Vatican area is surrounded by relatively open country as indicated in the sketch.

There are a few areas outside the Vatican which are under the control of the Holy See. For instance there is the summer residence of the Pope at Castelgandolfo about sixteen miles to the south-east of the Vatican. This comprises 100 acres. Adjoining the Vatican area on the south-west, but beyond the limits of the sketch, is an area of thirty-five acres which is reserved for new buildings which may become necessary. The Vatican receives food, goods, water and gas through Italy. It has, however, a large private radio station. The State is financed largely by the voluntary offerings (Peter's pence) which flow in from the 334 million Catholic souls who are directly linked with this tiny city of the Vatican.

Salt Lake City's dominance due to Mormon control

There is a particular interest in cities which develop far from extensive areas of settlement for they challenge the geographer to ascertain just what was the dominating factor which determined their origin. Often enough it is not possible to do this with cities in the midst of a closely settled area. However, Chauncy Harris has made an illuminating study of the environment at Salt Lake City in which the various environmental and other factors are discussed at some length.[1]

Salt Lake has a population of 150,000, and is the largest city between Denver and the Pacific Coast. In this journey of some 900 miles the central portion consists of the arid basin in part occupied with Salt Lake and the valley of the Humboldt river. Just short of half way to the west, the traveller must cross a long line of oases which extend in a north-south line at the foot of the Wahsatch Range. It is obvious that the regional centre would develop somewhere along this favoured belt of oases, and Harris devotes considerable space to making clear why Salt Lake City

[1] "Location of Salt Lake City," *Economic Geography*, April 1941.

gained a dominating position over Logan and Ogden to the north, or Provo to the south.

Water supply was the paramount factor in this semi-arid region. The Colorado and the Snake rivers have cut their way in deep gorges, and their basins are not suited for agriculture near the water. But in the Salt Lake area *arid* erosion, not river erosion, is the rule; and the basin has been levelling up with soil rather than robbed of its alluvial by juvenile streams. Along the Wahsatch Range and neighbour mountains the best streams are found in the northern portion where precipitation is greater; and where it is cooler, so that snow lies longer in the spring. The Weber, Provo, and Bear rivers are the best of these streams, though Salt Lake City is not watered by any of them. However, some of the best soils and lake terraces are to be found between the Weber and the Provo, where Salt Lake City was built. (For a time it seemed as if Ogden would dominate, for the first railway was built along Weber canyon to Ogden, and Salt Lake was bypassed.)

Let us now consider earlier history a little. The Mormons decided to move from Nauvoo in Illinois to some isolated spot where they would not be subjected to interference by intolerant neighbours. From journals by Fremont and others, they learnt that the *southern* oases were frequented by hostile Indians, so they made no move to go south to Provo near Utah Lake. It was also stated that the northern oases beyond Ogden were too cold, which again focussed attention on the region immediately east of Salt Lake.

Another important factor was ease of communication. There were several passes through the Wahsatch Mountains leading down to the plains of the Great Salt Lake. The lowest was the Weber Canyon, which passed through Ogden, but when the pioneer Mormons reached the region in 1847, this valley was blocked by deep water. Somewhat to the south was Emigration Canyon, and this led their wagons down to the terraces of the future capital. The pioneers found that the rich, easily worked loams of the eastern side of the great valley changed to salty soils just north of their temporary camp on City Creek. Hot springs in the vicinity were also felt to be an advantage. Consequently at an open-air meeting on July 28th, 1847, the pioneers voted to make their temporary camp the site of the new city. Temple Square, the centre of the settlement, was laid out between the two branches of City Creek.

In another publication[1] Harris describes details of the laying out of the city much as follows. The city plan had been drawn up at Kirtland, Ohio, by Joseph Smith. This plan called the 'plat of the city of Zion' is therefore

[1]Salt Lake City; University of Chicago, 1940.

much older than the city. After arrival at Salt Lake, no haphazard construction took place, and execution of the plan was entrusted to church officials. The general street pattern followed the checkerboard tradition of the plan which Penn had adopted in Philadelphia. The streets, however, were much wider, and the blocks were both larger and square. Indeed the streets follow almost exactly the plan of 'Zion'. They were oriented

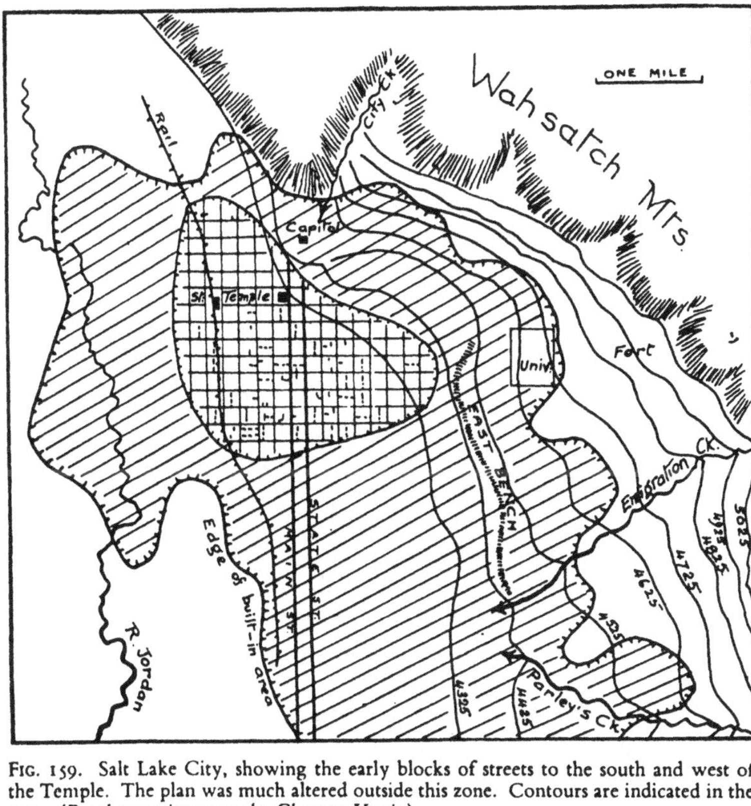

FIG. 159. Salt Lake City, showing the early blocks of streets to the south and west of the Temple. The plan was much altered outside this zone. Contours are indicated in the east. (*Based on various maps by Chauncy Harris.*)

in the cardinal directions, and carved the city into square blocks of ten acres each. The centre of the city was fixed at the south-east corner of the Temple Block. The four streets on the sides of this square were named East, South, West and North Temple Streets. However, East Temple was later changed to Main Street (Fig. 159).

Temple Square was the centre of the town, both physically and spiritually. Four other blocks were reserved for city squares, one being the location of the first night's camp on the site. This is now the corner of Fourth, South and State streets. Another was the site of the fort, built during the first winter. The ten acre blocks were divided into eight lots of one-and-a-quarter acres each. They were each to contain a single house placed twenty feet back from the street, and the front part was to be beautified with trees. Adjacent farm lands were included in the city; thus the 'Big Field' was divided into five, ten, and twenty acre farm plots. As a result the area of the city was over fifty-seven square miles, and in 1860 was exceeded in area by only four other cities in U.S.A.

It was soon found that some modifications of the original plan were advisable. In the steeper north-east part of the city smaller blocks of two-and-a-half acres were substituted, since here irrigation was needed. After 1877 the city extensions passed from the hands of the Mormon elders, and the new plans did not show much vision. Thus the distant north-west portion of the city was sold around 1890; and in the city itself lanes and courts were cut into the old blocks without much regulation. About 1900 Liberty Park was built up in bungalows, and served by street car lines. The higher lands on the eastern benches have been developed for the most part since 1915.

Some of the main features of Salt Lake City—based on the numerous detailed sketch maps by Chauncy Harris (*loc. cit.*) are charted in Fig. 159. Here we see the three streams debouching from the Wahsatch Mountains into the terraces and plains which border the Great Salt Lake. The Temple Square is shown, as well as the original blocks into which the city was divided. These are still found as the basis of the city to the north, south and south-east of the Temple. Some of the later lanes are indicated by broken lines. The city to-day extends four or five miles south of the Temple, and here the street plan is different, but is not indicated in the map.

The chief business section extends for about a mile to the south-east of the Temple along State Street. The railways cover a good deal of space (not indicated) in the west of the city between it and the River Jordan. The University has been built on the higher land to the east of the city near Fort Douglas. The best residences are found on these terraces south of the University. Second class residences fill the suburbs between the University and the Capitol. The lower land between Parleys Creek and the Jordan is occupied largely with third class residences, the poorest of all being found to the north-west of the station.

A few words may be given to other factors which decided whether Ogden or Salt Lake City should become the dominant settlement in the line of oases. Great mines developed to the south of Salt Lake City, especially at Bingham and Park City, and these were naturally in contact with the latter centre. Although the first railways favoured Ogden, the Denver line gave Salt Lake City an independent route to the east in 1883. Later the Western Pacific Railway built a direct route to the west (in 1910).

However, the deciding factor was the presence of the leaders of the Mormon Church in Salt Lake City rather than in Ogden. The Church leaders determined the location of the various villages in the region, the composition and leadership of the settlers in each village, and even the form of settlements in *villages* rather than on dispersed farmsteads. Furthermore the church was the most powerful economic and social agency in early days. Since Salt Lake City was the headquarters of the Church, this city since its very founding has been the principal centre of the oasis and the region (Harris).

Mormon Villages and small Towns

The early Mormons had three principles at the back of their settlements which lend a special pattern to their towns and villages. They believed that the Salt Lake Basin will be the 'gathering place' in the last days. They lived up to the tenet that each person is entitled to as much as he needs to support himself and his family, but no more. They tried to build their main city as a Utopia. A. L. Seeman has written an interesting account of the smaller settlements in the vicinity of Salt Lake City,[1] from which the following notes are borrowed.

Brigham Young always encouraged farming rather than mining in Mormon territory, because he believed that quick wealth tended to break down the sturdy character which he hoped to see widespread in his people. Each settlement was planned in the early days, and this wise practice obtains to this day. Thus near Malta in Idaho the Church quite recently bought a 4,000 acre ranch as a colony site. The migrants did not select farm sites, but they settled in the planned town. Indeed it has been observed that the church has created so many religious activities that it cannot carry out its programme except in organized communities. Hence there is little *isolated* rural life in Utah. Moreover individual efforts would be futile in an area depending on extensive irrigation.

Seeman describes various Mormon villages, such as Ephraim, 120 miles south of Salt Lake City. Such a community is composed of a main street with square blocks and side streets forming a grid. The houses are usually of brick, and surrounded by well-kept lawns. Behind the house

[1] "Communities in the Salt Lake Basin," *Econ. Geog.*, July 1938.

will be found the general farmyard with barns, etc. The church, social hall, and school are in the centre of the community. The holdings of the individual farmers are small, for the best land is divided among all the farmers as is the poorer land. As a result more than fifty per cent of the farmers have their holdings in more than one piece of land, and some farmers have as many as twelve pieces of land. The average size of an irrigated farm around Ephraim is about fifty acres.

Amana, in eastern Iowa: a composite religious community

Darrell H. Davis has given us a very interesting description of a group of villages in eastern Iowa, whose cultural pattern is very largely determined by their early communistic ideas.[1] It is of especial interest that the socialistic principles of the community have been much modified by the evolution of the surrounding population; so that in 1932, the group decided to turn themselves into a stock company of the capitalistic type.

Around 1714 a schism developed in the German Lutheran Church, and led to much friction among this sect in Germany. In 1843 five thousand acres of the Seneca Indians Reservation, near Buffalo, were purchased by the community. But in 1854 the growth of the nearby city decided the community to move into the west. They selected a site about twenty miles to the west of Iowa City, and to-day they hold about 26,000 acres on the Iowa River, about half way between Dubuque and Des Moines.

Members of the society renounced all personal property, and the seven villages constituted a sort of federation, which was virtually self-sufficient (Fig. 160). Each village (until the change in 1932) had its own meeting-house, school, general store and post office, and several 'kitchen houses'. Community barns and sheds were grouped at the east end of each village near the factories. The houses were surrounded by large gardens and orchards, and beyond these were the community fields. This pattern still persists.

By 1861 the Rock Island Railway reached Homestead, and this village was acquired as a means of controlling the shipping point of the community. To-day several paved and gravel roads link Amana to the surrounding villages. There was an adequate supply of timber on their territory, and they developed gravel pits and sandstone quarries. Agriculture is the main support of the colony. Maize, oats, barley and hay are fed to stock, while potatoes and wheat are also consumed locally by the human members. However, to-day cream is sold in neighbouring

[1] "Amana: a study of Occupance," *Economic Geography*, July, 1936.

villages, and pork is also distributed far beyond the colony. Wool from their sheep is made into blankets in their factories, and has a large outside sale.

The colony consists of seven villages at varying distances of one to five miles from each other. In planning these village groups two factors were taken into consideration: first the distribution of the farmlands, and second the social advantages of small communities. The seven

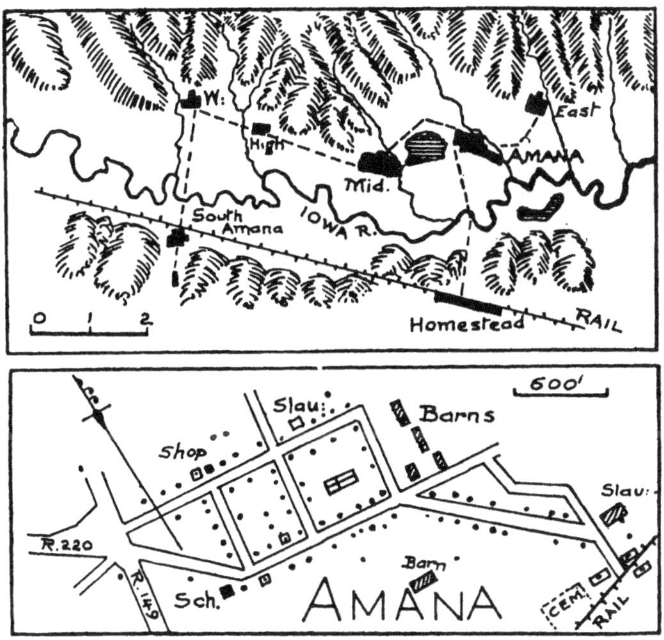

FIG. 160. The seven Lutheran Community Villages of Amana in East Iowa. The *lower* plan is of the main village. (*After D. H. Davis.*)

villages are arranged on each side of the Iowa River. On the north are West Amana, High Amana, Middle Amana, Amana, and East Amana; on the south side are South Amana and (in the south-east) Homestead (Fig. 160). Of these villages the largest is Amana with 437 inhabitants (1936); and it is the administrative centre. East Amana is the smallest of the group with 102 residents. Amana is also the focal point for the principal roads, and lies about sixteen miles south-west from the nearest town of Cedar Rapids.

Davis states that the village plan of these communities is strikingly different from that of most American farm villages. This is largely the result of the absence of any business section. Few of the shops have any different appearance from the homes, and this is an inheritance from the non-competitive organization of the past. There is a distinct tendency to the 'strassendorf' pattern, especially in Homestead and in Amana. There are many shady 'footpath' streets through the orchards, and the houses are shrouded in vines, but there are no shade trees or lawns along the main roads.

The total population in 1936 was 1,461 persons; but adults outnumbered minors in the ratio of 214 to 100. Until the first World War they all spoke German, and this is the case with the elders to-day. In addition to the woollen mills, there are woodworking and cabinet shops, flour and grist mills, bakeries and printing presses. The old age pensions are paid by the community, and there are many other features which make this experiment in corporate farming of more than transitory interest.

The Pilgrimage Town of Lourdes in the south-west of France

The site of Lourdes is charted in Fig. 143 on the river Gave de Pau about eighty miles from the Bay of Biscay. Here the river flows north from the rocky slopes of the Pyrenees, and makes a sharp turn to the west. On the outer side of the bend is a steep crag which was crowned with a castle, parts of which date back to the 14th century. This fortress commanded the valley of the Gave, and was the scene of many sieges in the early middle ages. The old town surrounds the castle, and was a walled city in early days, but not much is left of these fortifications.

The importance of Lourdes—which leads to its being visited by 600,000 pilgrims annually—dates back to 1858. In the bend of the Gave on the western bank is a considerable hill containing many grottoes. In some of these, prehistoric bones and artifacts have been found. On the north side of this hill a girl of fourteen named Bernadette Soubirous is alleged to have seen the Virgin Mary on several occasions. She was directed to build a shrine at this spot, near to a spring which issues from the hill (Fig. 161). Pilgrims soon flocked to this grotto, partly attracted by the cures reputed to be due to the waters of the spring. To-day the whole aspect of the little town of 7,000 inhabitants has been changed as the result of the visions of Bernadette.

In 1876 a Basilica was erected above the grotto, and in 1889 the Church of the Rosary in the Byzantine style was erected below the Basilica to the east. Nearby on the west bank is a large hospital, Notre Dame des Douleurs, and many hostels and apartments for the accommodation of the enormous numbers of pilgrims. The river has been diverted somewhat so as to give more space to the north of the grotto. A long Boulevard de la Grotte leads from the station to the grotto, and this is occupied largely with shops for the sale of religious articles and mementoes.

An interesting account of the site of the shrine before it rose to fame is to be read in *Lourdes*, E. Saunders, London, 1940. Massabieille is a steep rocky hill rising up from the river bank about half a mile to the west of the town across the river. It is green and wooded, but many bare patches of rock interrupt its grassy surface. At its base close to the river was a grotto, roughly circular in form; in it were fissures and openings, from

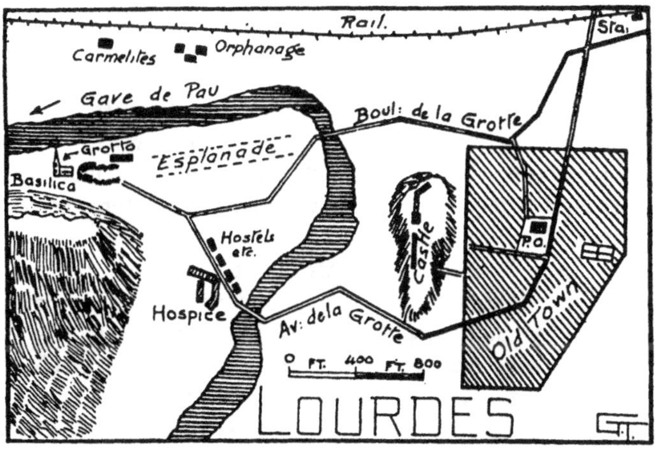

FIG. 161. Plan of Lourdes (somewhat simplified) to show the religious buildings (erected near the grotto), the old town, and the castle (see Fig. 143).

which narrow passages led into the rock to small interior cavities where stalactites had formed. Its sandy floor, covered with pebbles, was always damp. An underground spring was there, sometimes it overflowed and a narrow stream ran down to the river, but in dry weather the only evidence was a patch of liquid mud. The river occasionally flowed in after a heavy rain, but it was usually a few yards away. No one ever thought of entering the grotto except the village swineherd and a few fishermen. On the 11th February, 1858, a vision appeared to Bernadette in the form of a girl in a long white robe in a kind of oval niche in the grotto. A few days later the belief became general that the spring in the grotto possessed miraculous properties, and sick folk have visited Lourdes to bathe in the water ever since.

The Grotto in the rocks is a recess about fifteen feet deep and fifteen feet wide, and on a rock projecting above is a statue representing the Virgin as described by Bernadette. To the left is the miraculous spring, now confined by a wall garnished by taps through which the water flows into basins in which the pilgrims bathe. The Basilica is in the Gothic style with an elegant tower, and is built above the Grotto, while the Church of the Rosary stands in front of, but below, the Basilica. The Church has the form of a Rotunda in the Byzantine style.

Mennonite Strassendörfer in southern Manitoba

The first important agricultural settlement in modern days in the Canadian Prairies was that due to the Mennonites in 1875. These folk have a very interesting history, which largely accounts for the fact that they can still be distinguished among the millions of European immigrants to North America by certain cultural peculiarities. They are a religious sect whose tenets date back to 1525 in Zurich. The name derives from Menno Simons, a Dutch priest who expounded the new teaching around 1536. This may be summarized as a form of Lutheran worship marked by a refusal to take oaths or to fight.

In 1783-8 large communities of Mennonites migrated to Russia, but about 1870 the Russian authorities withdrew the special privileges granted to the Mennonites, and many of them left for the New World. Settlements on a large scale took place in Kansas and in southern Manitoba in 1875. The Canadian Government granted them exemption from military service, and allocated reserves of 720 square miles for their use. The main settlement was in the Municipality of Rhineland.[1]

In 1901 there were 15,246 Mennonites in Manitoba, and the bulk of this number dwelt in the region charted in Fig. 162. They have increased from the six thousand Mennonites who located on seventeen townships in the open prairie in 1875. In 1941 this total in Manitoba had increased to 39,336. An interesting account of these early settlements is given in the above-mentioned volume. The Dutch-German-Russian village was simply transferred to Canadian soil. Quaint houses with high-pitched thatched roofs, ancient flour mills with huge arms and sails, ungainly churches, arose, all higgledy-piggledy along the roadside. The village names are all German, and are usually somewhat romantic in origin. We notice in Fig. 162 Rosenort and Rosengart, Blumenort, Blumenfeld and Blumengart, Gnadenfeld and Gnadental, Schanzenfeld, Hochfeld, Schoenewiese, etc.

The earliest houses of these settlers were built of mud and sticks

[1] *Strangers within our Gates*, J. S. Woodsworth, Toronto, 1909.

thatched with straw or hay. Some of the oldest were still standing (around 1900) with walls of a delicate lilac, the window sashes a dull red, the shutters grey. A village of these houses, says the author, is more suggestive of Holland in the 16th century than Manitoba in the 19th century.

Each settler had a grant of 160 acres, and this was distributed according to their accustomed system. It was divided into three parts; a village plot, a strip of good land, and a portion of hay land more remote from the village. Each village had its own school and elected its own headman.

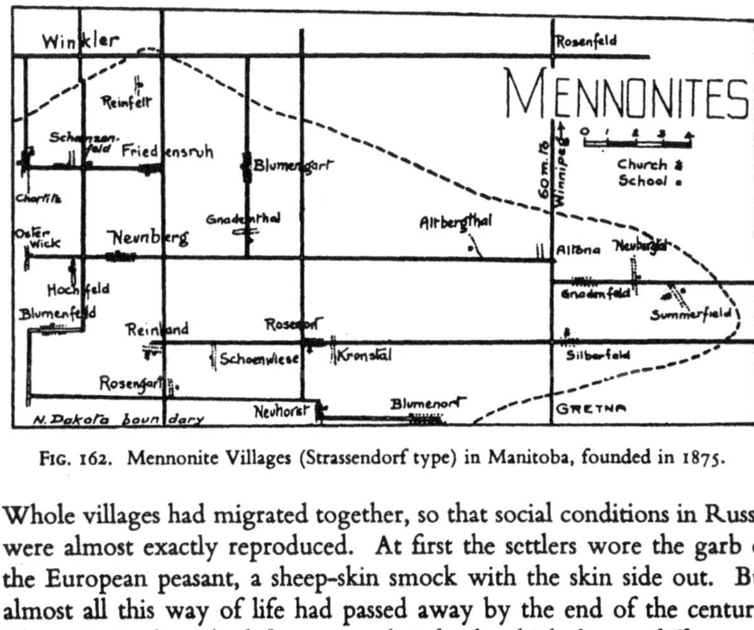

FIG. 162. Mennonite Villages (Strassendorf type) in Manitoba, founded in 1875.

Whole villages had migrated together, so that social conditions in Russia were almost exactly reproduced. At first the settlers wore the garb of the European peasant, a sheep-skin smock with the skin side out. But almost all this way of life had passed away by the end of the century. The majority have built homes on their land, which do not differ much from those of surrounding settlers of other cultures.

Large numbers of the younger generation have left the mother colony, and have taken up homesteads throughout the eastern prairies. New colonies have been founded at Rosthern and Hague. They still in many cases insist on the exemption from military service; but according to J. M. Gibbon (*New Canadian Loyalists*, Toronto, 1941) numbers have offered to enlist in the recent war as stretcher-bearers.

Reference to the map (Fig. 162) shows that the original strassendorf pattern is quite pronounced to-day in each of the twenty-two Mennonite communities there charted. In general there are about a dozen farmhouses built relatively close together on each side of the street running through the village. Half way along the village is the school house, which seems to be present in all but the smallest villages, while only about a third of the villages had a church according to the 1920 map.

PART II. RESORT TOWNS

The Swiss Health Resort of Davos

FEW mountain settlements are better known than Davos (*pron.* Dah-vòce) in the south-east corner of Switzerland. Here the upper Rhine and the upper Inn flow in parallel courses along fold-valleys from the south-west to the north-east. Just half way between, them in another parallel valley is a small stream called the Landwasser; which, however flows to the south-west (inset map in Fig. 163). No better example of a town illustrating the difference between the *adret* and the *ubach* sides of an Alpine valley is known to the writer.

The sketch (Fig. 163) shows that the town extends for about two miles above the broad flat floor of the alpine valley, but almost all the houses are confined to the sunny north-west side of the valley (i.e., the adret side). On the further slopes in the sketch are only a few chalets, which during a considerable portion of the day, even in summer, are in the shade of the great mountains which lie above them. The valley is nearly a mile wide, and owes its width mainly to bygone glacial erosion. The little Landwasser stream drains Davos Lake, but is a 'misfit' in the broad glacial valley.

There are two 'towns' here, Davos-Dorf at the north end near the lake, and Davos-Platz lower down the valley. However, both towns are over 5,000 feet above sea level; and it is this elevation, together with the sunny climate and absence of mist, which has led to the rapid growth of Davos in modern times. It is sheltered from the cold winds of the north by the high peaks of the Weissfluh (9,500 feet) and Strela (8,800 feet); and from those of the east by the Pischahorn (10,000 feet). Through the broad valleys of the south-east—which lead up to the high Fluela Pass or to the Sertig Pass—the mid-day sun shines strongly on the western slopes of the Davos valley, which in consequence is thickly covered with large pensions and dominated by an occasional huge sanatorium. I made a rapid survey of the distribution of the large boarding-houses (i.e., pensions), and the results are charted in the sketch. A light electric railway runs along the valley, with the station at Davos-Dorf. Here

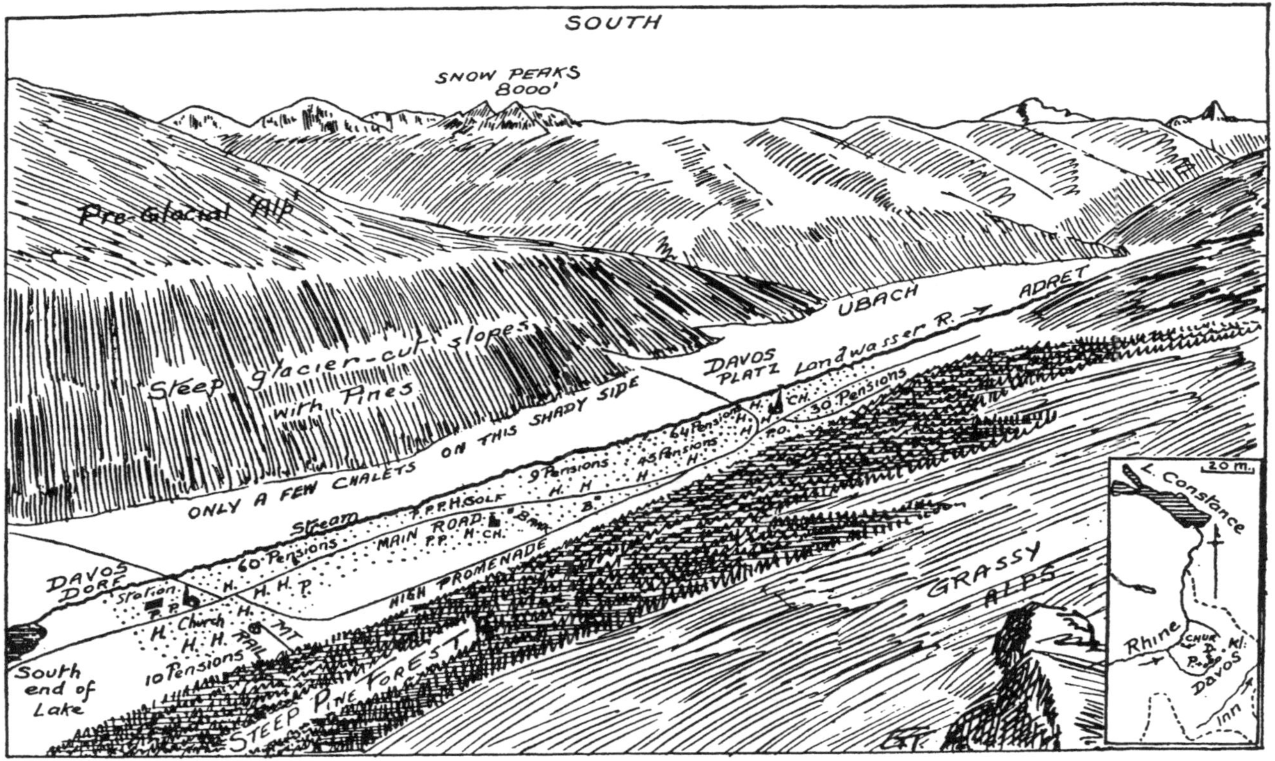

FIG. 163. Sketch of the Davos Valley (5,100 feet) looking to the south from the northern shoulder. The numerous hotels (H.) and pensions (P.) are strung along the main road on the sunny N.W. side of the valley. Dots are smaller houses. From the Lake to Davos Platz is about two miles. *Inset* is a sketch map showing the relation of the two towns (D. and P.) to the Rhine Valley. Note ADRET and UBACH aspects.

are clustered about seventy pensions, and six hotels, all of which are of recent date; but here also is the village church which dates back nearly to 1500. The little lake, whose end just appears in the sketch, is 5,144 feet above the sea; and its waters are led north over the low valley-divide to Klosters about five miles away, where they enter the turbines of a power-house.

Behind Davos-Dorf a mountain railways runs up the slopes to the flatter pre-glacial meadows on the shoulder of the main valley. In summer these appear as grassy 'Alps', but in winter this Schatzalp is the scene of some of the most popular ski-running in Switzerland. Ski huts and ski jumps are to be found in the vicinity; while a large sanatorium has been built here at a height of 6,112 feet.

The main road linking the two towns is now almost built-in with pensions or private houses, and a golf course and the bank are to be found half way between the two centres. Near Davos-Platz are about 150 pensions, and half a dozen hotels, mostly to the north of the old church built about 1480. Here also is a monument to the German doctor (A. Spengler) who made Davos a health resort, and founded the Kurhaus in 1867. In the year 1850 the population of Davos was only 1,680; but some seventy years later it had increased to 10,000. This is mainly due to its importance as a health resort through the year for patients afflicted with lung troubles, and to the tourism and winter-sports which attract the healthy visitor.

Winter Resorts in the Laurentians

The most popular ski-ways in North America are to be found to the north-west of Montreal. Here the North River and the Rouge River rise in the undulating plateau of the Shield whose highest point hereabout is 3,000 feet at Mont Tremblant (see the inset in Fig. 164). This mountain is close to the branch line which connects Montreal with Mont Laurier, and for forty miles the railway passes through the heart of 'Ski-Land'. In a single winter season the railways have received $300,000 for fares to this country, while the habitants have been paid $700,000 for board and lodging by ski enthusiasts. Four hundred special trains were run from Montreal, and there are hundreds who come to this area from Ottawa, Toronto, and from the United States.

To-day most of the farmhouses in this district have electric light and running water, and are advertising for the ski trade.[1] Shawbridge is the gate of entry from Montreal, and both the C.P.R. and the C.N.R

[1] "Ski-ing in the Province of Quebec," *Canad. Geog. Jnl.* n.d.

run trains to this point (Fig. 164). On the latter railway are prepared ski trails (as shown on the map) near Morin, and also just to the east at Christieville. But it is along the more northern line that the main ski-ing takes place. From here a continuous trail known as the 'Maple Leaf Ski Trail' leads for over forty miles to Mont Tremblant far to the north-west. (Fig. 164.) At Shawbridge is the Laurentian Ski Club Lodge, the first to be established in the area. To the north about six miles is St. Sauveur, where the village street is bordered with many ski cabins.

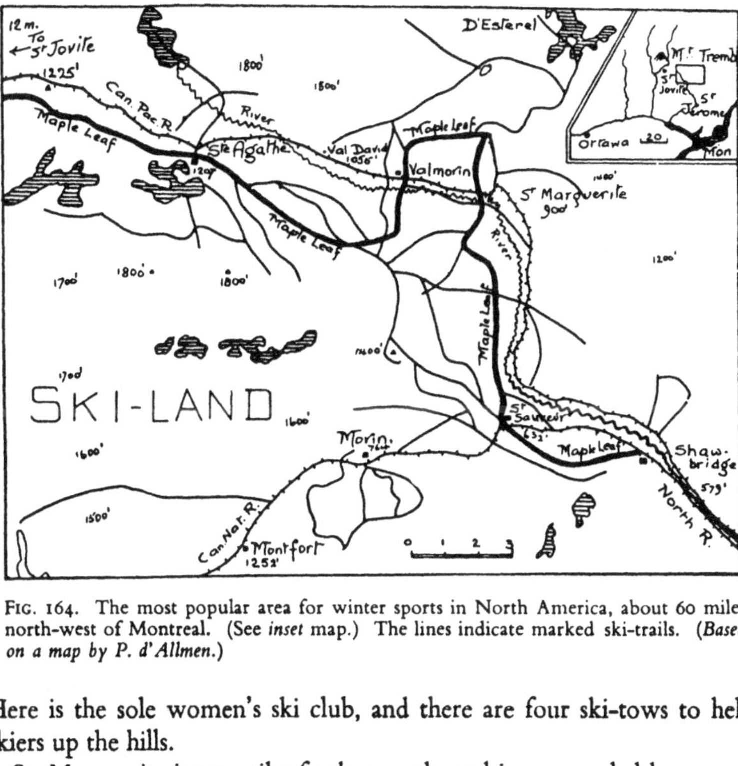

FIG. 164. The most popular area for winter sports in North America, about 60 miles north-west of Montreal. (See *inset* map.) The lines indicate marked ski-trails. (*Based on a map by P. d'Allmen.*)

Here is the sole women's ski club, and there are four ski-tows to help skiers up the hills.

St. Marguerite is ten miles further north, and is surrounded by a great number of well-marked and cleared ski trails. Here also are ski-tows, a large inn, and facilities for ski jumping.

A big establishment has recently been erected at the Domain d'Esterel, and there are two ski-tows and log cabins in addition. St. Agathe is about the highest of the ski settlements, being 1,200 feet above sea level. As the map shows it is surrounded by granite hills which rise about 600 feet above the general level of the plateau. There are many summer cottages on the lakes nearby, so that this district is attractive to tourists in summer as well as winter.

Twenty miles to the west along the Maple Leaf Trail brings us to the vicinity of St. Jovite (see inset map), where forty miles of trails radiate from the railway station. Near here are ski schools, and a pack of Siberian dogs. Mont Tremblant nearby is the scene of the Quebec-Kandahar race in the month of March. An American enthusiast has spent a million dollars on hotels and cottages at the foot of Mont Tremblant.

Each large centre of population in Quebec has its ski-land, though they are not so popular as the one just described. However, behind Quebec City is Lac Beauport, where ski-ing starts early in December and lasts until early May. Snow depths of six feet or more are common, and the resort is only nine miles from Quebec, and even in winter a bus service is available. Behind Trois Rivieres is Grandes-Piles with ski jumps and attractive trails. On the southern side of the St. Lawrence there are many smaller ski districts, as at Abercorn and Sutton. Further to the west in the Gatineau valley (north of Ottawa) there are a number of ski clubs which enjoy much popularity.

Resorts in the Rockies of Canada

Few better accounts of the way in which environment determines settlement in the scenic portions of a country have been written than that by Stephen B. Jones in the Bulletin of the Geographic Society of Philadelphia (July, 1936). Here all aspects of the problem are investigated, such as those based on structure, climate, accessibility, scenery, etc.

In the portion of the Rockies most visited by tourists, i.e., in the vicinity of Banff and Jasper, the mountains are divided into three longitudinal belts, each with special characters. The eastern belt is marked by westerly-dipping strata, the central belt contains the highest peaks and the most marked effects of glaciation. Here also are the typical pyramid mountains, which owe their shape to the horizontally-bedded rocks which compose them. In the western belt the mountains are lower, and the rocks are often much tilted. This western belt is much wetter, and consequently more densely covered with forest (Fig. 165).

Jones has divided the tourist resorts of the area into three classes. Headquarters resorts, objective resorts, and wayside resorts. The *Headquarters resorts* serve as residence centres, outfitting points, and trade foci for recreational regions. They are almost urban in character, and may be occupied throughout the year. They are all to be found in the eastern belt, for reasons which Jones considers in much detail. Banff and Jasper are typical examples. They have a milder climate than the Objective resorts, so that the evenings can be spent out of doors, which is not the case in the higher resorts. The summers are longer at the lower elevations, which enables the tourist season to be longer at Banff than at Lake Louise. In general the scenery at the Headquarters resort is restful rather than majestic; but perhaps the chief attraction is that at Headquarters there is a great diversity of recreational resources such as golf, tennis, swimming, dancing, and riding. Here only does it pay to have the elaborate carnivals such as those which take place at Banff. Almost inevitably there are more roads branching off from the Headquarters resort, and there are always plenty of paths suitable for short walks. Both Banff and Jasper are on the main railway lines, and the principal hotels are of course under the control of the railway authorities. Hence the cost of goods brought to such places is not much different from the cost at any other station on the main lines of Alberta. More important still, there are great numbers of other bungalows and houses where the tourist may obtain lodgings, and this tends to keep down prices and suit all pockets.

The *Objective resorts* have quite different characters. They are almost all to be found in the central belt of the Rockies. Often enough they are situated rather in a 'dead end' valley, with only one main objective in the vicinity. Hence they are suited for much shorter stays than are the Headquarters resorts. Jones points out that many Objective resorts have 'catchy' names, such as Twin Falls, The Natural Bridge, etc.; and this single definite feature is the sole attraction. Many are situated at the outlet point of the small lakes, where usually there is a small moraine. This gives a drier site for the log cabins, which are usual at Objective resorts. The weather is often unreliable in these rather elevated resorts of the central belt, and the bungalows must be artificially heated for a considerable portion of the day.

It is interesting to note that hardly any resorts have developed in the *Western* belt of the Rockies. Here the scenery on the whole is much less grand than in the central belt. Only one resort of note is available, that at Radium Springs in the Rocky Mountain Trench, where the Windermere Road from Banff reaches the great valley. This resort has some fine hot springs, and a picturesque canyon in the vicinity, but in Jones' opinion, it has a cramped site, a dearth of lakes and walks, and a lack of rail connection. These give it a far less favourable position than is the case with Banff or Jasper. It is to be noted that most of the tourists come to the Rockies from the east. The population of British Columbia, so largely concentrated at Vancouver, has available almost equally fine mountains and canyons anywhere along the coast, and so does not need to migrate to the Rockies for recreation.

Wayside resorts exist primarily to supply food and lodging to passing motorists. However they are usually sited near a canyon or picturesque bridge, though not usually on a picturesque lake; for the latter are found mainly in the tributary valleys away from the main valleys and main roads. A number of such resorts are indicated by 'W' on the map (Fig. 165).

Jones finishes his interesting study by an investigation as to why Field (Fig. 165) has slipped back as a tourist resort. He points out that the canyon is so deep here that few views are possible from the settlement. It is a railway 'division' town, and the busy railway traffic, and the noise of changing locomotives, etc., takes away from the restful character of Field as a resort. The good road makes it quite possible to reach Lake Louise in a short time, so that there is less temptation to make a stay at Field. For these and other reasons, Jones does not anticipate the development of further Headquarters resorts to rival Banff and Jasper. In a similar fashion it would seem clear that the number of coal towns all along the good road over the Crow's Nest Pass—150 miles to the south—will prevent that portion of the Rockies from rivalling the Kicking Horse Pass region as a tourist centre.

A few words must be given to the northern portions of the central belt of the Rockies. Extending along the eastern side of the divide in Alberta is the Banff National Park, which contains 2,585 square miles. At the Sunwapta Pass the Park is in contact with an even greater park to the north. This is the Jasper National Park, which comprises 4,200 square miles, and was established in 1907. In the last few years a beautiful road has been made available, linking Banff to Jasper through the middle of these parks, and this offers the tourist perhaps the finest mountain scenery in North America.

The road leaves Banff at 4,500 feet above sea level, and in thirty-seven miles reaches Lake Louise. Near this point it leaves the Kicking Horse Pass on the west, and soon crosses Bow Pass (6,878 feet) at the head of Bow River. Further north at Sunwapta Pass (6,675 feet) we enter Jasper Park. Many peaks flank the highway, including Mt. Athabasca which is 11,542 feet high. Here the road approaches within a few hundred feet of the Athabasca Glacier, and nearby is the Columbia Icefield which covers 150 square miles. Still farther north are the Athabasca Falls. After a journey of 186 miles the tourist reaches Jasper on the Canadian National Railway, which is another Headquarters resort.

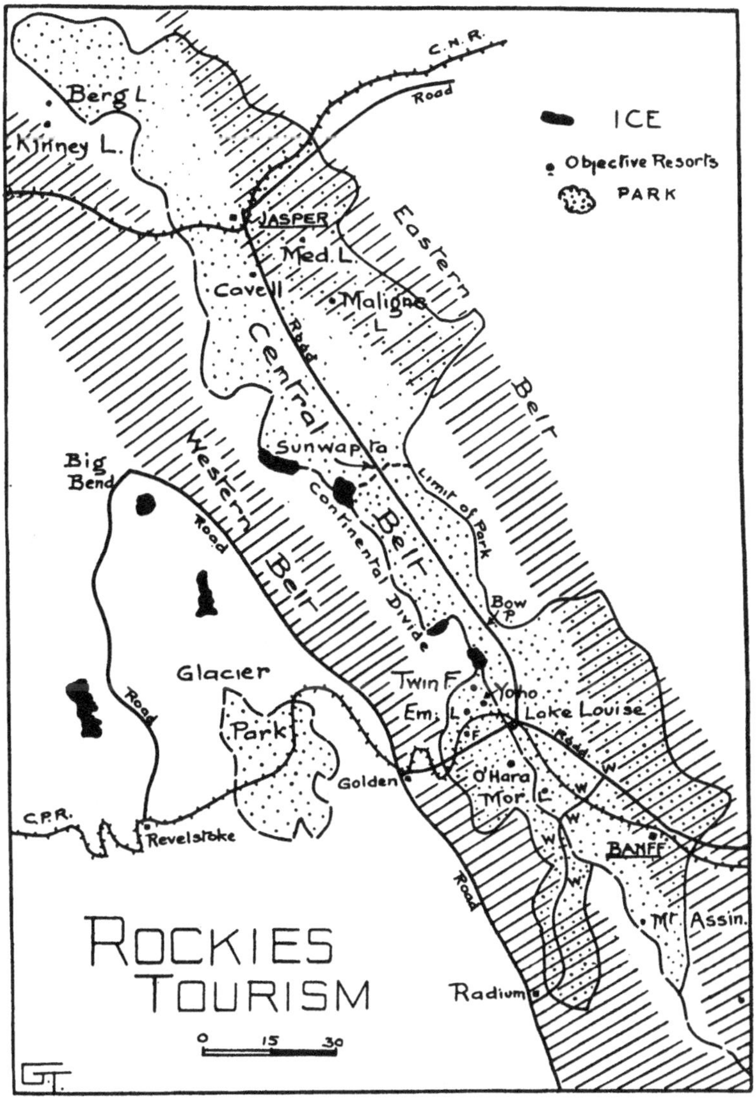

FIG. 165. The chief touring resorts in the Canadian Rockies
F. is Field.
(*Based on S. B. Jones.*)

Tourist Towns on the Blue Plateau, Australia

As explained in an earlier section (page 202) there is a large plateau composed of Triassic Sandstones some fifty miles to the west of the city of Sydney. This plateau rises gradually from near sea level at Penrith on the Nepean River (Fig. 94) to a height of almost 4,000 feet at Mount Victoria. It is one of the most remarkable 'monocline' folds which has been described in geographical literature, and the gigantic valleys are the direct result of this monoclinal structure. As seen from the large city of Sydney—with its million and a half inhabitants—the up-fold looks like a bluish range extending across the western horizon. Hence it received the name of the 'Blue Mountains'. Later investigation showed that this was a misnomer; and it is better called the Blue 'Plateau'.

For some twenty years in the earliest days of settlement the early explorers attempted in vain to find a passage to the west through the so-called range. This was because they rather naturally tried to advance along the valleys. These, however, open on to the coastal plain in narrow bottle-necks, such as that by which the Cox and Wollondilly-Warragamba system enter the Nepean south of Penrith. Here the rivers are sawing their way through the *inclined* layer of hard sandstone; whereas further to the west the rivers have cut through the *horizontal* sandstones (Fig. 166), and are sapping away at the soft coal measures below. Thus the Cox Valley (as shown in part in Fig. 94) is eight or ten miles wide; but it becomes narrower to the east, and its 'bottle-neck' entrance to the Nepean River is by a gorge only a few hundred yards wide.

The ravines which discharge their waters into the Cox valley occupy an area of about 1,200 square miles; and the valley is enclosed by sandstone cliffs increasing from about 1,000 feet near its outlet to about 2,500 feet near its western limits. Thus we find the anomaly that tourists who visit the famed Blue 'Mountains' for the first time see no mountains at all! For the railway has been built on the surface of the hard sandstones and all the scenery is down below in the remarkable valleys. As soon as the early explorers attempted to reach the west by way of the plateau surface they crossed the so-called mountains in a few days without any real difficulty—though a great deal of uncalled-for eulogy has been wasted on that first journey of Blaxland, Wentworth and Lawson in 1813.

As soon as the train has passed Penrith the climb begins. At Glenbrook forty miles from Sydney, the dissected valleys, though not very deep, contain the usual small waterfalls and fern gullies. Here the plateau is about 1,000 feet above sea level. At Lawson (fifty-seven miles) the surface of the plateau is 2,347 feet high, and the valleys are correspondingly deeper and wider. At Wentworth Falls (Fig. 166) we are sixty-one miles from Sydney, and are entering the best of the scenery. Here, a mile south of the station, are the chief falls, which have been described as follows: 'The streamlet precipitates itself over the precipice to the great boulders 1,000 feet below. Beyond these broken masses of stone and trees a second dreadful precipice and a second fall may be seen, so far below

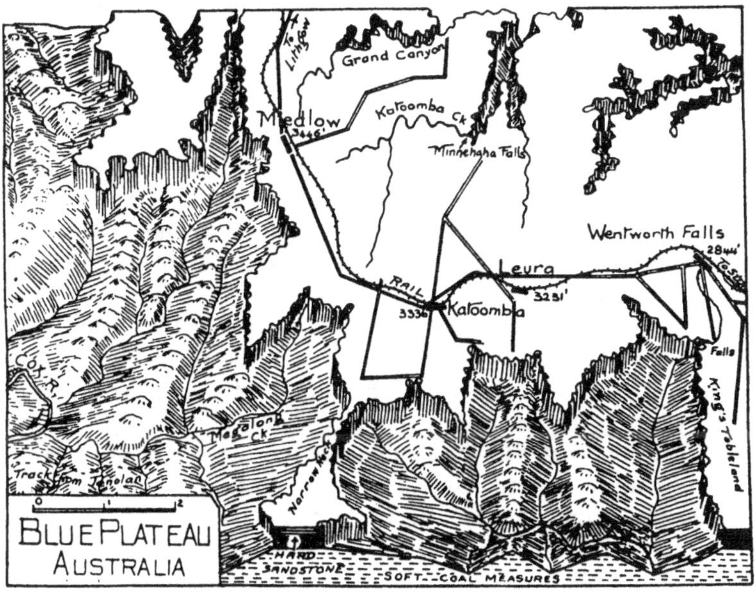

FIG. 166. Block diagram of a small portion of the sandstone plateau about 65 miles west of Sydney. Cox River is about 2,500 feet below the Plateau. Tourist houses are now almost continuous along the main road from Medlow to Wentworth Falls, but only the main roads are shown. A simplified geological section is given on the *front edge* of the diagram.

the dizzy altitude that no murmur from this or the next succeeding fall ascends to break the silence.'[1] As one reaches the lower levels of the valley the better soils of the coal measures are reached. Here the vegetation is quite different from the rather xerophytic scrubby eucalyptus forest of the plateau, and one passes into a jungle of semi-tropical trees, such as the coachwood and sassafras, with a beautiful underbrush of tree-ferns—the most attractive feature of Australia's eastern valleys.

From Westworth Falls the train proceeds west to Katoomba, which

[1]There is some poetic licence here! The total drop of the two falls is 646 feet.

is the largest of these plateau resorts, with a population of 7,000. There is no agriculture on the plateau, except to the north-east where some volcanic lavas have produced a rich soil now utilized by truck farmers. The timber is too poor for anything but fuel, and the dominant factor in settlement is the cool summer climate ten degrees below that of Sydney, and the magnificent scenery below the plateau level. The diagram in Fig. 166 shows that the railway has been constructed along the rather narrow divide between the Cox Valley on the south and the similar branches of the Grose Valley on the north. The latter valley also offers deep cliffed gorges containing small waterfalls and beautiful flora. Minnehaha Falls though not so high as Wentworth Falls contain more water, since the stream falling over the cliff is two miles long. The necessarily limited basins of the streams on this narrow plateau result in rather meagre amounts of water in the falls. However, the rains are fairly heavy, so that it is rarely that the streams cease altogether.

Medlow Bath is the next station, and its features are much the same as the others described. Further to the north-west are Blackheath (seventy-two miles) and Mount Victoria (seventy-six miles) where the railway is 3,424 feet above sea level. Here the Victoria Pass road leads down into the Cox Valley, and then divides to go west to the much-visited Jenolan Caves, and north to Lithgow and Bathurst. Of course the best route from Sydney to Bathurst and the west would be along the Cox River itself. This would obviate the climb of 500 feet up the plateau to Mount Victoria, and the corresponding descent of 500 feet to Lithgow, which is on the same Cox River at a height of 3,000 feet above sea level. However, as far as the writer is aware, not even a track for pedestrians has been cut along the difficult 'bottleneck', where the combined Cox-Warragamba River finally enters the Nepean (Fig. 94).

CHAPTER XVIII

CITIES BUILT TO OTHER THAN CHECKERBOARD PLANS

Introduction

By the time a village has grown to the size of a city there must inevitably be some sort of a plan in the design of the streets and house blocks. But, as we have noticed in many of the later chapters, this is usually merely an extension of the checkerboard pattern of the real estate agent. It serves to cut the land up into sections which are easy to survey, and which can be readily subdivided into lesser blocks. Such rectangular blocks lend themselves to our conventional rectangular houses. But they do not harmonize with the topography, and they are very wasteful of the traffic-time, since they entail much 'going round corners'. They pay little attention to the prevalent winds, or to the dominant solar rays; nor do they lend themselves to the display of fine buildings or monuments.

In preceding chapters we have on many occasions seen some indications of planning, right from the dawn of history. The checkerboard was honoured in the prehistoric terramara settlements (Fig. 52), as well as in some of the earliest Greek cities, such as Selinus (Fig. 50). The Roman castrum crystallized this pattern, so that not only such places as Pompeii (Fig. 53) and Timgad (Fig. 55), but also cities of much later date like Aosta (Fig. 54), Chester (Fig. 60) and Cologne (Fig. 62), were strait-jacketed by the Roman 'grid'.

In the Middle Ages the necessity for defence led to the building of rectangular walls, later of circular walls, and finally of star-shaped walls. These naturally considerably affected the pattern of the streets and house-blocks in the interior of the walls. For examples, see the various town plans in Chapters IX and X. With the close of Medieval Times when cities were built without confining walls we may perhaps place the dawn of modern city planning.

It is worth remarking that enlightened citizens in these later times realized that the checkerboard had little to recommend it. For instance after the great fire of London in 1666 there were two plans put forward to ensure that the new London should be an improvement on the old. That of Sir Christopher Wren is sketched in Fig. 167; and shows that the advantage of 'radial' streets, and 'round points' linked to notable buildings, was quite apparent to city planners as far back as the middle of the 17th century. Karlsruhe and Versailles with their plans have been described at some length in Chapter X, and they but carry a stage further the ideas of Sir Christopher Wren. L'Enfant's plans for Washington are also based on the radial street and the 'round point', but in this case we have to do with the design of a large city. Perhaps this may be taken as the beginning of modern town planning.

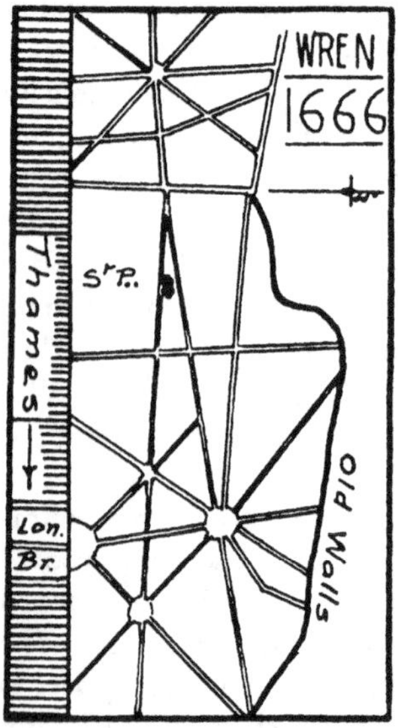

FIG. 167. Wren's proposed plan for rebuilding London after the Great Fire. It was not adopted.

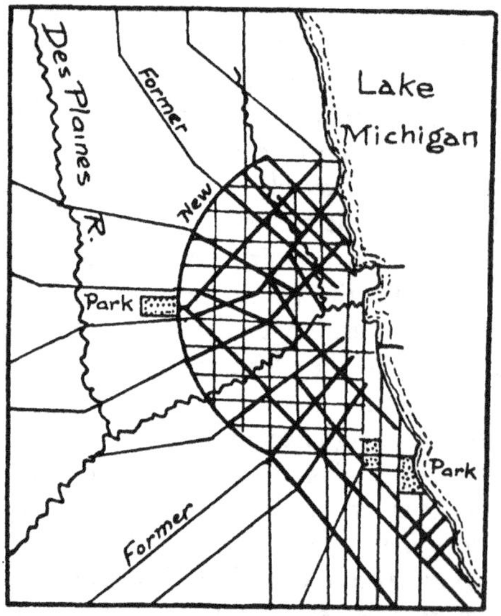

Fig. 168. The main features of Burnham's plan (1909) for Chicago. The diagonal streets (heavy lines) are to be added to the conventional grid (fine lines).

About the same time—the end of the 18th century or a little later—some small towns were being laid out in the Dominion to the north. For instance, Charlottetown makes use of small parks to diversify the checkerboard. Goderich and Guelph use the radial streets in the centre of the town as shown in Fig. 22 and Fig. 21. Haussmann in Paris towards the end of the 19th century showed that boulevards cut through the slums vastly enhanced the amenities of the city (Fig. 77). All these aspects of a city began to be integrated in a regular study only in the present century—giving rise to that branch of the arts which is known as Town Planning.

Geography is pre-eminently the liaison science, and in the field of town-planning the geographer joins forces with the architect and the engineer. However, in questions of site and approach, and the relation to environment generally, the problems of town planning belong equally to the field of geography. Hence some early paragraphs of the present chapter may well be given to discussing briefly some glaring examples of 'plans which do not fit the topography'.

The first plans given in this chapter deal with attempts to improve the checkerboard, and make the plan fit city needs rather than those of the real estate agent. Thus Fig. 167 shows us the very large area of London which was destroyed in the Great Fire of 1666. Almost everything south of the ancient walls (which are indicated in the map) was burnt, including the 'Gothic' St. Paul's. Wren made the approach to London Bridge the centre of his plan, with one round point just north of the Bridge, and another about the position of the Bank. A third round point was to the west, outside the Roman walls. Radial streets linked these round points, and Wren's new St. Paul's was to be bounded by two of the main radials. No part of this excellent plan was carried out; but, as stated, it includes many of the major ideas of later planners.

The second map shows that the evils of the checkerboard were recognized in young American cities. Chicago was barely ninety years old when the Burnham plan called for the superposition of a network of radial streets—leading towards the centre of the city—upon the primitive checkerboard of the early designers. These radial streets are shown by heavy lines in Fig. 168, while the original design is in faint lines. A number of these radial streets have been cut through in Chicago, but most of the design has not yet been carried out.

A few pages are given to a discussion of the interesting town of Williamsburg in Virginia. It illustrates a simple town plan laid out with some regard for topography, and for vistas of fine buildings. The approaches to the centre of the city are by radial streets, so that this plan of 1699 shows that early planners were not wholly obsessed by the checkerboard. The plan of St. John's (Fig. 126) should also be consulted. The capital of South Australia—Adelaide—illustrates some interesting aspects of city planning. Here as early as 1837 Colonel Light surrounded his embryo city with a broad belt of parklands, as is sketched in Fig. 170.

Our next example is of a small port in the far west of Canada, where the Canadian National Railway reaches the Pacific at Prince Rupert. Here around 1910 the design for a small city, to be built on a rocky and hilly coast, was carefully thought out. The plan has been utilized (Fig. 171), but the city has not yet grown near the limits expected by the sanguine town planners. Somewhat similar principles are illustrated in the 'company town' of Grand Falls (Fig. 114). A rather complete study of the evolution of Canberra—the capital of Australia—follows. Here the association of round points and radial streets to form a series of 'cobweb units' has been carried out to the full.

Most of the preceding cities have been of the western industrial type, but the evolution of Dakar in semi-tropical West Africa shows that there is need for design in building a city which is to include folk of very different cultures (Fig. 174). Two 'Garden cities' are next described, Letchworth and Romerstadt, which illustrate the very newest ideas of the last few decades. Finally some brief discussion is given of the latest capital of all—that which it was proposed to build for the United Nations some thirty miles to the north-west of New York (Fig. 177).

City Planning and Topography

Folk in the middle of the 20th century are often somewhat unjust in their criticism of the founders of modern cities. It is true that the 'grid-iron plan' has the support of the real estate agent, since it ensures simple surveys and the maximum amount of lots to the acre; and it has ruined many a city. But it must be admitted that our forefathers had no idea that their villages were to grow into cities of fifty or a hundred thousand inhabitants. Many villages have never grown to occupy one-tenth of the area anticipated by their founders (e.g., Whitby, Fig. 20); and the simple grid is suitable for a cross-roads village—however unsuitable it may be for a New York or a Toronto.

Charles Zueblin in his book *American Municipal Progress* (New York, 1916) gives many examples of shortsighted planning. Thus New York lacks thoroughfares to get traffic uptown, because Randall planned the earliest streets to give short cuts to the waterfront. To-day the city must tunnel waterfronts to connect its scattered boroughs. San Francisco has utilized its waterfront admirably, but its streets clamber over the hills instead of circling them. New Orleans is another city 'utterly dependent on a river, but with a plan weird in its violation of the laws of nature'. The old French town radiated from a reasonable centre near the market and other important buildings. The new American town radiates *against* the traffic (Fig. 107), and the streets 'lead from nowhere toward the river, instead of from the river to somewhere'. Seattle has tried to accommodate the hills to its plan, by washing them down to make a level business centre! Los Angeles, built on a level plain, has dealt most unwisely with its additions and subdivisions. There are not half a dozen satisfactory thoroughfares in Los Angeles, because each subdivision was built without regard to its predecessor; every long street is full of jogs.

Williamsburg in Virginia, planned in 1699

One of the most interesting small towns in North America is Williamsburg in Virginia, about forty-eight miles to the south-east of Richmond. It has recently been restored so as to give the best picture of an early Colonial town in the United States. It was founded as early as 1632 on the peninsula between the York and James Rivers, and was at first called the 'Middle Plantation' from this position. As usual in those days

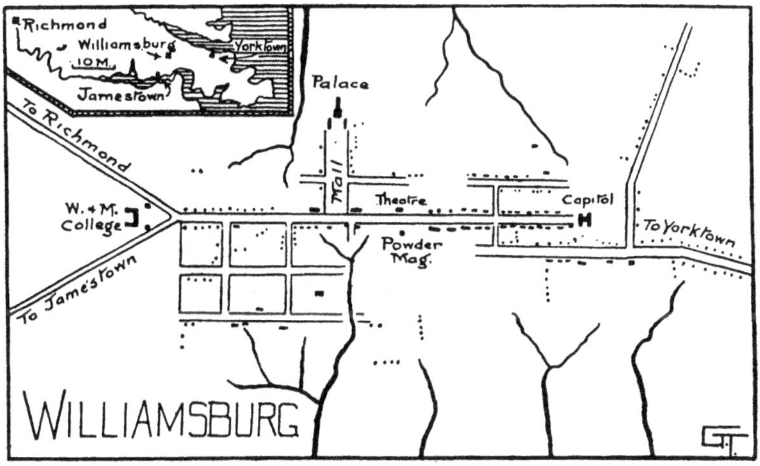

FIG. 169. Williamsburg in eastern Virginia was laid out in 1699, and in some ways anticipates the plan of Washington, D.C. (*After T. Adams.*)

a stockade was built around the settlement, as a protection from the Indians, but no sign of this remains. In 1698 it became the capital of Virginia, and next year the name was changed to Williamsburg in honour of William III. However, in 1779, it was decided to make Richmond the capital, and Williamsburg has not changed much since that date, for its population is only about 2,500.

The essential features of this early example of town planning can be gathered from the map on page 351. Some of the features remind one of the plan of Washington, which was laid out a century later. Thus there is a main street with imposing buildings at each end. The famous College of William and Mary, founded in 1693, has the position of the

Lincoln Memorial at Washington; while the capitol has much the same position in both plans. In Williamsburg the Governor's Palace (like the White House) has been built on one side, with a broad avenue leading from the main street called the Mall.

It is to be noticed that the design incorporated diagonal streets approaching the capital from the four corners. The roads in the west led to

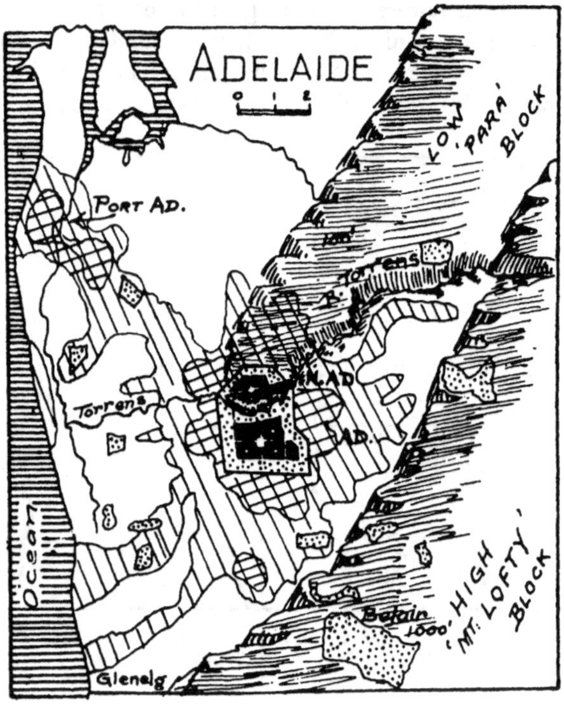

FIG. 170. The city of Adelaide with its surrounding Park-lands (dotted). Cross ruling shows dense houses, diagonal ruling scattered houses. (*Based on C. Fenner.*)

Richmond and the early settlement at Jamestown (founded 1607). The latter has vanished, while the former, as stated, has displaced Williamsburg as the capital. The eastern roads lead to the famous town of Yorktown, where Cornwallis surrendered in 1781. The planners paid attention to the topography, and it is obvious that the main street is laid out along a well-marked flat ridge, from which streams drain to north and to south.

Adelaide, South Australia: a capital planned in 1837

When the colony of South Australia was first settled, late in 1836, the Surveyor General, Colonel Light, was given the duty of picking a site for the capital and laying it out. The general character of the coastland was of course known by this date. A site on Kangaroo Island was favoured by some early settlers. Others supported the claims of the fine harbour at Port Lincoln (across Spencer Gulf from Adelaide). Governor Hindmarsh wished it to be placed at the mouth of the Murray, on the open coast of Encounter Bay. However, Light moved north along the western side of the Mount Lofty Horst, which separates the Gulf of St. Vincent from the lower Murray River, and was greatly taken with the broad plains near the mouth of the little river Torrens.

Light looked for the following features in the proposed site. First, shelter from most gales, which inclined him to the east shore of the Gulf of St. Vincent. Second, wide fertile spaces; which he found at the mouth of the Torrens. Third, a secure harbour, as in the estuary of the Torrens, now Port Adelaide. He also required a supply of fresh water, as in the Torrens; and finally an area free from danger by floods, which led him to place the city on the low Para block (Fig. 170), as this rose well above the lower plains of the Torrens. On 11th February, 1837, the survey of Adelaide was begun, at a slight rise of about forty feet known as Newmarket Hill.[1]

It must be remembered that Light was a trained surveyor, and it is interesting to see that he modified his main plan in many details to suit the topography of the site. The first important decision was to build the city in two parts, using the high level land (140 feet above sea level) on both banks of the Torrens. North Adelaide was built on the southern tip of the Para Block, and the streets were twisted a little from west-east so as to lie parallel to the river.

We do not know what gave Light the idea of a broad belt of Park Lands completely enclosing the two parts of his city. These open areas are shown by the dotted portion which surrounds the black patches in Fig. 170. It has been pointed out that the modern 'garden cities' of Letchworth and Welwyn use the same idea of a belt of agricultural land effectively shutting in the building areas. In each of the two parts of his city Light adopted the central large square. This of course was not unusual in America; for instance in the case of Charlottetown (Fig. 25) planned some sixty years earlier. So also in the case of the main southern part of Adelaide he arranged for four small squares, which are suggested

[1]Charles Fenner, *South Australia*, Melbourne, 1931.

in Fig. 170. In both north and south Adelaide the eastern sides of the plan were definitely modified to suit the slope of the ground. Colonel Light had travelled a good deal in Mediterranean regions, and it has been suggested that the plan of Catania in Sicily gave him some of his ideas.

To-day the city ot Adelaide has spread far beyond the belt of Park Lands, as may be learnt from the map (Fig. 170). The close-set population (shown by cross ruling) is nearly as dense as that within the Park Lands. The diagonal ruling shows how the suburbs have grown out towards the sea coasts at Glenelg in the south and at Port Adelaide to the north. The high scarp of the Mount Lofty Horst effectively controls the spread of settlement on the east, but houses have been built thickly in the broad valley between the Para Block and Mount Lofty. Many small parks have been alienated in the plains near Adelaide, and there is a large park near Belair, where the railway climbs the scarp by a wide loop to the south of the city. The population of Adelaide was about 313,000 in 1933.

Prince Rupert: a small planned port

In the summer of 1941 the writer made the second of his traverses in British Columbia. On this occasion he travelled by sea from Vancouver to Alaska, and then on the return voyage made a survey of Prince Rupert. It was of great interest to find that this little town of 5,000 had been laid out according to a definite town plan. This differed very advantageously from the checkerboard plan beloved of real estate folk in America, and so it may well be described in this chapter on town planning.

When the Grand Trunk line to the coast was decided on, a site on Kaien Island (Fig. 171) was found to be most advantageous for the terminal. It lay near the mouth of the Skeena River, the obvious corridor to the coast; it was well protected from the ocean by Digby Island; the water was deep close to the island and free from ice in the winter. Chief advantage of all, along this fiord coast, there was a considerable area of lowland on the west side of the mountainous island, which was suitable for the layout of a city. It was, of course, hoped that the new port would grow as Vancouver had, and it was pointed out that its distance along the Great Circle Route to Japan was about seven per cent shorter.

Clearing began in 1906, and a Boston firm designed a city plan suitable for the peculiar topography of the site (Fig. 171). The development of a pioneer city according to plan is rather rare in North America;

and although Prince Rupert has by no means grown to the assumed population of 50,000, the little town of 6,000 or 7,000 has benefited from its adherence to the original plan.

Except for Home Defence Forces. which around 1945 filled Prince Rupert to excess, there has been little change in the population for many years. The town shows no sign of ever competing with Vancouver as a world port.[1] The halibut and salmon fisheries are the chief local activities.

A rocky ridge runs parallel to the coast in four 'levels'. The first, or shore, platform, which carries the railway, has been levelled and broad-

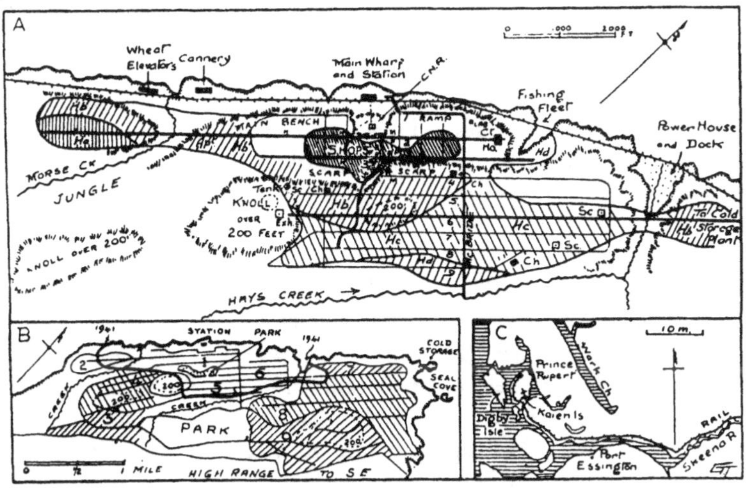

FIG. 171. Prince Rupert.
A is a functional map; *B* shows the original plan; *C* is a locality plan.

ened, especially at the mouth of Hays Creek, where a power-house and small shipbuilding dock have been erected. A fish cannery is operating south of the railway station, and some large wheat elevators have a few busy weeks in the autumn. A small rocky 'step' accommodates the offices of the Canadian National Railway. The main road to the town rises by a ramp around this step. The second main 'step' is about 100 feet above sea level and carries the two main streets (2 and 3), with the shops and hotels (H), the local museum (Mus.) and the post office (P.O.). A steep, rocky scarp rises about sixty feet higher to the main backbone of the ridge, along which streets 4, 5, and 6 have been laid out. In the

[1]The economics of Prince Rupert have been described by J. Q. Adams in "Prince Rupert, British Columbia," *Econ. Geog.*, Vol. 14, 1938, pp. 167-183. See also W. S. Tower, "Western Canada and the Pacific," *Geog. Rev.*, Vol. 4, 1917, pp. 284-296.

south-west this ridge rises to a rounded knoll about 250 feet above the sea, on which stand a large water tank and the wooden Exhibition Building. Beyond, the ground falls away to the broad valley drained by Hays Creek. The valley is now wooded with small second-growth timber, in which a few clearings show an acre or two under cultivation. To the east is the high central ridge of Kaien Island.

The original plan is given in B. In each of the nine sub-divisions the street plans were in accord with the topography. A large park was arranged for directly behind the centre of the city, but it cannot now be distinguished in the wooded valley. The limit of the built-up area to-day is shown by the heavy line. The private houses are classified in four groups: Ha, the largest houses, with considerable gardens around them, chiefly at the extreme south-west; Hb, pleasant homes in small gardens, occupying the 'backbone' and in the northern suburb; Hc, smaller houses of artisans; Hd, poorer houses and shacks near Hays Creek and the fishing port below the Court House. At Seal Cove, two miles to the north, is a large cold-storage (fish) plant.

Canberra—the capital of Australia—built to a plan

One of the most interesting of towns is Canberra, the capital of the Commonwealth of Australia. Most of the world's chief capitals have been fairly large towns before they became of great political importance. For instance, there was a considerable population at Ottawa before it was chosen in 1858. Washington on the other hand seems to have been of little or no importance before 1790, when Hamilton, it is said in return for political favours by Jefferson, agreed to the latter's desire to place the capital on the Potomac. However Canberra belongs to this century's history, and every detail of its development is ascertainable. Moreover the writer had a fairly close personal association with the capital in its earliest years, which may add to the interest of this brief record.

In 1900 it was decreed that the capital should be built in the state of New South Wales. But partly due to the jealousy of the other states it was ruled that the city should not be less than 100 miles from Sydney (the state capital). Since the western part of New South Wales is so hot and dry that it is not suited for agriculture, it is clear that the new city would develop in the rather elevated and wetter areas in the east of the state. Various commissions visited suggested sites, and the first inspection placed Bombala ahead, with Yass and Orange second (Fig. 172). In 1903 another commission favoured Albury, with Tumut, Orange, and Armidale following. Then Parliament voted, and chose Tumut in 1903, and

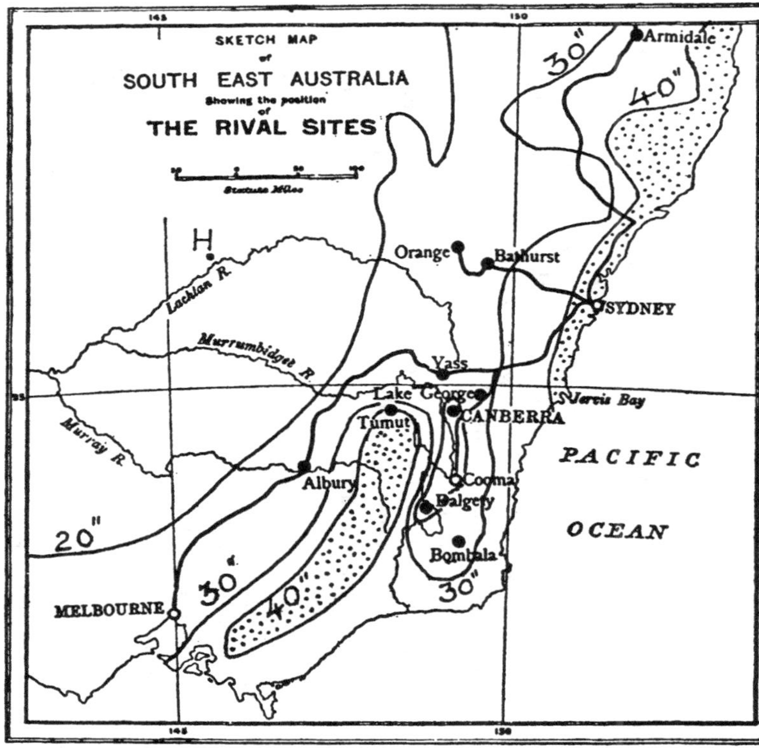

FIG. 172. Map of Eastern New South Wales to show the rival sites for the capital 1902–1908. The main isohyets are shown, and regions with over 40 inches are dotted. Note the 'dry loop' from Dalgety to Canberra. 'H' is Hillston.

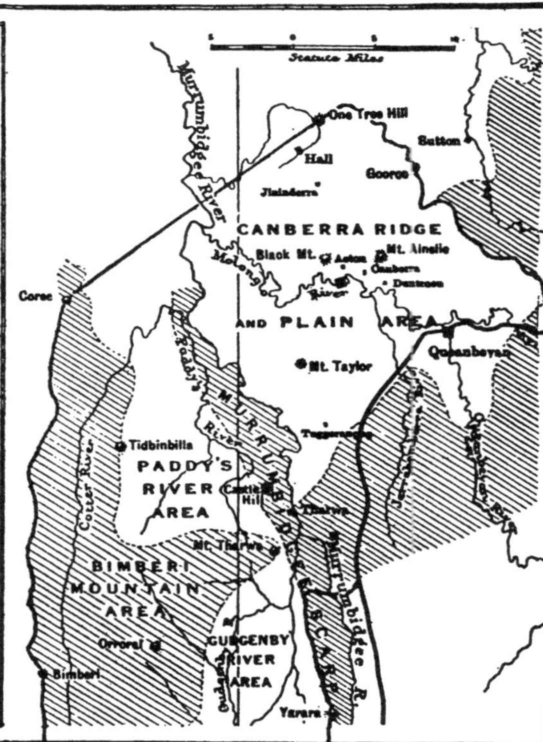

FIG. 172a. Map of most of the Federal Capital territory in the south-east of New South Wales. The ruled areas are rather high and rugged. The small black square shows the centre of the city near Acton. (*See Geog. Jnl.* (*London*), *April–May*, 1914.)

Dalgety in 1904. In 1908 at the final discussion, Yass-Canberra was elected, through the union of the parties supporting Yass, Lake George, and Canberra, which are all in the same region. The territory around Canberra, comprising 900 square miles, was vested in the Commonwealth on 1st January, 1911. (See Frontispiece.)

Probably any of the ten places shown on the map would have been fairly suitable. Canberra has, however, some advantages. It lies close to the route linking the two chief cities of Sydney and Melbourne. It is 1,900 feet above the sea, which gives a cooler climate than that of most of the sites. It is close to a long plateau (Bimberi) which rises over 6,000 feet, and here the Cotter River acts as a valuable collecting ground for the water supply of the city.

Canberra is on the line joining the centre of population of Australia (at Hillston, H on Fig. 171) to the east coast. It is not too far from the Federal port at Jervis Bay, and yet it cannot be shelled from the sea. On the other hand the rainfall is low (twenty-three inches), since the site is at the north end of the Cooma rain-shadow strip (Fig. 171). This, however, meant that the land was used primarily for grazing, and was not in private hands. It is in the faulted area of south-east Australia; but earth-tremors, though not uncommon in the district, have never been alarming.

The two first pastoralists settled in the district about 1823; and one of these used the name Canberry for the district which surrounded his ranch, which he called Acton. The other built the house named Duntroon on the adjacent ranch to the east. The same squatter also erected the church at Canberra in 1826. For a time Acton was in the hands of the Brasseys —a well-known naval family in England. Further west was Yarralumla, now used by the Governor General. The whole Federal Territory was occupied by a score of stations (ranches) mostly raising sheep, but in the very rugged south-west section beyond Tharwa, cattle are grazed.

By 1909 there were a few tiny settlements serving these stations, such as Canberra, Tharwa, Hall, and Uriarra, some of which are shown on the map (Fig. 172). The total population was about 1,500. The writer had described the topography of the upper Molonglo River as early as 1907 (in his paper on Lake George), but did not visit the actual site until July, 1910. At this time the Federal Survey was housed in tents and in two fabric huts just north of the Red Hill. To-day the Prime Minister's Lodge occupies almost the same spot. (P.M.L. in Fig. 173.) A preliminary paper on the topography of the Territory appeared in December, 1910 (*Meteorol. Bulletin* 6); and I gave a full discussion of the topography and history in the *Geographical Journal* (London) of 1914.

There were only two large houses on the city site in those early days. One was Acton used by the Surveyor General, and the other was the Rectory about a mile to the north-east. A little house near Ainslie was the official post office. Duntroon to the east soon became the nucleus of the Military College; but it is still well outside the city, though a flourishing suburb of the latter. The main road led from Queanbeyan to Yass and ran just to the east of Ainslie. Here another road came up from Tharwa in the south, while less important roads led to Acton and Yarralumla. All these are shown by broken lines on Fig. 173.

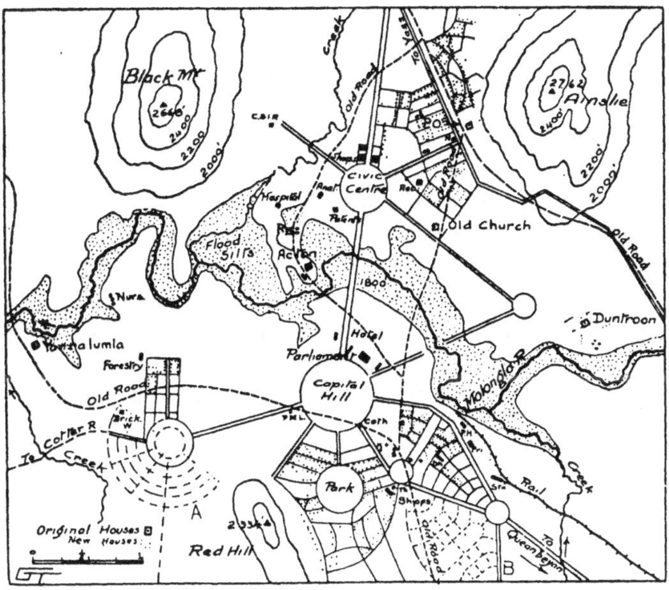

FIG. 173. Map of the new city of Canberra, the capital of the Australian Commonwealth. Note the Molonglo River running west to the Murrumbidgee River. The half dozen original houses, and the early roads (broken lines) are indicated. The 'cobwebs' in 'broken' lines are part of the design, but have not yet been built. (*From the Official Map*, 1933.) (See also the Frontispiece.)

After my return from Antarctica I was engaged on the geological survey of the Territory. Our chief object was to find brick-clays and building stones, which unfortunately are not altogether satisfactory in the Territory itself. By 1912 there were several wooden cottages near Acton for the surveyors and clerks. A post office, drawing office, and bank had also been erected in the vicinity, and a large house (R. in Fig. 173) was built for the Administrator. Soon a short railway was run in from Queanbeyan eight miles to the east; while the Power House (P.H.) was perhaps the most ambitious building in those early days.

Meanwhile plans were invited for the future capital, and those of a Chicago architect (W. B. Griffin) were chosen. The actual street-plan follows his design, which gave much prominence to the local environment. The city has developed in a broad flat plain between three conical hills, which rise respectively 400, 700, and 800 feet above the Molonglo plain (at about 1,900 feet). The Molonglo River winds in its flood-silts across this plain, running westward to join the Murrumbidgee River (Fig. 172). The main vista is the view northwards between Black Mountain and Mount Ainslie. Here Commonwealth Avenue was made the axis of the chief corridor; and it leads northwards from a low hillock, called Capital Hill, to the Civic Centre across the Molonglo River.

The cobweb pattern was adopted, with two main cobwebs at Capital Hill and the Civic Centre. But other cobwebs were also laid out, most of which are still only in the embryo stage. However, in my sketch map I indicate a number of these lesser cobwebs; i.e., one north-west of Red Hill and three between the latter and the railway station. Another lies near Duntroon, while the largest cobweb of all is quite outside the sketch map to the north. This last is not yet bordered by any houses. It will be noticed that there is not much similarity between the simple early road-pattern (shown by heavy broken lines) and the rather complicated cobwebs of to-day.

The actual arrangement of modern houses is indicated in Fig. 173 though every small street is not included. It will be seen that there are three or four scattered quarters separated by wide spaces where the streets have not been developed yet. South of Capital Hill are the largest houses, with the Prime Minister's Lodge (P.M.L.) in the north. Another quarter served by a cluster of shops and a theatre is to the east. Still further to the east are the Railway Station, Power House and Printing Office with adjacent houses. Parliament House and the chief Government buildings are about a mile north of these residential quarters.

The Civic Centre is three miles to the north across the river, and the other main sections of the town lie between it and Mount Ainslie. The latter sections occupy the once empty plain between the old Rectory and the post office, where the writer investigated the claims of certain water-diviners way back in 1920.[1]

In the sketch map (Fig. 173) the streets in use are indicated, while some

[1] "Water Divining," Griffith Taylor (*Royal. Soc., Victoria*, May, 1921). A map of the empty site may be found of interest.

of them not yet laid out (near Red Hill and the railway station) are suggested by broken lines. Presumably similar streets will in the future fill all the empty portions of the map, as well as the flat plain north of the Civic Centre. A dozen small parks are scattered among the street blocks, but they are omitted in my sketch plan. (See Frontispiece.) The splendid plantations in the west of the city are an attractive feature, and a small hamlet has developed near the Forestry Building to care for them.

Clearly we have here a departure from the plans of most cities of to-day. The experiment of splitting up a small town of some 7,000 people into civic and political units is very interesting. The rather wide distance between the southern and northern clusters seems at present to waste a good deal of the citizens' time. But in Canberra the designer is commendably looking forward to a city of three or four times the present size. The siting of the individual cobwebs has been wisely influenced by the relation of the river to the striking conical hills. Hence, as usual, in city evolution, we must agree that nature and man both play important parts.

Dakar, a planned capital in Tropical Africa

Few tropical towns have been planned almost completely during the last forty or fifty years, but Dakar in latitude 14° 40′ N., on the extreme western point of Africa, is one such town. We are fortunate in having a full description by a competent geographer of this most interesting combination of European and African activities from the pen of Derwent Whittlesey.[1] It also illustrates the transfer of commercial interests from the safe island-harbour of Goree to the nearby rocky cape on the mainland, as soon as political conditions made the latter settlement available for Europeans.

In the early days of the Portuguese navigators they pushed south along the coast of the Sahara until they reached the relatively green rocky cape which they called Cape Verde (v. in inset, Fig. 174). Here, in the time of the summer rains, the baobab trees and scanty crops of the natives give a green tinge to the arid landscape, which was delightful to the eyes of the explorers. The cape is part of a once rocky island now tied to the mainland by a sandy tombolo some twenty-five miles long. This is shown in the inset map in Fig. 174. The southern point of the rocky peninsula is called Cape Manuel; and it is in the lee of this headland of basalt crags that the port of Dakar has been developed.

The map given in Fig. 174 shows the topography in the vicinity of

[1] "Dakar," *Geographical Review*, October, 1941, 20 maps and photos.

Dakar. Cape Manuel rises to forty metres (130 feet), and just to the north a ridge about the same height runs west-east for about a mile, on which the chief portion of the town has been built. To the north (the right in the map) the ground drops to a flat lowland built of calcareous sand, which rarely rises more than twenty feet above the sea. Most of the natives live in this portion of the town. The headland protected ships in early days from the north and west if they anchored to the north of Dakar Point, where the original tiny breakwater (B.W.) was erected. But in those early days France had no control over the mainland natives,

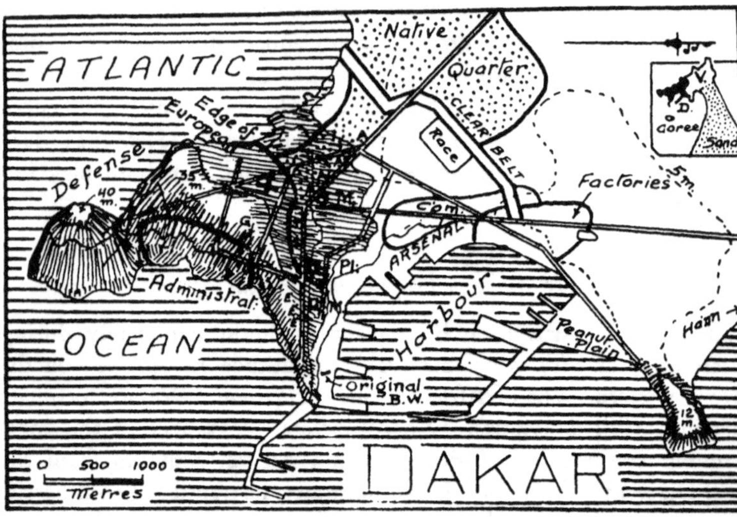

FIG. 174. Dakar, the capital of French West Africa, built on a former island (130 feet high) which is linked by a sand-bar to the mainland 12 miles to the east (see D. in *inset* map). European residences are on the low hill, negro huts are shown dotted. The Plaza (Pl.), Markets (M.), European (E.), Syrian (S.), and African (A.) shops are indicated. (*Based on Derwent Whittlesey*).

and the French port of call was the little island of Goree about two miles to the east (see inset).

About 1857 the French built a fort on Dakar Point, and a few French merchants settled nearby. But for the rest of the century Goree was still the sole settlement of note, with a population of 3,000. Peanuts were grown in the dunes to the north of Dakar, and quite a port developed here for a time at Rufisque. However, about 1866, a pier 330 metres long was constructed near Dakar Point, which, however, gave little protection from north-east winds.

Around 1860 a plan for a town was drawn up in which due notice was taken of the healthy north and south sea breezes. The main street ran west-east between the two markets (M. in the map), and a plaza (Pl.) was laid out about half way. The town was built on the 'grid' plan to the north of this street at about fifteen metres above sea level. In 1885 a railway was constructed for about 250 miles to the north linking Dakar with St. Louis, the early capital of this part of Africa.

About the beginning of this century the superior claims of Dakar as a port were recognized, and the capital was transferred from St. Louis. The French now developed their plans for a great West African Empire using Dakar as their base. By 1912 a jetty 2,000 metres long was built thus enclosing a spacious harbour on the north. Much of the shore was filled in, and the new land was devoted to docks and a large Arsenal.

Water supply was a great problem, but this difficulty was solved in time by utilizing the rain stored in the dunes to the north of the town. Marshes were drained, and eucalyptus trees were planted to lower the groundwater level and so prevent mosquitoes from breeding. As the town spread the new suburbs developed along radial roads, and the 'grid' plan was wisely abandoned. For a time the thatched huts of the natives alternated with the residences of the French, but this condition is not so apparent to-day. More and more water was needed for the town, and deep supplies were found at Hann (Fig. 174) (and even in dunes thirteen miles away) which were collected and brought to reservoirs in Dakar.

One aspect of urban life in Dakar of great interest is the segregation of the negroes. This was necessary because many of the natives came in from the bush and had no knowledge of town life or of sanitation. A special district was laid out in the north-west quarter, called Medina; and here land was given to the negroes to induce them to leave Dakar proper. To-day there are 20,000 negroes in the town, and the majority are to be found in the Medina district. A 'clear belt' 900 metres wide has been left between Dakar and the negro quarter.

To-day 6,500 French, i.e., half the Europeans in French West Africa, live in the southern part of Dakar. Here are fine veranda-bordered bungalows often shaded by bougainvillea creepers, where the French resident lives for the three or four years he spends in the colony. The numerous Administration offices and Government House (G. in map) are on the rocky south-east shore of the ridge. The southern cape and much of the west side of the promontory is reserved for defence purposes.

The European shops are found in the eastern part of the town near the older market (M.); while the western market is used by the negro citizens exclusively. Other shops catering to the negroes extend along the road towards Medina, and are labelled A, A, A, on the map. Just to the west of the Plaza (Pl.) are many shops owned by Syrians, who sell textiles and clothes to the negro population. There is very little manufacturing now at Dakar. The old handicrafts have largely been abandoned since the natives prefer European articles. However some refrigerating plants, a brewery, etc., have been erected near the coal wharf at the north end of Dakar.

The chief business of the port is fuelling ships in transit. Many boats bring coal to the stock piles and take away cargoes of peanuts from the new wharves on Peanut Plain (Fig. 174). Thus coal, oil, and rice are the chief imports, while peanuts, gum arabic, and sisal fibre are the exports. The French have spent great sums in fortifying Dakar, and there was an important submarine base here before the last war. Dakar is of great significance in commerce since here the Atlantic is only about 1,650 miles wide. The first transatlantic air service linked Dakar to Natal in Brazil; and during the war this route was one of the most vital corridors for supplying our African armies. Several railways besides that to St. Louis have been built through French Sudan from the port of Dakar. It has clearly many advantages as a port of call, and as an entrepôt to Africa, which must lead to considerable growth in the future.

Garden Cities: Letchworth

Early reformers such as Owen, Henry George and Wakefield had criticized the lack of planning in the founding of cities, but it was Ebenezer Howard in his book *To-morrow* (London, 1898) who led the way to the creation of towns which would combine the advantages of city and rural life. He realized that the increasing population of these days was being drawn to the larger cities from the country environment. In the cities there was no provision for these huge increases, and without a plan, there was every inducement for folk to be packed closer and closer in worse and worse surroundings. Howard was not much enamoured of the 'Company Town', mainly because no alternative to the one form of industry was available in such communities. Also organized labour has always resented the control that such a company—however benignant—must exercise over the home life of the worker.

Howard's new type of settlement—to quote Mumford—would combine the hygienic advantages of the open suburbs with the social advantages of the big city, in short would offer a balanced environment. To this new urban nucleus Howard gave the name of 'Garden City'.

In such settlements the land is not parcelled out into individual ownership; it must be held by the common authority under which it is developed. Such increments as may arise through the growth of the garden city must be reserved for the community. In this way Howard eliminated the temptation to increase density in order to raise land values.

Howard also recognized that there was an easily reached optimum of population for such a garden city. He proposed to surround every city by a permanent reserve of open country, either for agriculture, or (as had been shown in the case of Adelaide) for recreation. His idea of a maximum was 30,000 inhabitants, which is not far from that generally accepted to-day, which is about 40,000 citizens. Howard also insisted on a balanced environment, so that home, industry, and market, were all to be considered in the total plan.

Letchworth was the first of such Garden Cities to be built in England. A private company was formed in 1903 known as the 'First Garden City Limited' with a capital of £300,000 in five pound shares. They purchased an area of 3,800 acres about thirty-four miles north of London (Fig. 175 inset), and this estate was divided into a central urban area of 1,200 and a surrounding agricultural belt of 2,600 acres. Parker and Unwin laid out the city, and the main features can be ascertained from Fig. 175. It will be noticed that there is no hint of the checkerboard in this plan. A straight avenue leads from the station to the town square, and is prolonged to the south-west. But for the most part there are irregular segments of curved streets, with ample room for recreation in the form of large sports grounds and commons.

It is important to notice that a number of minor industries have moved to this site, such as printing and engineering establishments. These are found mainly in the east of the city, near the railway line as might be expected. Now that power may be obtained readily almost anywhere near the railway, there are many advantages in the decentralization of industry. Labour costs are often lower in the country, and it is certainly much more attractive for the workers themselves. Letchworth allows for houses of various classes, such as small houses for the lower-paid worker, and larger houses for professional residents. In this respect it differs from some of the other 'Garden Cities' such as the Hampstead Heath Development in the north of London, where there is mainly provision for better middle-class homes, and no factories are permitted. Each house at Letchworth has a garden, and a considerable area is allotted to each home. This had happened often enough in *early* English towns, but as time elapsed the former gardens were filled with smaller buildings, so that ultimately slums have developed. This will never be possible in a planned town like Letchworth, where no one is allowed to crowd the land. In 1933 there were 15,000 people living in the Garden City.

The Greater London Plan, 1944

Brief mention must be made of a far-reaching plan proposed, mainly by Professor Patrick Abercrombie, for the 2,600 square miles surrounding the County of London.[1] It is suggested that 1,720 factories should be

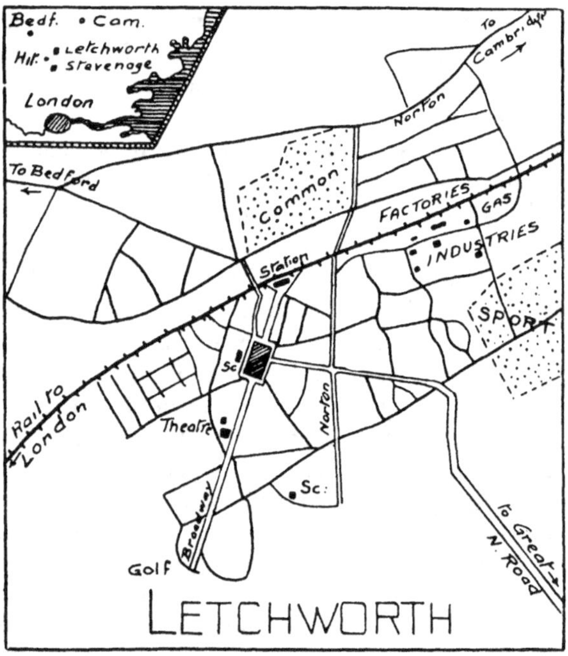

FIG. 175. Letchworth Garden City, 34 miles north of London (near Hitchin). The map is two miles wide. (*Based on T. Adams.*)

decentralized, together with quarter of a million employees and three times as many dependents. The new industries would be located mainly in the 'Outer County Ring', which would contain eight 'satellite towns' with a population limit of 60,000 apiece. Two present-day villages,

[1] A very good summary of such views, entitled "Changing City Patterns," by M. D. Meyerson and R. B. Mitchell, may be consulted in the *Annals of the American Society of Political and Social Science* (Philadelphia), for November 1945. A number of sketch maps illustrate this paper.

which are to receive vast increments of citizens according to this plan, are Stevenage about thirty miles north of London (Fig. 175) and Crawley in Sussex. It is interesting to note that here, as in a number of other town-planning suggestions, a population around 50,000 is felt to be about the optimum in one unit.

Romerstadt—a novel garden suburb near Frankfurt

Very different from the usual modern suburb (with its adherence to the Gothic type of architecture, and to the belief that every house

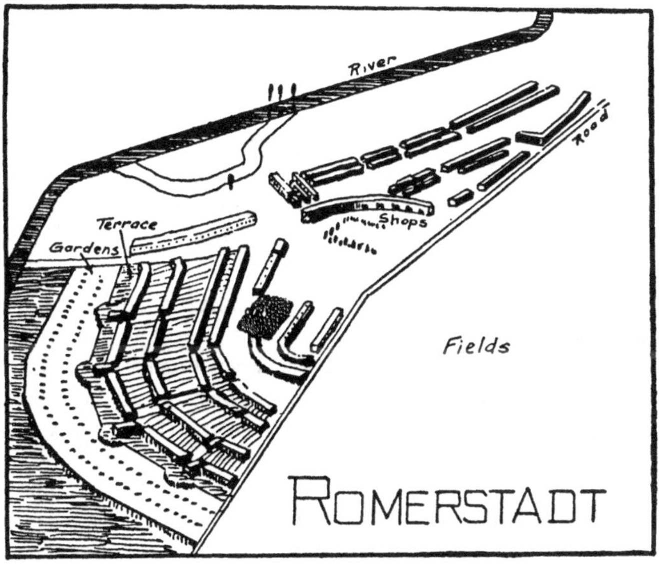

FIG. 176. A sketch of the modern plan developed at Romerstadt, near Frankfurt. (*Based on an air-photo in Mumford's volume.*)

should have its own garden) is the lately-built suburb of Romerstadt, near Frankfurt, in Germany. Here, in 1926, the erection of a great number of homes was commenced, which, in Mumford's opinion, incorporate many of the best ideas of the modern town planner. The following description is based on his comments.

The general plan may be gathered from the sketch given in Fig. 176, which is based upon an air photo of Romerstadt. Here we see parallel rows of houses in long blocks unbroken by traffic streets: narrow roads bordered with verdure for local service; large open spaces with ample lawns, flower beds and dining terraces behind the houses. The curving unit in the middle of the air view is a four-storey block containing shops, and it can be readily reached from any part of the settlement. In the case of the elongated blocks of apartments, some are several storeys higher than others, especially near the radiating alleys, thus lending much variety to the town-plan.

In the new suburb consideration was given to the direction of the sun's rays and also to the advantage of sweeping views across the landscape to the River Nidda. The gardens form a belt on the lower side of the main terrace behind the houses. On the other side there is a broad belt of land, reserved from building, which will serve as a 'green belt' on this side also.

The proposed United Nations Capital, 25 miles N.E. of New York

It is of considerable interest to study the early plans for the United Nations Capital in the vicinity of New York, even though they were not carried through. Moreover a very complete study has recently appeared,[1] from which the following data have been borrowed. In October, 1945, the Executive Committee voted (9 to 3) that the capital should be in the United States. Certain conditions were laid down as to the four areas to be inspected. The chosen regions were (a) within sixty miles of Boston; (b) between twenty-five and eighty miles of New York in the Hudson valley; (c) east of the Hudson River in New York or Connecticut; (d) in the Princeton (New Jersey) district.

The following special conditions were to be met in the site chosen. Accessibility to a large metropolis; agreeable climatic conditions, especially in summer; attractive natural features; possibility of constructing a city; education; accessibility to mountains and sea-coast. Fourteen sites were examined, five being near Boston, two near Providence, seven north of New York, and one at Princeton. The final choice was North Greenwich—North Stamford, on the borders of New York State and Connecticut and about twenty-five miles to the north-east of New York City. The area to be held by the city is forty-two square miles, and may be extended later to 172 square miles. A majority of the residents in the chosen region were averse to the loss of ownership implied in this transfer.

The environment of the proposed capital is shown in Fig. 177. The circle to the north-east of Manhattan (the centre of the city of New York) is on the borders of Westchester County (W.) in New York and

[1] C. B. Hitchcock, "Westchester-Fairfield," *Geog. Review*, July 1946.

Fairfield County (F.) in Connecticut. The general elevation is about 300 feet above sea level, rising to 600 feet to the north-west. In the vicinity is the great Croton Reservoir which supplies New York, and the smaller Kensico Reservoir (K.), and one of these presumably will be used for the new city.

There is very little settlement in these hills, though a fine automobile road known as the 'Parkway' runs through the area. Three railways,

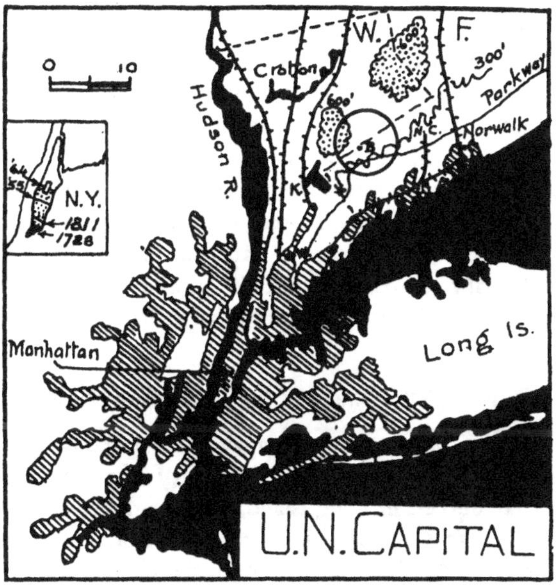

FIG. 177. Capital was the circle about 25 miles north-east of New York. Diagonal rulings show present residential areas around Manhattan (N.Y.). *B.* is Banksville; *K.* is Kensico Reservoir; *N.C.* is New Canaan; *W.* is White Plains. *Inset* map shows New York in 1728, 1811, 1855 and 1864. (*After C. B. Hitchcock.*) The arrow shows 47th Street.

as shown in the sketch map, run along the east side of Hudson River, while the coastal railway through Norwalk sends a branch to New Canaan, which may be the nearest station. A village called Banksville (B.) seems to be quite close to the proposed centre of the U.N. Territory, but the total population within the circle (which does not accurately represent the area) is only about 1,000. The diagonal ruling in the map shows the limits of residential areas around New York as given by Hitchcock in the paper quoted.

Originally much of Westchester County—over 86,000 acres—was owned by Stephanus Van Cortlandt, and for many decades the farmers were merely tenants of the landowner. At the beginning of the 19th century Poundridge (near New Canaan) had about the same population as Yonkers (south of White Plains, W. in the map). To-day Poundridge has a population of 800, while Yonkers is part of the huge New York Conurbation, and has a population of over 142,000. All through this region—town and country alike—there are about eighteen per cent of the folk who are foreign-born. These are chiefly Italians, but there are large minorities of Germans, Poles, and Irish.

It is worth noting that the Parkway is bordered by lands which are publicly owned. Cross-roads burrow under the Parkway, and do not hinder the rapid flow of automobiles, for which such a Parkway is intended. There are three airports in the vicinity, near White Plains, Bridgeport and Danbury respectively; but no doubt a special airport will be built near the capital. To-day there are 573,000 folk living in Westchester County and 418,000 in Fairfield. At Bedford Hills (near Croton) the January temperature is 18.6° F. and the July temperature 72.2° F. The total rainfall is about forty-five inches.

Further Development of the U.N. Capital Site

While the Site Committee was engaged in selecting the final site, the personnel of U.N.O. was allotted temporary offices in some industrial buildings at Lake Success in Long Island, about eighteen miles to the east of New York. Here was assembled a secretariat of some 2,700 persons (*Harpers*, April, 1947). In October, 1946, the choice of a Westchester site seemed certain, but new offers of sites were entertained during the closing months of 1946, from Philadelphia, San Francisco and elsewhere. On December 7th it was disclosed at Lake Success that the world capital of the United Nations may be established as a huge international skyscraper centre in New York City (*New York Times*, Dec. 8th, 1946). This shift of opinion was due to an offer from John D. Rockefeller, Jr., of a block of land adjacent to East River in the heart of the City of New York. On the evening of December 11th, 1946, the committee voted 33 to 7 in favour of the New York City site, at the east end of 47th Street.

This shift from an outlying special city-site, removed from the influence of any other city, to the centre of the world's outstanding megalopolis seems to the urban geographer very much of a change for the worse. It must add to the intolerable congestion of New York, and it seems to be contrary to all the modern ideas in favour of decentralizing cities rather than increasing their complexity, which the author has tried to demonstrate in the preceding pages. The matter is debatable, however; and the early steps leading to the (apparently abandoned) Westchester plan are well worth a place in this study, as illustrating the many factors which enter into the field of city planning.

CHAPTER XIX

REGIONAL DISTRIBUTION AND REGIONAL SURVEYS

Introduction

ONE interesting section of Urban Geography is concerned with the relation of cities and towns to each other. What determines the distance between adjacent towns? What part does the city play in the administration of a large region or smaller district? How large is the district which can be satisfactorily controlled from a large town? What 'tentacles' in the way of aqueducts, power-lines, and the like does a city send out into the surrounding countryside? What area should be set apart for large recreation reserves to serve a metropolis? An attempt will be made to give actual examples, which will help the reader to answer these questions. Inevitably the sections in this present chapter must be somewhat disconnected, since so diverse problems are indicated in the preceding lines.

The first section attempts to show the effect of rainfall on the distribution of large towns in the United States. Next a discussion of various papers by English geographers, such as Gilbert, Fawcett and Dickinson, will give us some idea of the purposes of sub-dividing the national area into more readily administered regions. Similar problems have been tackled by the Germans since 1933. Some notes on the greatest 'Conurbation' in Europe, i.e., that of the Ruhr, follow on this section.

Turning to America, we have in the Tennessee Valley the most elaborate attempt at regional planning. This is described in the following pages. New York State at an earlier date attempted to carry out a rational plan for the development of the State, and it is illustrated herewith with several maps. One of these shows how the great Highways leading from the Metropolis of New York should be planned. The chapter ends with a description of the small but interesting area—the Ganaraska Basin—near Toronto, where a beginning has been made with regional planning in Canada. A few notes of a tentative plan for southern Ontario are added, and two maps help to explain the suggestions of the author.

Climatic Control of Town Distribution

If we assume that the population of the United States is moderately stabilized, i.e., that there are no large areas of unused good land, and that the density of population is adjusted to the natural controls, then it is possible to gain a rough idea of the controls which govern the density of towns. In Fig. 178 I have made a tentative graph to bring out such controls. by the use of the following technique.

If we ignore all towns having less than 17,000 citizens, we find that some states have many larger towns, while others have only a few. For instance Nevada, with an area of 109,000 square miles, has only one town with over 17,000 inhabitants. This is in large part due to the climate of Nevada, which does not permit of large scale agriculture. The

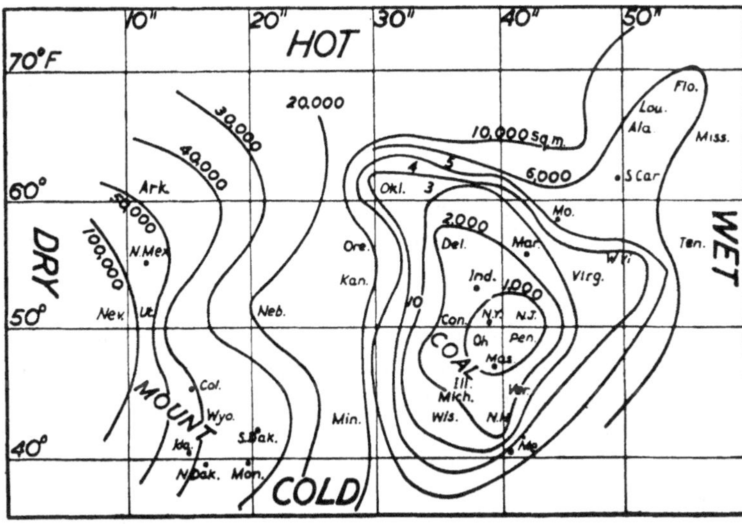

FIG 178. A tentative isopract graph showing how the distribution of towns (exceeding 17,000 folk) varies in the United States with climatic controls. Figures represent square miles allotted to each town.

average temperature of Nevada is about 50° F., while the average rainfall is around ten inches. In the isopract graph shown in Fig. 178 we place Nevada at the dry side of the graph in the position indicated above, and write against it the figure 100,000 square miles—which is about the area 'occupied' (in the broadest possible sense) by this one town. The state of New York has forty-four such towns in an area of 47,000 square miles; and according to our arbitrary relation, we may state that each town is concerned with an average of about 1,000 square miles. Accordingly we place New York in its proper position—where the 50 degree and 40 inch lines cross, and then label this isopleth with the figure 1,000 square miles. Carrying out this procedure with the other states we obtain the isopract graph which forms Fig. 178; and from this we may draw some interesting conclusions as to the characteristics of the settlement in U.S.A.

It is obvious that there is a strong tendency for the closely clustered cities to keep to an environment expressed by a temperature around 50° F., and a rainfall between thirty-three and forty-three inches a year. But other factors must not be ignored altogether, though we cannot give them consideration in this simple isopract diagram. For instance the presence of coal is of great importance, and this is suggested by writing the word COAL against the appropriate states in the diagram. So also the rugged topography of some states prevents the clustering of cities, and this is suggested by inserting the word MOUNT in the diagram.

From a study of the arrangement of the isopleths it seems obvious that the rainfall control is more important than the temperature control. Indeed if we make a traverse from the New England states across to arid Nevada, we can say that the 'area for each town' increases by about 3,000 square miles for each diminution of an inch of rainfall. However there would seem to be an 'optimum' of rainfall near forty inches, for there is also an increase in the distance between the towns as we move south from New England into the hot wet states of the south-east. (The state of California is so diverse in its environment that it is omitted.)

If we ignore the temperature factor in the above relationship, then we can express the effect of rainfall in terms of the distance apart of the towns. We can assume that the towns of a state are equally spaced, each in its own 'square of influence'. Then the square root of the area will give us a measure of the distance apart of the towns. This is suggested in the following table for a traverse from Nevada to New England.

RAINFALL AND TOWNS

Rainfall in inches ..	10	15	20	25	30	35	40
Distance apart (miles)	320	187	141	124	100	45	14

Regionalism

In this chapter certain interrelations between the city and its surroundings will be considered in a somewhat broader sense than the discussion of *umlands* given on page 216. The civilized world is demanding many of the benefits of city life, and at the same time wishes to retain some of the special advantages of the older rural way of life. It is logical that the administration of a region shall be centred in the 'hub' of that region, to wit the main city. For these and other reasons a somewhat new chapter of geographic research is developing, which is sometimes called Regionalism.

Many of the main problems of Regional Planning were considered at a symposium of the Royal Geographical Society in London in January 1942. A brief discussion of the resulting paper (*Geog. Jnl.*, Feb., 1942) will serve to introduce this topic.

E. G. R. Taylor put forward the map which appears as Fig. 179 to show the relationship of various parts of England to the largest cities. The shaded areas in this map are not connected with any obvious great centre, yet when it comes to planning, it was pointed out that they must have their centre of leadership. Parts of the north-west (Cumbria), of Wales, of the south-west, and of East Anglia must therefore be somewhat arbitrarily subdivided; and the heavy lines in the map show how Professor Taylor proposed to do this. The heavy *broken* lines show less important subdivisions.

Dudley Stamp suggested that it was important to get a balance between town and country, so that if Birmingham became the regional capital, the voice of rural Herefordshire shall also be heard and considered.

E. W. Gilbert has written at great length on this topic, especially in his lengthy study (with twenty-five maps) in the *Geographical Journal* for July, 1939. He is strongly in favour of a set of divisions which break away as little as may be from the accepted *county* boundaries.

In 1941 a Civil Defence grouping of the various sections of England was authorized, which was of profound significance during the last years of the war. Gilbert tried to devise a sounder grouping of the counties than that used in the Civil Defence map, and produced a map which differed from that shown in Fig. 179 mainly in the eastern part of England. His map appears on page 76 of the *Geographical Journal* for February, 1942; and in almost all cases he has managed to make the greater part of the limits of the region agree with the ancient county boundaries. The number of divisions is about the same as that given in Fig. 179; i.e., fourteen for England and three for Wales.

Gilbert's earlier paper (*Geog. Jnl.*, July, 1939) must certainly be consulted by those readers who wish to go further in this problem of regionalism. Here he points out that such subdivisions have been arranged for either statistical, administrative or private reasons. His first two maps reproduce those originally prepared by C. B. Fawcett, in 1917 and 1919. This latter writer chose a dozen regional capitals, and then allotted territory to each, much as suggested later in Fig. 179. A later map given by Gilbert shows the divisions proposed for housing purposes in 1920 —again a dozen in number. The divisions used by the census and by the Meteorological Service follow. The religious divisions, i.e., the forty odd dioceses, depart in an interesting fashion from the forty odd counties at times. These are charted in his Fig. 21. Later maps to show divisions used in earlier wars—e.g., those of Cromwell and Napoleon—are reproduced.

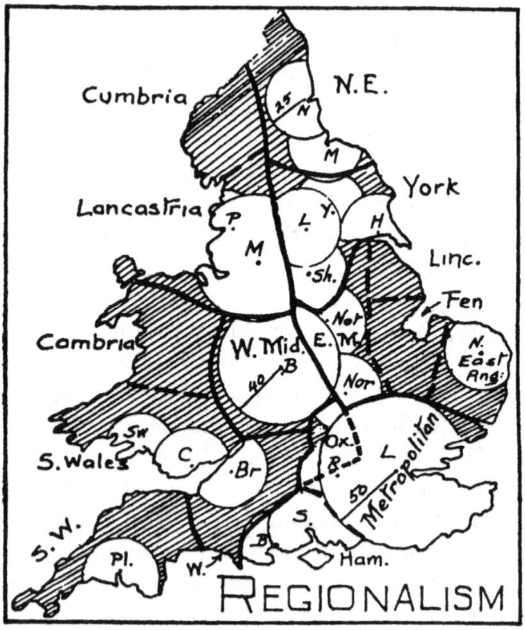

FIG. 179. England as subdivided by E. G. R. Taylor to illustrate problems of Regionalism. The shaded areas are not much influenced by the vicinity of large cities.

Perhaps the most interesting section of his paper has to do with the popularity of the regional capitals in the many maps charted. Gilbert shows that in the nine most typical regional maps, London is always made a regional capital; so also are Birmingham, Manchester, Leeds, Bristol and Cardiff. Newcastle, Cambridge and Southampton are the next most often chosen as regional capitals.

R. E. Dickinson summarizes recent work by German geographers in the same field in a valuable paper 'Economic Regions of Germany' in the *Geographic Review* for October, 1938. Ten maps are given in this paper of which three are especially interesting. The first shows the 'trade areas of the ports'; and the second the Economic Regions, which are conceived as centred on economic capitals. Some of these capitals are Konigsberg, Stettin, Berlin, Breslau, Leipzig, Hamburg, Hanover, Nurnberg, Munich, Stuttgart, Frankfurt, and Essen. The regions surround these capitals, and in general agree pretty well with the political areas surrounding the respective cities.

The third map shows the new Regional Planning Districts established by the Nazi Party in 1936. Reichsminister Kerrl stated that it was their endeavour . . . 'to direct all changes in the German State, whether effected by settlement, commercial developments or the erection of industrial establishments or through other demands on the land, on a basis of planning and foresight in the common interests of the people and the State'. The Reich was divided into twenty-three Planning Regions which generally coincide with the provinces of Prussia and the *Länder*. In each region the chief planning authority is the supreme representative of the Reich, and the actual work of planning is in the hands of the Regional Planner. A monthly journal was published, entitled *Raumforschung und Raumordnung*. Here the Germans clearly recognized the virtue of keeping to old-established boundaries, and the various Regional Capitals are as follows: Kiel, Oldenburg, Schwerin, Dresden, Magdeburg, Weimar, Kassel, Darmstadt, Saarbruck, Munster, etc.

Dickinson points out that the major economic units find expression in the new planning regions. It was a tenet of the Nazi philosophy to preserve the major cultural entities in the Reich, and this was kept in mind in deciding the limits of the various Planning Regions (*Plannungsraume*).

Conurbations and 'Megalopolis' in the Ruhr

The growth of the vast factory cities is one of the main characteristics of the Industrial Revolution. But even more striking is the development of the phenomenon which Geddes long ago called the '*conurbation*'. It is well to dwell on such examples of human agglomeration, for they emphatically prove the control exercised by the environment on human affairs. In a sense London could be called a conurbation since the City of London joined the City of Westminster in one solid block of built-in streets about the end of the 17th century. Since that date, as already noted, London has swallowed a dozen independent towns now all included in the 'Greater London' of over eight million inhabitants. But a conurbation more strictly means the welding together of several more or less equally important towns, and for this the vicinity of Manchester or of Leeds give us better English examples.

However, the best sample of a conurbation in Europe has developed in the vicinity of Essen in the Ruhr region in north-west Germany. The key to this unequalled collection of factories is given in the geological section inset in Fig. 180. Here we see that a thick bed of very valuable black coal lies on top of the Oldermass and is capped by the Youngermass. The coal layers dip to the north, so that the coal is right at the surface near Essen, but needs deeper and deeper shafts in the pits as we proceed northwards. The River Ruhr flows along the southern outcropping edge of the coal, and here a string of coal towns developed in the 19th century, such as Essen, Oberhausen, Duisberg, Dortmund, etc.

About a dozen miles to the south is the rather deep narrow valley of the Wuppeı, and valuable iron deposits were found in the vicinity. Here a series of steel works developed, based on the nearby coal and iron ore. To-day of course the main bulk of the iron ore comes from foreign sources, such as Lorraine, Sweden and Spain. But the factories are still tied to the coal; and some forty towns have developed in this 'black country' of the Ruhr.

The growth of many of these towns can be learnt from the following table which shows how they have increased during the later years of the Industrial Age.

TOWNS IN THE RUHR COALFIELD AND VICINITY

Populations in thousands

Town	1939	1910	1871	Ratio 1939/1871
Essen	659	295	52	12.5
Dusseldorf	539	359	69	7.7
Dortmund	537	214	44	12
Duisburg Hamborn	431	330	33	13
Wuppertal	398	339	145	3
Gelsenkirchen	313	170	8	40
Bochum	303	137	21	14
Oberhausen	191	90	13	14
Krefeld	169	129	57	3
Hagen	151	89	20	7.5
Solingen	138	49[1]		
Remscheld	103	72	22	4.5

[1]In 1905.

Thus the towns of the Ruhr region have increased about ten times in their population since 1871. Since the nuclei of these towns were only about five miles apart at the start of the Industrial Age, it is obvious that they must have now begun to crowd each other in many areas. From Duisburg to the north-east as far as Herne, along the famous Rhine-Ems Canal, there is now practically one 'Conurbation'. So also along the southern belt of factories, based originally on the local iron ore, there is almost another 'conurbation' extending from Dusseldorf

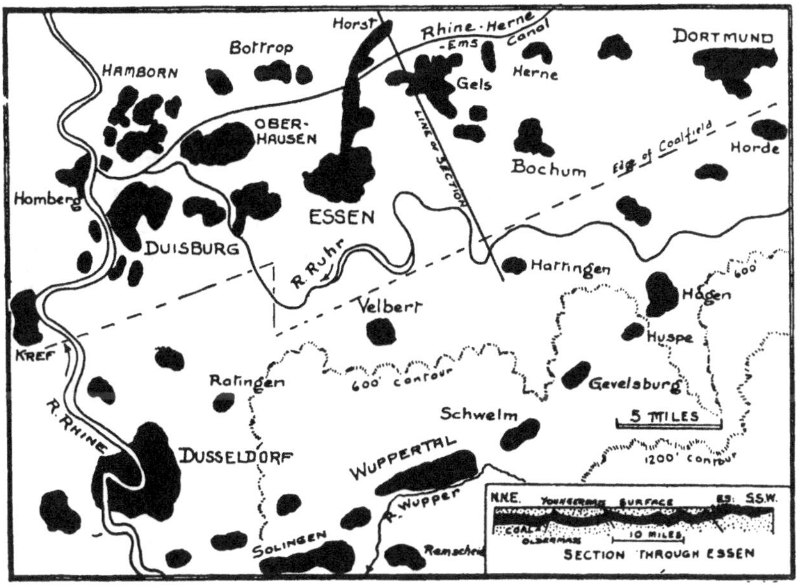

FIG. 180. The Conurbation of the Ruhr Valley in western Germany, centred on Essen. A second incipient conurbation is developing along the Wupper Valley and linking to Dusseldorf.

to Hagen up the rather narrow Wupper valley in part (Fig. 180). At the eastern end of these two conurbations is the huge centre of Dortmund and this city will soon link with Horde and Hagen. Hence we may see ultimately produced, if the R.A.F. have not destroyed it too completely, a somewhat ring-shaped conurbation linking these two present groups already specified.

There is little if anything to be said for these conurbations as a feature in our civilization. They consist of square miles of smoky factories, usually destroying the adjacent countryside through the sooty pall which settles thereon. The workers live in monotonous blocks of mean streets, and too often have neither the time, energy, or capital which would enable them to improve their environment. It is quite unsatisfactory to depend on the benevolence of the factory owner for parks, museums, art galleries, and other amenities. In a better social system the profits of the industrial age would produce these amenities as a first charge. However it is not for a geographer to dwell on these purely sociological aspects of the city problem. (See also map of Essen on p. 319.)

Mumford's conclusions as to the seeds of decay inherent in a giant city should be read by everyone interested in our evolving civilization. He points out in his *Culture of Cities* that the city rose as a special kind of environment favourable to nurture and education, because it was a *protected* environment. In the small city, co-operative actions prevailed over the more predatory modes of life. But the giant city tends to produce many of the evils of 'crowd associations'. Most forms of crime are more prevalent. Many activities are only pursued *vicariously*, such as the games and boxing matches. In time of war the 'megalopolis' is particularly vulnerable from the air. Its huge populations are readily brought to the verge of starvation when outside supplies are merely interrupted. High land values, and high costs of transport, all increase with the congestion. Recreation spaces, at the rate of one acre for 300 persons, can be provided in reasonably-sized cities. But New York has only one acre of space for each 1,234 persons. However, Mumford sees the end of this unhealthy growth in the 'cutting off' of the supply of new citizens from the countryside or elsewhere. After a city grows beyond forty or fifty thousand it fails to reproduce the full quota of its citizens. The decline of the birthrate will cut down all immigration; and this is perhaps the chief reason why in the next stage of our civilization we shall see much smaller cities than to-day.

Regional Planning by the T.V.A.

As an example of planning on a very large scale few projects are of more interest than that carried out in recent years by the Tennessee Valley Authority. The area involved is four-fifths of that of England and comprises 40,600 square miles; and the outlay involved in planning was estimated to be 185 million dollars by the year 1940. Almost 5,000 professional employees and 11,000 workmen were involved in this giant attempt to raise land values in the basin of the Tennessee River. During the work more drastic changes in the environment of the inhabitants were carried out than perhaps have been seen in any comparable period of time elsewhere. Hence this change in the environment of cities, towns, and villages, and especially in the farm life of the poor whites and negroes, well merits our brief attention.

The best short account known to the writer, which especially stresses the structural features, is that given by George B. Barbour in the *Geographical Journal* for May, 1937. The general topography can be gathered from the block diagram given in Fig. 181. The Tennessee River flows through two regions of quite different structure. At its headwaters the Powell and Clinch tributaries are cuesta streams, flowing along the eastern edge of the Cumberland Plateau. This is composed

FIG. 181. A block diagram of the Tennessee River Basin, showing the huge dams built or controlled by the Tennessee Valley Authority (T.V.A.). (*Partly after G. B. Barbour.*)

of Carboniferous strata, which are fairly level-bedded. To the west of the Cumberland Plateau the dominant feature is the 'Nashville Dome', formed of Lower Carboniferous rocks round the rim, with the underlying Ordovician beds exposed in the eroded centre of the Dome.

It is clear from the diagram that the Tennessee River flows round this Dome much as the River Trent (in England) flows round the elevated Pennines at their southern end. The big river does not receive many tributaries in its lower section, except from the southern portion of the Nashville Dome. As Barbour points out, there is a definite narrowing of the basin near Chattanooga, where the river cuts through the extension of the Cumberland Plateau. Here the Hales Bar Dam was built many years ago. Further down the river at Florence the fall is the steepest of all, and here (near Muscle Shoals) the Wilson Dam was built about the time of the First Great War. In all, the river falls about 500 feet between Knoxville and Paducah, where it enters the Ohio River not far from the Mississippi. The positions of Norris Dam (on the Clinch),

Coulter Shoals near Knoxville, Chickamauga Dam near Chattanooga, Guntersville Dam at its most southern point, and the Gilbertsville Dam at its outlet, are all indicated in the block diagram (Fig. 181).

It will be noted that the north-eastern tributaries rise in or near the Blue Ridge, and then cut right across the highest part of the Appalachians. Just north of Ashville is Black Dome, which is 6,700 feet above sea level. At Guntersville Dam the river enters the lowland area, and most of the land is below 600 feet west of this point. It is anticipated that hydro-electric power will be available from all the dams tc the extent of 600,000 kilowatts, or one-third more than the output of the American plant at Niagara Falls.

Navigation and flood-control are also to be greatly benefited by the numerous dams on the river. But the most interesting features concern the rehabilitation of the farms in the whole river basin. Two-thirds of the area is owned by farmers, who constitute one-half of the total population of two-and-a-half million in the T.V.A. region. The agricultural experts are introducing crop rotation, preventing soil erosion, re-afforesting, and instructing the farmers in the use of fertilizers, especially of phosphates. The latter are found locally, and are being made into fertilizer on a very large scale at the Muscle Shoals plant. In conclusion it may be noted that professional geographers were found to be of special value in this project. No less than fifteen geographers (with Ph.D. rank) were employed in the geographical section at Knoxville; and the writer may be pardoned for mentioning that half of these were formerly students at his classes in Chicago.

Regional Planning for New York State

Much as New York City is the pioneer in preparing a complete plan for the future development of a *metropolis*, so also New York State has led the way in the planning of a much larger region, i.e., the *whole state* itself. The Regional Planning Commission was set up in 1923, and Lewis Mumford has based one of the most instructive sections in his book *The Culture of Cities* (New York, 1938) on the work of this Commission. The diagrams given in Fig. 182 are taken from Mumford's book.

We may consider that the settlement pattern in the state passed through two major phases. The first was during the period from 1840 to 1880. The pattern depended on the use of the forests, and water-mills for sawing wood and grinding grain and running machinery. Many small factories were in operation, and were situated mainly in farming villages and towns. The second period was one of industrial and financial concentration; it probably reached its apex between 1910 and 1920; although the inordinate growth of Buffalo, and still more New York, continued by inertia, even after the main supply of immigrants from Europe had been cut off.

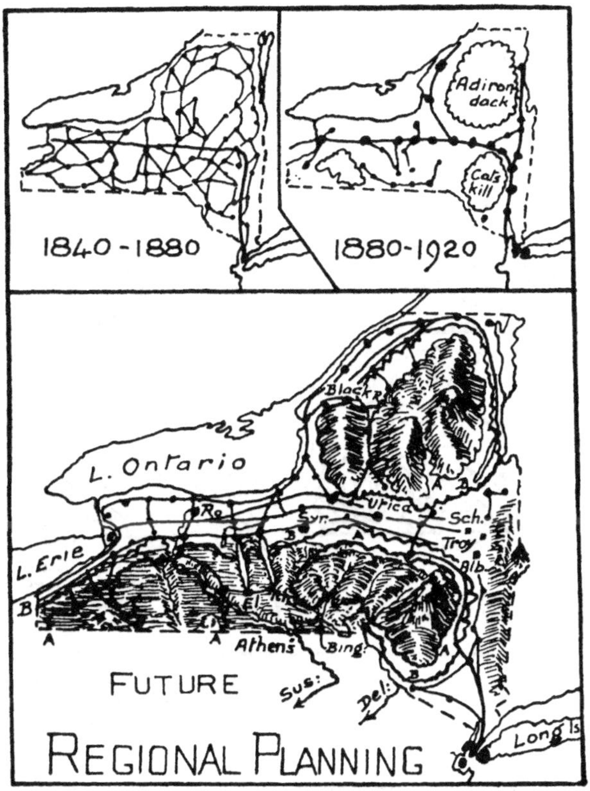

FIG. 182. Mumford's analysis of future Regional Planning for New York State, based on two earlier epochs shown in two smaller maps.

The present and future phase is suggested by Mumford's third diagram (i.e., the larger map in Fig. 182). Now new technical facilities and revised social aims will it is hoped improve upon the undiscriminating diffusion of the early period and the congestion of the second phase. The research on the soils, the spread of electric power, motor highways and the reservation of great park areas are all tools giving us an opportunity for better regional planning. Instead of encouraging the further building up of metropolitan areas, Mumford is convinced that a rational policy would lead to new settlement in Green-Belt Towns.

New York is more fortunate than many states in having a large portion of the land area of the state unsuited for farming,[1] but well suited to serve under public ownership the essential purposes of watershed protection, timber production, and recreation. If all this land were level and fertile it would be very expensive or perhaps quite impractical to secure an adequate amount of common land to conserve water, soil, timber and

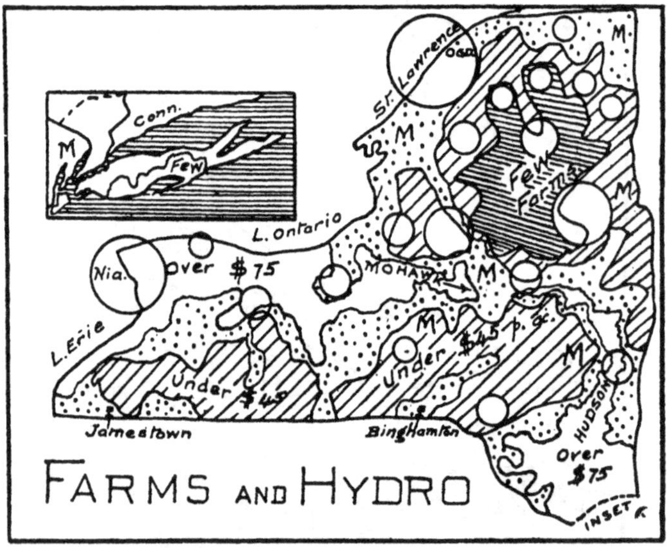

FIG. 183. Regional Planning, in part based on the value of Farm-lands and potential Hydro-electric Power (Circles). 'M' indicates the chief Milk areas for New York City. (*From maps by the New York State Planning Board.*)

human resources. The Compendium states further that nine million acres of land should be included in the enlarged public domain of the state. About one-third of this is state-owned in forest preserve. They suggest that all the Adirondacks become a Park and most of the Catskills, while large areas of idle or submarginal farm lands—mostly along the south side of the Mohawk-Erie route (Fig. 183)—should also be acquired and administered as similar public domain.

A later map (page 28) shows, as one might expect, that the more rugged the land the less productive is the soil. The best soils are all

[1] *Graphic Compendium of Planning Studies, Div. of State Planning*, Albany, 1939. Contains over 100 excellent maps.

along the Mohawk-Erie corridor, as well as along the Black River valley. The Hudson valley and the actual shore of the lakes seem to be second-class lands. Since the land has long been settled, we find that the value of the farm lands (Fig. 183) agrees moderately well with the preceding map. However, the most valuable farms are in the lower Hudson (over $100 per acre) and along the shores of the two big lakes. The Mohawk Corridor contains farms worth from $55 to $100 per acre, and this is also the case with the farms along the slopes in the Buffalo-Syracuse belt.

Water conservation is very important in this rugged state, so close to the largest city in the world. Existing reservoirs in 1934 totalled seventy-three billion cubic feet, and plans are ready for 379 billion more (Fig. 183). The best areas are in the Adirondacks, especially on the Black River, but there are good sites in the Catskills, and on the Genessee river above Rochester.

New York State ranks second in the nation in water power resources. The estimated total is three-and-a-half million horse power, of which one-and-a-half millions are already in use. Half-a-million units are developed at Niagara Falls, while about the same amount can be obtained from the St. Lawrence when the power houses are built in the new International scheme (Fig. 186). The upper Hudson (near Schenectady) and the Black River area already produce more than half of the supply possible in those areas. In 1932 New York City contained 2,220,000 consumers of electricity, while outside the huge city the number of consumers was only 1.4 million. The principal generating stations are interconnected by a system of high tension transmission lines. These extend all along the Hudson-Mohawk-Erie Corridor to Buffalo and Jamestown. Branches reach Binghamton in the south, and Ogdensburg in the extreme north of the state. About twenty-eight per cent of the farms in the state are using electric power.

New York draws on all the farm lands of the state for its milk supply (M. in Fig. 183), especially in the Mohawk Gate around Utica, and along the St. Lawrence River in the north of the state. The map showing Highway Traffic (page 107) shows how heavy is the traffic along the Hudson-Erie Corridor. But there is very heavy traffic up the Delaware valley, and it is surprisingly heavy in the hilly country between Syracuse, Binghamton and Elmira. The roads around the Adirondack Dome are much less utilized by highway traffic. There are seven metropolitan districts in the state; New York, Albany, Utica, Syracuse, Rochester, Buffalo, and Binghamton. In 1930 these seven cities contained eighty-one per cent of the population of the state.

Bearing in mind these major features of the landscape Mumford and other planners would divide the state into three main divisions (Fig. 182). The Adirondacks, the Catskills and the roughest country in the south-west would be mainly forest, park, game, and water-collecting areas, i.e., all included within the isopleth A. A. A. The marginal areas between isopleths A. A. A. and B. B. B. consist primarily of the less profitable farmlands and would be developed accordingly. Indeed much may be added to the public domain. The broad belts from New York to Schenectady (Fig. 182) and from Schenectady to Lake Erie should be traversed by several parallel highways, linked by numerous cross-roads, with the cities at the intersections. Indeed regional planning consists essentially in classifying the lands as a whole, and considering the developmental proposals also for the region as a whole. Such planning will in time replace the 'hit-or-miss' methods of settlement, which, however, are almost inevitable during the early stages of a democratic nation.

Regional Highways near New York

Linking the city to the surrounding region are the highways, which have often grown up in a haphazard way by linking irregular portions of short roads between neighbouring villages. In the vicinity of a metropolis it is necessary to create a plan which has a measure of regularity, and provides roads which will serve as main arteries for the countryside, as well as for the local *village to village* traffic. In the unequalled planning carried out for New York, and published in over a dozen volumes in 1929, there are several chapters dealing with the regional highways. The final pattern is sketched in Fig. 184.

The New York region roughly resembles three-quarters of a circle, the south-east quadrant being occupied by the Atlantic Ocean. The main key of the proposed road system is the Metropolitan Loop or belt-line highway, which circles the most intensively developed areas in the region. It is labelled 'A' in Fig. 184, and forms an oval, about thirty miles across with Manhattan Island in the centre.

Instead of a 'Main Street' the centre of traffic is an area already intersected with a fairly complete system of rectangular highway routes, connected at the edges of that area with many arterial highways.

Outside the Metropolitan Loop are the Radial Routes which are numbered from 1 to 20. They lead directly to the arterial roads far from the metropolis, and also serve to connect the outer circumferential routes 'B' and 'C', which are more or less concentric with the Metropolitan Loop. It is to be noted that these routes, both radial and circumferential, are carried to the east so as to include Long Island.

According to the Plan the best bypass route, whereby vehicles can pass—say—from Philadelphia to Boston without entering the congested city highways, makes use of parts of circumferential routes 'B' and 'C'.

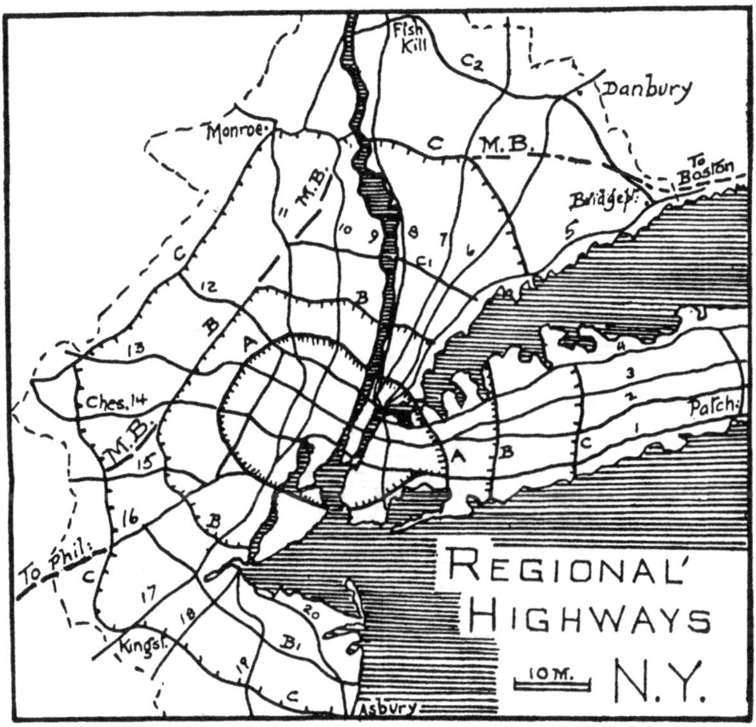

FIG. 184. The regional highways leading out of New York, according to the 1929 Graphic Regional Plan. At 'A' is the proposed 'Metropolitan Loop'. 'M.B.' is the proposed Metropolitan By-Pass around New York.

The chief connecting links are shown by heavy broken lines labelled 'M.B.' in Fig. 184. In many of these routes existing roads may be utilized in part, but many sections and some bridges will need to be built, other sections must be straightened or made wider, before the 1929 plan is actually put into practice.

The Ganaraska Regional Survey

We have in the Dominion no such elaborate an organization as T.V.A. but several smaller surveys made with the same purpose show that Canada realizes the value of intelligent plans for the development of the countryside. A few notes on the Ganaraska Plan will show what should be done in much of the older-settled portions of the Dominion.

The Ganaraska River is quite a small one, consisting of a cluster of streams which enter Lake Ontario at Port Hope, about fifty miles east

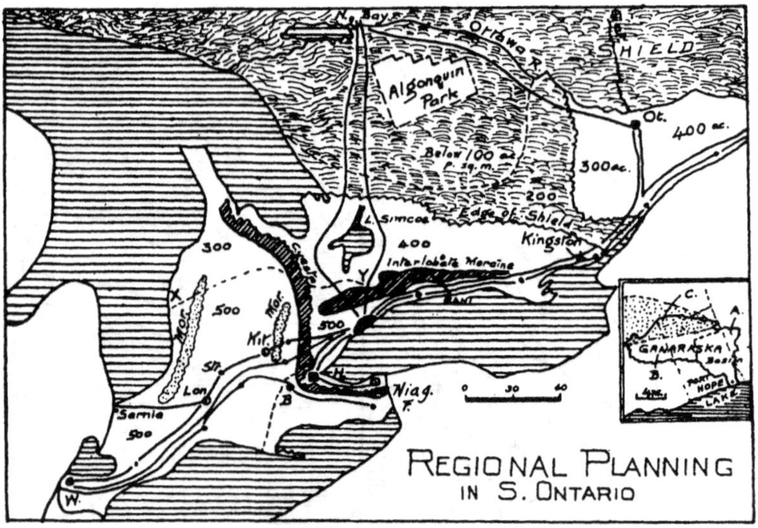

FIG. 185. Some environmental factors which determine Regional Planning in southern Ontario. 'XY' separates lands where 500 acres (out of 640) are farmed, from lands where 300 (out of 640) are farmed. *Inset* is a map of the Ganaraska Basin.

of Toronto (GAN. in Fig. 185). The river basin has an area of 103 square miles, and is charted in the inset in Fig. 185. The headwaters rise in the very striking Interlobate Moraine, which is an irregular set of ridges of gravel running west-east across Southern Ontario. This portion of the basin is shown dotted in the small inset map. No rock outcrops in the basin, most of the surface consisting of fluvio-glacial sandy loams.

The region was surveyed in 1791, and for a time the first settlement was known as 'Toronto', but the name Port Hope was adopted in 1819. The town grew slowly till 1836, when there were 1,000 folk in the settlement. Then progress was rapid, since there was a fine market for the lumber of the basin across the lake in U.S.A. The peak population was 5,585 in 1881, but by this time the best timber had gone, and tariffs in U.S.A. prevented the sale of Canadian timber there. To-day the population is about 5,000 in Port Hope, and there are half a dozen villages scattered through the basin. Some half a dozen other villages have been abandoned since the decline of the lumber trade, and much of the soil has lost its fertility through the crude farming of early days.

Three railways traverse the region; the C.P.R. passes along the coast to Toronto; the C.N.R. runs to the north-west along the Ganaraska and serves Canton and Osaca; while another branch (to Peterborough) serves Quays, Campbellcroft and Rosemount within the basin. The rivers supplied many small mills for grist and grain, and in 1860 there were sixty of these mills. To-day there are only ten of them in operation, for the water-supply has greatly diminished owing to the changes in the forest and run-off. In 1933 the pitchblende ores from the Great Bear Lake were brought to Port Hope to the refinery which extracts the radium content of the ore.

In Southern Ontario it is estimated that there are 8,500 square miles of land which are marginal or submarginal for agriculture. In 1941 the regional survey was initiated in the typical area of Ganaraska, and the report 'The Ganaraska Watershed' was published in Toronto in 1944. It covers many phases of rehabilitation, such as agriculture, springs, silting of Port Hope Harbour, soil surveys, floods in Port Hope, and especially forest preservation and extension.

The soil surveys are the best guide to future development. The best soils are found in the east of the basin (A. in inset map). Second-class soils cover the south-west (B.), while the interlobate moraine (C.) is too poor for extended agriculture. The Survey suggests that almost all the latter area be devoted to forest, amounting to 20,000 acres; though perhaps a quarter of this may grow some crops. They advise the building of three large dams to prevent floods in Port Hope. There have been forty extensive floods since the town started, and that of 1929 resulted in damage amounting to 250,000 dollars, a large amount for a small town. Several areas in the north of the basin are to be set apart as recreation reserves. Three per cent of the area suffers from extensive soil erosion, and the report includes many other recommendations of value to all interested in the scientific development of Southern Ontario.

Regional Survey of Southern Ontario

On the basis of earlier surveys in U.S.A. and at Ganaraska, Ontario, it is now possible to suggest some of the main environmental factors which must be considered in any plan for the rational development of Southern Ontario. Topographic Surveys have been completed of most of the area shown in Figs. 185 and 186. Soil Surveys have been carried out in most of the region south of the Shield, and a brief summary of these (based on work by Putnam and Chapman) can be consulted in the author's recent book, *Canada* (London, 1947). The structure of the region is also quite well understood in this early-settled portion of the Dominion.

The chief topographic features are indicated in Fig. 185. In the north is the Pre-cambrian Shield, which sends a tongue across the St. Lawrence just to the east of Kingston. The younger rocks overlie the Shield, and all dip gently to the west or south-west. Most of the rocks between Ottawa and the Hamilton Cuesta are of Ordovician age. From Kingston to Lake Simcoe is mostly Trenton Limestone, while the area from Ganaraska to Hamilton is mostly shale, and therefore less rich in lime. The high Hamilton Cuesta is formed of hard Silurian beds dipping to the west, while younger Devonian rocks fill the triangle west of Stratford. The main deposits of Glacial Moraine (Pleistocene) are also indicated on the map.

In the Regional Survey of the State of New York three maps seemed of the first importance. These were the farm-value maps, hydro-electric distribution, and water-supply. In the latter case the sole large cities of Toronto and Hamilton are supplied from one of the greatest bodies of freshwater in the world, i.e., Lake Ontario. Hydro-electricity is supplied mainly from Niagara, or from the Gatineau region near Ottawa. The value of the farms is indicated by the number of acres of improved land in every square mile (640 acres). These figures appear on Fig. 185, where '500' means that 500 *out of each* 640 *acres* in a given county is farmland. This best portion of the region is south of the broken line XY. Near Lake Simcoe, and again far to the east near Ottawa, about 400 acres out of 640 are farmlands. On the Shield the proportion drops to 200 in the Frontenac Axis, while the northern portion around Algonquin Park has only a small portion of each square mile developed into farms.

Turning now to the map given as Fig. 186 the distribution of the towns in Southern Ontario is charted. At the 1931 census there were four large towns in the region with over 100,000 inhabitants; four more with more than 25,000; and about a further score with more than 10,000. These are plotted in Fig. 186, and one quite striking feature is emphasized there. The region which contains these large towns is extremely elongated, and serves to accentuate the importance of the water-borne traffic of olden days. It is shown by the close diagonal ruling in Fig. 186, and extends from Cornwall in the north-east to Sarnia in the west. The two isolated patches at Ottawa and Niagara Falls are also obviously determined by water-transport in part.

The most interesting exception to the general rule mentioned above is found in the region just to the west of Hamilton. Here three of the four third-class towns (shown by rings) are not particularly close to the great waterways of the St. Lawrence or Ottawa. They are, however, situated on the banks of the two main rivers of this part of Ontario, that is the

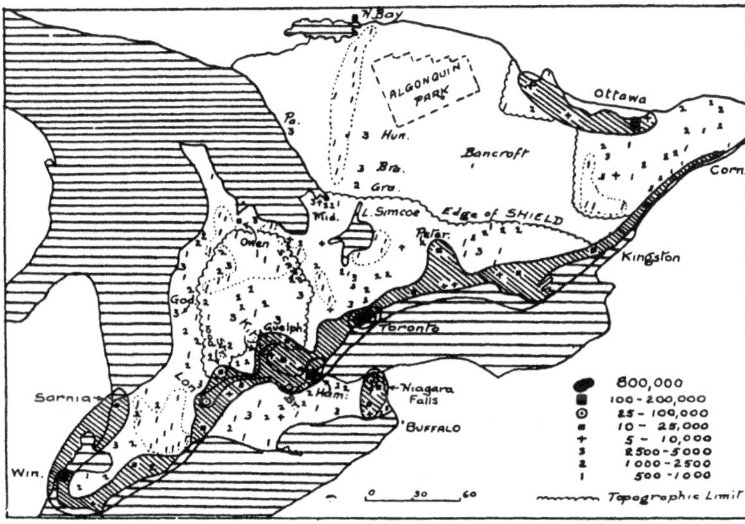

FIG. 186. The distribution of the towns in southern Ontario, classified according to population in 1931. Note the belt of large towns (over 10,000) extending from Cornwall to Sarnia. Goderich, Brantford and Stratford are indicated. (*Partly based on Schott.*)

Grand and the Thames; and this in part explains their origin. In the early days of settlement a good site for a grist or saw mill often decided where the settlement should develop. Hence arose Brantford, Galt, Preston, Paris and Fergus. However, Kitchener is the largest town of this 'constellation', and it is not on the streams mentioned. But it was early served by a railway, and this was the second factor in determining growth. Thirdly all these little towns benefited because they were relatively close to the giant supply of hydro-electric power at Niagara Falls.

It is interesting to compare York (later 'Toronto'), Port Credit, and Whitby, which started about the same time in the same environments.

However the choice of Toronto as the capital led to its rapid growth; so that to-day it is the centre of a built-in area, where live just about one million people. The other two towns have hardly grown in the last half century.

Along the 'strip of cities' from Cornwall to Sarnia the big towns are strung out at rather regular intervals. Thirteen cities are dotted along this belt, and the average distance between them is forty-one miles. We may consider for a short time the meaning of the empty areas in Fig. 186. They are of course readily explained by the structure, since they all lie on the sterile Shield. Only near the great north route from Toronto to North Bay are there any towns of note. Gravenhurst, Bracebridge, and Huntsville all depend very considerably on the tourist traffic in summer. The numerous lakes and the relatively untouched forest have naturally great attractions for tourists from the closely settled areas to the south. These cool summer resorts draw many visitors from the United States. It is worthy of note that the thriving farming district to the north of the Hamilton-London 'constellation' contains no large towns, but supports thirty-three small towns (3,000 or less) in an area of 5,000 square miles. This gives us one small town for 130 square miles of farm land, i.e., the towns are about eleven miles apart.

Now that we are familiar with the main environmental and human factors in the problem, it is possible to make a tentative regional plan using New York State as a pattern (Fig. 185). The total population of Ontario is 3.8 millions, of which total about half a million live on the Shield, and the remaining 3.3 millions are in the small area left white in Fig. 185. If it be an advantage to New York State to have a large proportion either mountainous or sterile (as stated in the *Graphic Compendium*, *passim*) then I suppose some folk would rejoice that such a large portion of Ontario is also sterile, and can never be closely settled! It does mean that within a hundred miles of the million living near the metropolis of Toronto there is an unlimited extent of unexploited territory, i.e., most of Shield shown in Fig. 185. Already the authorities have wisely reserved Algonquin Park as a place of recreation and a game preserve. All the rest of the Shield is also submarginal, and should be under government control so as to prevent the ruin of the native trees and the spreading of soil erosion.

Nearer the centres of large population are other areas which will produce nothing more valuable than the original forest cover. This forest was laboriously cleared by the ignorant pioneers, and should be planted again with young trees. The Interlobate Moraines extend right alongside the main traffic routes and should come under public control. The great

Niagara Cuesta is covered with very thin soil for the most part, and a grand scenic driveway could be constructed in the future along this scarp, which rises three or four hundred feet quite abruptly from the farmlands to the east. In the western division of Fig. 185 are the best farmlands. Here recent soil surveys show long patches of gravelly moraine second in extent only to the Interlobate Moraines. These could also be alienated, and converted into Government Forests and Public Recreation Reserves.

I have suggested where the main lines of traffic occur. Between Toronto and Hamilton there already stretch three excellent roadways, the best being named Queen Elizabeth Highway. We shall see these prolonged to the east to Montreal and to the west to Windsor. To the north a similar pair of highways will extend to North Bay, while two or three 'parkways' may be expected to connect Hamilton with Niagara and Buffalo. The whole subject of Regional Planning is so new in Canada, that this is perhaps about as far as it is expedient to carry the subject at present.

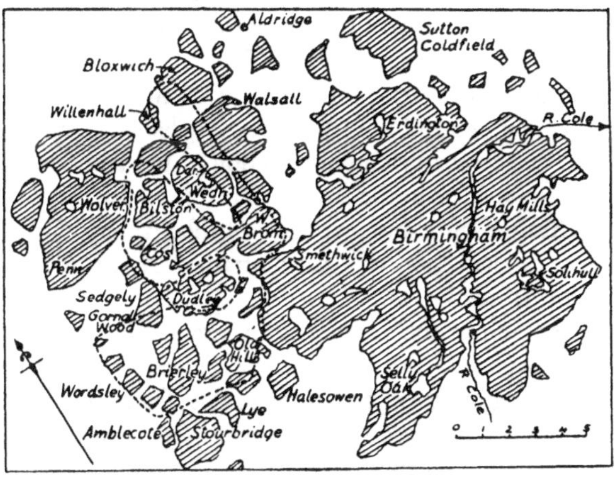

FIG. 186A. Birmingham Conurbation. The plan for its re-development in the 'green setting' of open spaces. Many of the once independent industrial areas are indicated. The broken line encloses the area where there is now much wasted (derelict) land.

Conurbation, A Survey of Birmingham and the Black Country

This remarkable volume dealing with the conditions in and near the city of Birmingham was published in 1948 by the Architectural Press, London. It is concerned with post-war reconstruction and planning. Many committees were concerned, the geographical section including Dr. Priestley, R. H. Kinvig and K. M. Buchanan. The volume includes 60 large maps. In Fig. 186A is indicated the plan proposed by the Group for the re-development of the Conurbation; the blank spaces representing green belts enclosing each town unit.

CHAPTER XX

CONTROL AND CLASSIFICATION OF CITIES

Introduction

IN this final chapter it seems advisable to discuss some of the political aspects of towns and cities, even though it be granted that these belong to fields rather far removed from the main content of this volume. A few paragraphs will give the reader some knowledge of the great diversity of control exercised not only in different countries such as U.S.A. and Britain, but also as between State and State in U.S.A. Great changes have developed in city administration as they have grown larger and larger, and these also are summarized. Finally it is very interesting to compare the different methods in use in the most important urban centres of Germany, Britain, and U.S.A. Each country has its special advantages, each its special defects; so that a combination of the best features of each nation is to be looked for in the future. To this section may be added a reference to Jefferson's interesting concept of the 'Primate City', in which he shows some of the results of a concentration of power and privilege in the great cities of the world.

The second section in this chapter is an attempt to produce a logical classification of large cities based on their *evolution*. In an earlier section (in Chapter V) a logical classification for towns, from their birth to their attainment of what we may call the 'climax' population (about 50,000) has been worked out; but quite a different set of conditions affects the growth of a huge city, say from 50,000 up to 500,000 people. A fairly full picture is given of the salient features in the development of the greatest of cities, New York, as well as a briefer discussion of London. On the basis of these data a very tentative classification of cities is attempted. So far as the writer is aware no attempt has been made to produce classes based on common stages of evolution, though several classifications based on differing functions—regardless of the size and development of the city—are described.

PART I. CONTROL OF CITIES

Villages, Boroughs and Counties in U.S.A.

Usually the procedure for incorporation requires a petition from the inhabitants of the proposed village, and a popular vote on the proposition.[1] The statute may establish a minimum population such as 100 in Alabama or 200 or 300. In New York the area is a square mile, but two square miles in Michigan. In New York and Pennsylvania the minimum population for cities is 10,000; and all municipal corporations below this figure are villages or boroughs. In the New England states the usual maximum for villages and boroughs is 5,000. In Missouri and Alabama the maximum is 3,000.

NUMBER OF SMALL MUNICIPALITIES WITH VARIOUS POPULATIONS IN 1906

	4000–8000	1000–4000	Under 1000
New England	29	43	33
Mid Atlantic	123	513	1354
N. Central	231	1305	3581
S. Atlantic	52	272	1080

Nearly half of the village and borough incorporations in New England were established between 1890 and 1900. The principal authority is the Village Council. The chief officer is called the Mayor.

The New England towns are irregular in form, and usually contain from twenty to forty square miles. The larger places have been incorporated as cities, and five towns in Massachusetts had each over 12,000 in 1900. Most of these towns have no separate government for the villages, and the town performs both the functions of 'township' and 'village' in states further west.

Before the Civil War the states were divided into counties, and the county was subdivided into smaller districts; but these varied from the New England *town* through the *township* of the Middle West to the *precincts* in the South. Later the northern 'township' was transplanted to the southern states, so that in a number of states the name township has replaced the former 'precinct'.

[1]J. A. Fairlie, *Local Government*, New York, 1906.

In U.S.A. the major district for local administration is the county (page 57, Fairlie), which is created by the legislature. The usual limit is about 400 square miles, but in Texas it varies from 700 to 900 square miles. In Ohio a county with over 100,000 inhabitants may be divided, while in Michigan any city of 100,000 may be made into a separate county. There is of course a great difference in the number of counties in a state, e.g., Delaware has three counties and Texas 243. The most usual areas are between 400 and 650 square miles. The average population is 26,646, and the median about 18,000. The counties are naturally rural in character.

In most of the central states the area is divided into civil townships. In Illinois seventeen counties out of 102 do not have township organizations; but in Missouri only twenty out of 114 counties have township organizations. Generally the cities are entirely independent of the townships.

Political Organization in Modern U.S. Cities

In other paragraphs some of the features of a large American administration have been described. Metropolitan New York is composed of five boroughs, five counties and a regular constellation of corporate units (Fig. 187).[1] Chicago has 390 distinct units of local government if we consider all the boards which control sanitation, libraries, forest preserves, etc. The Boston Metropolitan Region contains 14 cities, 26 towns and 5 counties among its many subdivisions. Los Angeles consists of a county, 38 municipalities, 187 school districts, 34 lighting and 33 road districts. These complexities have gradually developed; and have become more and more numerous as the populations have increased. It should be possible to correlate the accretion of more and more governing boards with the growth of population. It seems to the author that here we may find the 'evolutionary key' to the classification of great cities, which obviously cannot be classified by those determiners which were satisfactory in the classes 'sub-infantile' to 'late mature' which have already been briefly discussed in connection with towns.

One of the peculiarly American ways of expressing democracy was found in the opposition to special privilege, and this thought became more effective under the national leadership of Roosevelt.[2] The public utilities were now seen to be by nature essentially monopolistic. Safeguards over monopoly grants were introduced in Houston in 1905, and the

[1]Gist and Halbert, *Urban Society*, New York, 1933.

[2]*City Government*, E. S. Griffith, Oxford Press, 1927.

Wisconsin Law of 1907 recognized the jurisdiction of the state over private corporations.

'Home rule' amendments of various types placed control of many phases of city life outside the pale of state legislative action, and vested them in the cities. Thus Oregon in 1906 conferred charter-making power on her cities, and Michigan soon followed suit. The methods of government were simplified, so that by 1915 only nine of the largest fifty

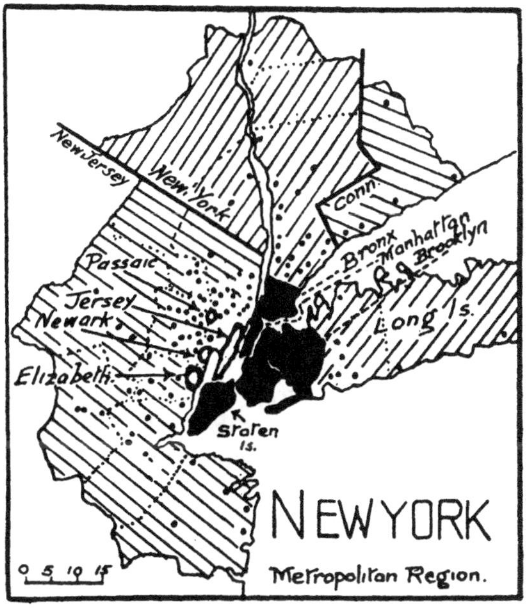

FIG. 187. The New York Metropolitan Region cuts across three states, and includes nearly a score of counties, as well as the five boroughs of New York, which are shown Black. The four cities named are not included in New York proper. Dots are satellite towns, each with over 3,000 citizens.

cities were under bi-cameral councils, and in 1923 only Kansas City and Richmond survived. Commission government began after the floods in Galveston of 1901; where the irresponsible mayor and council government seemed pitifully inadequate. Hence all authority was vested in a board of five commissioners. The five jointly confirmed all appointments, and prepared the general annual budget. In 1913 no less than 371 cities were governed by commission.

A movement to approach the plan used in Swiss municipal government developed after 1900. By 1900 South Dakota and Nebraska had made 'initiative' and 'referendum' available for their cities. In 1903 Los Angeles added power to 'recall officials'. Another feature 'the direct primary' was adopted in Wisconsin in 1903. The study of city government revealed the superiority of the German method, and in 1908 Staunton (Va.) hired a 'city manager', who should take entire charge of administration. In 1913 the Dayton flood caused that city to adopt the 'city manager' method of government, and by 1923 there were 310 cities under city manager government. Cincinnati and Cleveland followed in 1924-6. Boards of Municipal Research began to function in 1906 in New York, Philadelphia in 1908; and many other cities soon instituted such bodies. According to Gist and Halbert (*op. cit.*) there were in 1931 no less than 429 cities in the United States where city managers had been appointed. This comprises seven per cent of the whole number, and eleven of these cities had populations exceeding 100,000.

Control in Britain

We have examined elsewhere some of the main features which distinguish village from town and city. In England by the Act of 1894 a *parish council* was created for every rural parish having a population of 300 or more. This may be taken as the smallest official agglomeration in that country. There were at this time about 15,000 civil parishes in England and Wales. Another political unit in England is the *County Borough*, which is an urban district separated from the rest of the county for administrative purposes and acting much like a town in this respect. In 1894 all the local boards became *urban district councils*. Where the urban district is a borough or city the mayor and his council form the authority to deal with sanitary and other matter of common interest.

English counties average 1,000 square miles and contain about 300,000 population. French Departments average 2,000 square miles, and have about 400,000 population. Much more local authority rests with the locally elected officers in U.S., while in Europe the central officials have much power in the county.

In Ireland there are thirty-three County Councils and six County Boroughs who are responsible for roads, hospitals and education. Dealing with smaller areas are ninety urban district councils, twenty-nine small town and 213 rural districts each with a district council. They in part control local roads and health regulations.

A comparison of American, British, and German Cities

For an interesting account of the differences in the controls and ideals which affect occidental cities the book *European Cities at Work* by F. C. Howe (London, 1913) should be read with care. There have been some changes since Howe wrote, but the differences are still to a considerable degree present in the three nations.

Let us first of all see how the German City is governed. In Germany young men prepare themselves for city administration as they do for law, medicine or any other profession. In time they hope to become burgomasters, and if they make a success of their city, their reputation is known all over Germany. The burgomaster, however, does not hold as much legal power as the American mayor. Associated with the burgomaster are a large number of expert advisers, who form the Magistrat and exercise much of the municipal power. In Dusseldorf for instance in 1898 there was a population of 200,000, and Dr. W. Marx was called to be burgomaster. Under his administration the municipal debt was increased to $100 per capita. The city engaged in the purchase of land, a great harbour was constructed; and in a dozen years Dusseldorf acquired the name of the garden-city of Germany. There is keen rivalry among German cities, and the professional burgomaster has much to do with developing arts, beauty, and comfort, in the city.

The German city is free to own almost anything; free to borrow and free to experiment. In America in most cases the city can only enter on the smallest undertaking after it has secured permission from a reluctant legislature. The American city 'lives within the most carefully prescribed rubrics, designed, not to promote efficiency, but to protect property' (Howe). Monarchical Germany reverses this principle, and assumes that a city should be as powerful as a private individual. It is almost a little republic, like the *freistadte* of the middle ages, which owed allegiance to none. The commercial classes have never been powerful enough to control politics, as in America; nor have they united with the landed aristocracy as in Great Britain. The factory appeared in Germany almost one hundred years after it came to England. In Germany the ruling political class (in the early years of this century) was the landed aristocracy—an aristocracy almost as feudal as it was at the end of the 18th century. But the junker had one virtue. He had no city possessions to protect as in England, and having no privileges to protect, the junker gave the cities the right to own things and to control property.

Howe's opinions of the British city are not very favourable. He discusses the tenements of Glasgow and the slums of Whitechapel, and blames the system of land-ownership for many of these blots on civilization. The extent of land monopoly in England is almost incredible. One fourth of the land of the United Kingdom (around 1910) was owned by 1,200 persons, and another quarter held by 6,200 more. There were twelve landlords who owned four-and-a-half million acres between them. The land underlying London with its seven million people is owned in large part by nine estates. The city of Huddersfield with its 95,000 people is owned by another noble lord; and so are large areas of the cities of Sheffield and Bury.

It is the universal custom in England to lease land to tenants with the provision in the contract that all the improvements made by the tenant shall *return to the landlord* without compensation on the expiration of the lease. Cities have no general rights of compulsory purchase as in U.S.A.; hence although London is largely dependent on Covent Garden Market the latter is completely in the power of the family of Bedford. The cost of this class control of cities is colossal. It is seen in the indebtedness of the towns, and in the burdens imposed on municipal enterprises. Most of all it is seen in the poverty of the cities (F. C. Howe).

In England the *taxes* are based on the rent paid by the tenant. Parliament has ruled that if the property is not occupied it pays no taxes at all, for there is no one from whom they can be collected. It may be in the heart of the business district and be worth millions of dollars. Howe states that it would be difficult to exaggerate the evils which flow from the fact that in England the control of the city is by Parliament, a control which is in effect lodged in the hands of those who own the things which most vitally affect the city's life. Just as the American city it cramped and confined by class legislation inspired for the most part by public-service corporations, etc., so the British city is in chains to the landed aristocracy. This group refuses to endow it with home rule, and watchfully supervises every grant of power to insure that its own property and privileges will be protected. Most British cities are ugly, for they have none of the charm that makes men love cities. The appearance of the British city is traceable to the fact that there is no pressure behind the landowner to improve his land, while the tenant has no incentive to do so.

The evils of American city growth are largely traceable to individual license. To quote Howe, 'We awake in the morning to find a noisy factory in course of erection, and at once all values in the neighbourhood go tumbling, and an exodus to a new region begins'. The British city has suffered from the same license, but there it has been sacrificed to the unrestrained privileges of the landowning classes. The Germans on the other hand think in terms of public service, while Americans and Britishers think in terms of private right. Yet there are many things in American cities which other nations envy. The American local taxation is more just than in any other country; while the schools, parks and libraries are viewed as models. 'The cry of the almost disfranchised German is for democracy; of the British rate-payer for the American system of local taxation, while the cry of the American city is for the things the Germans and English—in spite of their handicaps—have already achieved."

In the British Isles, while the American cities were acquiring home rule, the cities were being legislated *for* by Parliament, and by departments that thought pre-eminently along national lines. Broadly speaking England trusted most her Parliament, less her cities, and least of all her people. America trusted least her legislature, more her cities; and most of all her people. In the former humanitarianism inevitably became paternalism; in the latter it became equality of opportunity (E. S. Griffith).

Rank among Cities—the Primate City

We owe to Mark Jefferson an interesting discussion as to the relation of the first two or three cities in a nation to each other. In an article entitled 'The Law of the Primate City' (*Geog. Rev.*; Apr., 1939) he gives various lists of the great towns in each county, and discusses their characteristics. He points out that in twenty-eight of the leading countries of the world the largest city is more than twice as large as the next, in eighteen more it is more than three times as large—a constancy of recurrence that gives this relation some claim to the status of law. In this main list he gives the value 100 to the largest city, and corresponding values to the second and third in each country. The table on p. 404 shows some of the results of this treatment, though only the first two cities in each country are listed.

Jefferson in his discussion points out that the primate city expresses the national disposition more completely than any other city. 'Paris is the front office of France', and as such contributes far more to the unification of the nation than any other city. He is of the opinion that *small* countries suffer loss by attraction of their vigorous youth to adjacent world primate cities. Amsterdam would be much larger, in his opinion, if London, Paris, and Berlin were not within a relatively short distance. He instances the case of Professor Brunhes, who wrote the valuable text

"PRIMATE CITIES (Mark Jefferson)"

Marked "Primacy"				Very Little "Primacy"			
Nation	Primate	Second	Ratio	Nation	Primate	Second	Ratio
Austria	Vienna	Graz	100 : 8	Italy	Rome	Naples	100 : 96
Denmark	Copenhagen	Aarhus	100 : 11	Spain	Barcelona	Madrid	100 : 91
Hungary	Budapest	Szeged	100 : 13	Australia	Sydney	Melbourne	100 : 80
England	London	Liverpool	100 : 14	South Africa	Jo'berg	Capetown	100 : 80
Mexico	Mexico	Guadalajara	100 : 18	India	Calcutta	Bombay	100 : 79
Rumania	Bucharest	Chisineff	100 : 18	Canada	Montreal	Toronto	100 : 77
Peru	Lima	Callao	100 : 20	U.S.S.R.	Moscow	Leningrad	100 : 76
Argentina	Buenos Ayres	Rosario	100 : 22	Netherlands	Amsterdam	Rotterdam	100 : 76
Turkey	Istanbul	Izmir	100 : 23	Afghanistan	Kabul	Kandahar	100 : 75
Cuba	Habana	Holguin	100 : 25	New Zealand	Auckland	Wellington	100 : 70
Bolivia	La Paz	Cochabamba	100 : 26	Yugoslavia	Belgrade	Zagreb	100 : 70
Finland	Helsinki	Viipuri	100 : 26	Brazil	Rio	San Paulo	100 : 66
Chile	Santiago	Valparaiso	100 : 30	Iran	Tehran	Tabriz	100 : 61
Belgium	Brussels	Antwerp	100 · 30	Switzerland	Zurich	Basel	100 : 59

Geographie Humaine in Fribourg, Switzerland. Paris called him, and he was lost to Switzerland. So also Bruckner of Bern went to Austrian Vienna. This is largely because Zurich is not sufficiently ahead of Basel to attract young Swiss, and Basel and similar small Swiss cities are too near other primate cities.

It is obvious that the primate city develops best in a long established nation. In general the low figures (i.e., 'Very Little Primacy') belong to young countries like Italy, Australia, South Africa, etc., though there are some exceptions like Spain. Jefferson points out that in this country there is no real union between the Castilian of Madrid and the Catalan of Barcelona. In Italy it seems probable that Rome will rapidly leave Milan behind, now that the nation is approaching a century of real union (1870 to 1970).

The average figures for the ratios given in the table by Jefferson are as follows:—Primate, 100; second, 30; and third, 20. Most of the nations not given in the above list come somewhere near these average figures. Brussels, Antwerp and Ghent in Belgium have ratios 100:30:18; so also the leading cities of France and Germany agree quite well with the expression of Jefferson's ratio of 100:30:20 as stated above.

PART II. A BASIS FOR CLASSIFICATION

Stages in the Evolution of Megalopolis: New York

Since New York is to-day the largest city in the world, it will be well to summarize the main stages in the evolution of this huge human agglomeration. In Fig. 188 I have made an attempt to show how the city has progressed in its growth to a total population of nearly eight millions. The graph on the left shows in some detail the features characterizing the city's development until (about 1850) it attained a population of 500,000. By this time New York was a very large city, and had long passed through all the characteristics of the 'Seven Ages of a Town' which are discussed elsewhere (page 77). The most notable feature in this graph is the marked change in the slope around 1800. Before that date the town grew quite slowly to a population of 60,000; soon after its increase was much more rapid. This date marked the onset of the Industrial Evolution; so that before 1800 New York did not exhibit the characteristic features of a *modern* occidental city, but was more akin to the older cities of Europe in the same condition. Before 1800 the city was confined to the southern two miles of the elongated island of Manhattan, which is nearly thirteen miles long (Fig. 177 inset).

Let us rapidly traverse the history of 'Old New York'. It seems likely that Dutchmen had visited Manhattan before Henry Hudson explored the Hudson River in 1609. A lucrative fur trade soon developed, and Christensen built some houses at the tip of Manhattan in 1612, while in 1614 a yacht was built here. In 1621 the region was given to the Dutch West India Company, and in 1623 thirty families emigrated to New Amsterdam. Sawmills furnished timber, and two-room huts, built of wood instead of bark, soon sprang up. In 1626 Peter Minuit was appointed Director General, and he purchased the whole of Manhattan Island from the Indians for twenty-four dollars.

In the middle column of Fig. 188 we see the order in which the various urban amenities came to the young city. First the fort with its four bastions was built at the south-west corner of the cape, and the Broadway ran from this corner due north along Manhattan (see inset at right of Fig. 188). The 'Castello plan' of New Amsterdam (dated 1660) shows that a definite wall had been erected about half a mile from the tip of the island with six bastions and a gate at Broadway. This line of defence occupied the site of the famous financial centre of to-day, i.e., Wall Street (inset, Fig. 188).[1] The first Trinity Church was built here in 1697.

Meanwhile we can trace on the graph the order in which the first civic buildings were erected. After the Dutch church, a prison and post office followed. The first newspaper did not appear till about 1723, and a theatre was built about this time. The first library of note was opened in the fifties. By this time, as the inset map in Fig. 188 shows, the city had spread north to Frankfurt Street, i.e., the southernmost mile of Manhattan Island was covered with streets. The city had been subdivided into six wards in 1683, and already a number of families had moved far to the north-east and founded the village of Bronx. Ferries had been in operation for some time linking the city to Long Island and New Jersey, while a bridge crossed Harlem River in the north as early as 1693. By 1737 Wall Street was the fashionable quarter of the town with many large houses (Ha). A seventh ward was added in 1763.

After 1760 the streets were lighted with oil lamps, and the old water supply (derived from wells) was improved by a reservoir. A steam pump was in operation by 1799, and six miles of wooden pipes were laid.

[1] *History of New York*, W. L. Stone, New York, 1872.

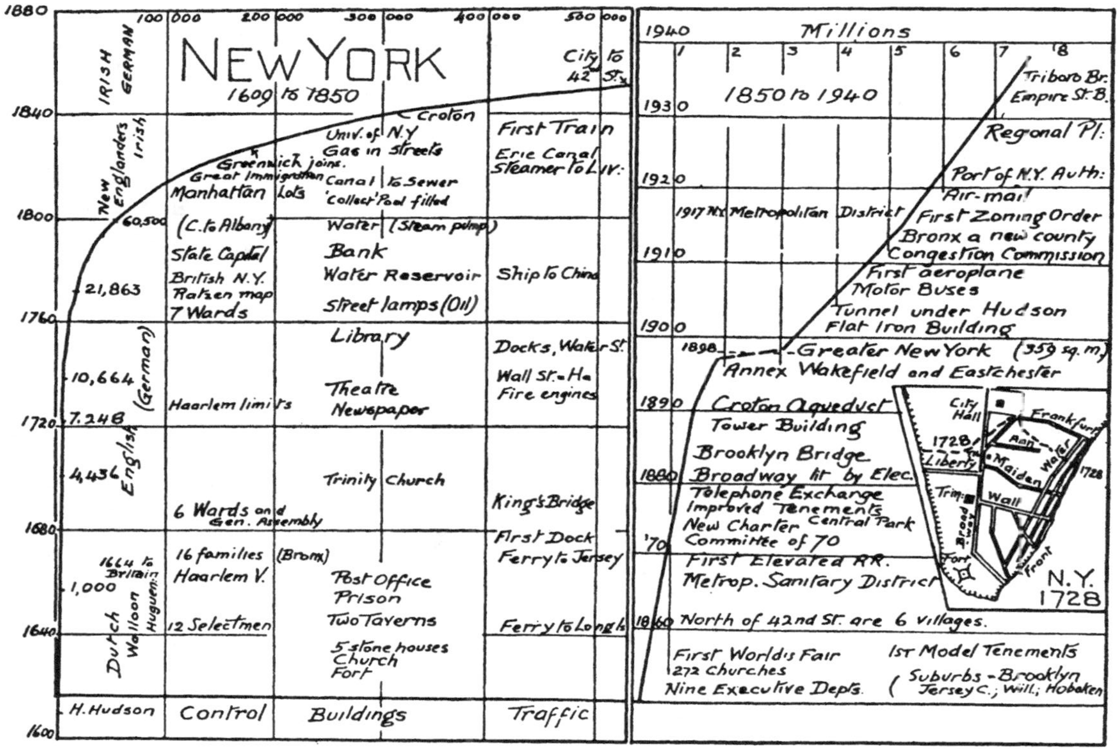

FIG. 188. Historical Population Charts for New York. (*Left*, 1609–1850; *Right*, 1850–1940.) *Inset* on the right is a map showing the area occupied by New York City in 1728. (See also *Inset*, Fig. 177.)

In 1784 New York became the state capital, but this honour was transferred to Albany in 1796. The Bank of New York began operations in 1784. During the Revolution the British held New York from September, 1776, till November, 1783. The first census of the city, taken in 1790, showed a population of 33,131. About 1807 J. Randel and others subdivided the whole of Manhattan Island into lots, and laid out the streets of the future city. It is worthwhile quoting from their report on the merits of the 'grid' which they adopted.

They discussed 'whether they should adopt some of those supposed improvements, by circles, ovals and stars, which certainly embellish a plan, whatever may be their effects as to convenience and utility . . . they could not but bear in mind that strait-sided and right-angled houses are the most cheap to build. It may be a matter of surprise that so few vacant spaces have been left . . . for the benefit of fresh air . . . but those large arms of the sea which embrace Manhattan render its situation peculiarly felicitous. . . . It is improbable that (for centuries to come) the ground north of Haerlem Flat will be covered with houses'.[1]

Modern New York

We may date the beginning of the modern city from about 1825, when the village of Greenwich—which lay just to the north-west of the City Hall (inset, Fig. 188)—was linked by houses to the city. Gas was in use in the streets about this time, and Washington Square was laid out on the site of the old 'Potters Field'. From 1837-42 new sources of water were being developed, and a large reservoir was built at the northern edge of the city on 42nd Street, which was supplied from Croton Reservoir some thirty miles to the north-east. The first steam train was used in the city in 1834, and the city reached the half million about 1849. (See inset to map on page 371.)

About this time the city was governed by nine executive departments, and extended to 45th Street or thereabouts. There were farms in the northern part of Manhattan, with half a dozen villages such as Harlem and Bloomingdale (about half way along the island), Manhattanville, Harsenville and Yorkville. More distant suburbs were Brooklyn, Jersey City, Williamsburgh and Hoboken. The erection of model tenements was exercising the authorities as early as 1855. From 1857 to 1876 plans for the creation of Central Park were under consideration. This remarkable area of open parkland is the finest feature in Manhattan Island, and

[1]*New York Past and Present*, I. N. Phelps Stokes, New York, World's Fair, 1930.

compensates to a considerable degree for the strait-jacket pattern of the Randel plan of 1807. The park is over one-and-a-half miles long and nearly a mile wide, and I know of no other city which has devoted so large an area to a 'lung' in the middle of such a congested area as Manhattan Island.

Telephones and electric lighting followed about 1880, but the greatest episode in the history occurred when Greater New York was created in 1897. This charter united Brooklyn, Queens, and Staten Island, with Manhattan and the Bronx; and gave Greater New York an area of 359 square miles and a population of more than 3,100,000. The addition of this large area accounts for the sudden bend in the population curve shown for 1898 in Fig. 188. The first decade of the new century saw the beginning of sky-scrapers, the use of motor buses and the coming of the aeroplane. Congestion was rife, and a Commission to consider this problem was authorized in 1910. In 1913 Bronx was separated from New York County and became a separate county.

One obvious sign of the difference between a small town and a large city, is the subdivision of the city into wards or boroughs or other convenient districts. Thus New York City soon expanded greatly, and was at first administered in five boroughs. These are Manhattan (the nucleus); Brooklyn, Bronx, and Queens to the east, and Richmond (occupying Staten Island). They are shown as solid black in Fig. 187. In the New York Metropolitan Region instituted in 1917 portions of three states and nearly a score of counties are involved, as is clear from Fig. 187. Much the larger portion of the region is in the state of New Jersey, while a considerable portion of Connecticut is incorporated in the north-east. Moreover in the Metropolitan Region of New York are quite large cities; such as Passaic with 62,000 folk, Jersey City (302,000), Elizabeth (110,000) and Newark (430,000). Many independent municipalities, each over 50,000 population, are also included. All of these are in New Jersey, and therefore outside the state in which the city of New York originated. Thus geographical association overpowers early political divisions.

The first Zoning Order was promulgated in 1916, and a special Commission was instituted to supervise the Port of New York in 1921. Of great interest to town planners was the publication of the unrivalled *Regional Survey of New York* in over a dozen volumes in 1929. The World's Fair of 1939, and many improvements in the slums and tenements of New York were accomplishments of the late thirties.

Immigration as a factor in early New York

Theodore Roosevelt in his volume *New York* (New York, 1891) comments in an interesting fashion on the changes in the cultural composition of the city. 'At its founding the Dutch were dominant, but with a considerable Walloon (French Belgian), element, which was soon absorbed by the Hollanders. There was a large element of French Huguenots who were absorbed more slowly, as well as many English and a few Germans. After the final English conquest there was a fair amount of immigration from England and Scotland; the Huguenots also continued to come in for a little while, and there was a large German and a considerable Scotch-Irish immigration.

'At the end of the Revolution all these peoples had grown to use the English tongue, and were fast being welded together; but the great majority of the citizens were non-English by ancestry. There then began a great inrush of New Englanders; and for the first time the citizens of English "blood" grew to outnumber those of any other strain. All, however, were soon fused together and became purely American. The immense immigration between 1820 and 1860 changed this. By this latter date the men of Irish birth had become more numerous than any other; the Germans at some distance off, next; while the native Americans, who still controlled the others, were a close third. Of course, however, the older "races" of the city made a mould into which the newer were poured.'

There was considerable immigration from abroad even before the war of 1812, but it soon swelled very rapidly. In 1818 and 1819 over 20,000 immigrants arrived in New York, and most of them were very poor and ignorant. They housed in sheds, cellars, and rookeries, of all kinds; since they preferred to huddle in the large cities rather than go into the country. After the potato famine in Ireland and the revolutions of 1848 in continental Europe, fugitives came over by hundreds of thousands. Probably five-sixths of those who settled in New York City during the half century before the close of the Civil War were Irish and Germans, the former being at this time much in the lead (Roosevelt).

Social evolution in modern Cities

Characteristic of the development of the larger occidental cities is the increasing cost of community services. These indeed determine in large measure the superior attractiveness of life in a large city as contrasted with rural or 'small town' life. Since they are of great importance in the evolution of the large city we shall do well to consider them, though they belong to the sociological as much as to the geographical aspect of city life.

Consider for instance the character of the *streets*. We are told that the modern paving of streets only became at all general after 1840 (Gist and Halbert). Until 1866 Glasgow had no streets in the modern sense, but only wynds, vennels and closes, which were crowded with congested tenements. Elsewhere we have shown that Paris blazed the trail with her broad streets cut through the heart of the slums. The automobile age commencing with the new century has done much to improve the width and direction of city streets. *Per contra* it has vastly added to the danger to life and limb.

Water supply is a costly service in the great city; and it has been stated that the ultimate size of cities may be fixed by the difficulty in obtaining further possible water supplies for the large populations. Aqueducts were used in medieval days, but pumps replaced these in part in 1652 in London. New York had a pumping system by 1800, and to-day it uses twenty-four reservoirs—mostly to the north of the city, which by 1930 had cost 193 million dollars. Los Angeles—which increased its population twenty-five per cent in the period 1920 to 1940—has built reservoirs 250 miles away in the Sierra Nevadas.

Sewage disposal on a grand scale is characteristic of the great city. London installed great sewage tanks as early as 1859 and to-day dumps the sludge into the sea. Berlin has sewage *farms*, and it has been stated that about half the local vegetables are grown with the aid of sewage manures. Chicago built a huge canal about 1890 to turn the purified sewage into the Mississippi river system. New York has 2,800 miles of sewers, and it has been calculated she could conserve three million dollars worth of fats annually, if her sewage were scientifically treated. *Lighting* of cities is costly, and the modern system of electric lighting started in Cleveland, Ohio, in 1879. All these services add greatly to the cost as well as to the amenities of life in a large city.

The later evolution of some great American Cities

From 1900 to 1930 the population of the United States increased about seventeen per cent each decade; while the cities increased about thirty-seven per cent.[1] To quote Mark Jefferson: The earlier period still fell in that area of great industrial expansion when the people shifted their

[1] "The Great Cities of the United States," Mark Jefferson, *Geog. Rev.*, July, 1941.

main occupations from farm to factory, and created cities everywhere. But with the onset of the 'depression' in the decade 1930 to 1940 we saw people straggling away. Twenty-six great cities, all in the industrial north-east quarter, have fewer people than they had ten years ago; something that has not happened before. The following table shows these changes for New York, Washington and the Republic as a whole.

DECENNIAL PERCENTAGES OF GROWTH 1840 TO 1940

	1840–50	1850–60	1860–70	1870–80	1880–90	1890–00	00–10	10–20	20–30	30–40
U.S.A.	36	36	33	29	26	20	21	15	16	7
New York	78	73	32	31	33	37	39	17	33	6
Washington	122	52	64	47	28	48	19	32	11	36

Jefferson points out the anomalous position of Washington, 'our contrary political capital', which accelerates its growth when the country is in trouble, and slows down in periods of national prosperity! Thus, during the Civil War, New York dropped to thirty-two, and the nation to thirty-three; but Washington increased its rate of growth to sixty-four per cent. In the first World War the same thing happened in all three cases. Washington doubled its population, while the other two decreased greatly. In the depression (1930, etc.), Washington, 'the job-hunter's paradise', jumped up, while the nation's rate decreased to six per cent.

The same authority gives us interesting data as to the fusing of satellites to form the City of New York. As in most other huge cities there are a number of contradictory boundaries, depending on the function under discussion. In 1917 New York Metropolitan District was created, and this includes 3,768 square miles. It comprises no less than 373 independent self-governing municipalities; and in 1926 had a total population of nine-and-a-half millions.

New York had only 200 citizens in 1626, and this had grown to 1,000 by 1656. A century later the total was only 14,000. It was in 1871 that it reached a million. (See the graph in Fig. 188.) As stated, in 1898 the five boroughs of Manhattan, Brooklyn, Bronx, Queens and Richmond, united to form Greater New York. The addition of other units is indicated in the table herewith. (Figures are thousands.)

	1840	1850	1860	1870	1880	1890	1900	1910	1920	1930	1940
New York	391	696	1175	1478	1912	2507	3437	4767	5620	6930	7455
Jersey City		7	29	82	121	163	206	268	298	317	301
Hoboken			10	20	31	44	59	70	58	59	50
Yonkers				13	19	32	48	80	100	135	143
Mt. Vernon				3	5	11	21	31	43	61	67
Bayonne				4	9	19	33	56	77	89	79
Union City				5	6	11	15	21	21	59	56
New Rochelle						8	15	29	36	54	58
W.N. York City							5	14	30	37	39
N. Bergen									23	41	40
New York	391	703	1214	1605	2103	2795	3839	5336	6306	7782	8288
Percentage of growth		80	73	33	31	33	37	39	17	23	6

Jefferson asks the question: Why not include Paterson and Newark in the New York composite? The criterion suggested is that the margin of a city is reached when one comes to open country. However, before reaching Paterson, when driving from New York, one certainly passes through open fields and woods, not merely open spaces in parks and recreation areas that belong with a city. Hence Paterson is not yet merged in 'Megalopolis'.

Some Characteristics of London

London has passed through somewhat the same phases as New York in its development during the industrial expansion of the last two centuries (Fig. 189). The City proper is quite a small area of less than 700 acres, comprising the lands on the north side of the Thames between the Tower and Temple Bar, while Finsbury borough bounds it on the north. It is not very different from the city of late Middles Ages or indeed from the city within the Roman Walls shown in Fig. 85. Within this area the 'sleeping population' (on which the census is based) is quite small, and is decreasing. Thus in 1861 this population was 112,000, whereas by 1901 it had fallen to 27,000. This city area is still the financial centre of the British Empire. The shipping offices are hereabouts still, and the publishers frequent Paternoster Row near St. Paul's. But many of the other centres of trades and professions, which necessarily started in the city, have now moved to other parts of Greater London.

The County of London was established in 1888, and covers about 117 square miles. It is much the same as the area over which the Metropolitan Board of Works had jurisdiction in earlier days. The County of London is divided into twenty-eight boroughs, which are shown in Fig. 189. Each of these has its special interests, and it will be worthwhile to consider these briefly in our study of the great city. Ten of the boroughs lie to the south of the Thames in the counties of Surrey and Kent, while those on the northern bank are mostly carved out of the county of Middlesex, and do not extend east of the River Lea into the adjacent county of Essex. (Fig. 83.)

We may commence this brief description of the boroughs of London at the eastern end. Here Woolwich is the site of the Royal Arsenal on the river. Greenwich contains engineering, telegraph, and chemical works. It is noted for Greenwich Hospital, and for the Observatory from which longitude is reckoned. Deptford is famous for marine engineering and large cattle markets. Bermondsey—as already suggested—is very low; and has many tanyards. The river is here fringed with wharves, and the Docks of Rotherhithe import grain and timber. Southwark is filled with factories and small mean streets. Lambeth is a peculiarly shaped borough extending from the river, near the famous palace of the Archbishop of Canterbury, to the Crystal Palace at the extreme south end of the county. Battersea contains large parks; while Wandsworth, the south-west borough, is the largest of all, and like two other large boroughs in the south (Camberwell and Lewisham) is full of pleasant residences, especially in those portions furthest away from the heart of the city (Fig. 189).

The northern boroughs are rather high on the west, rising to 450 feet, but in general become lower on the east, where the River Lea with the broad, marshy meadows of Hackney forms the boundary of the county. Hampstead is the north-west borough, and is noted for its large heath, a park of 240 acres adjoining a 'garden city'. St. Pancras contains three great railway stations at its southern end, and has many mean streets in adjacent areas. Islington contains many charitable institutions, and a great Exhibition Hall where many shows are held. Hackney includes many areas of small houses, but has open spaces on the east along the River Lea. Shoreditch and Bethnal Green are also filled with fourth-class houses and small factories. Finsbury is noted for printing and watch-making establishments. Holborn occupies the site of the ancient Hole Burn, and here are many of the famous Halls of the lawyers, such as Lincoln's Inn. The British Museum, and Bloomsbury, with the University of London, are also in Holborn (Fig. 189).

The borough of Marylebone was formerly known as Tyburn, and formerly that little stream flowed through the district near the church of 'St. Mary on the burn'. Edgeware Road (the former Watling Street) runs along its western boundary, and 'Tyburn Tree' (i.e., the gallows) stood at the southern end. In this borough are Regent's Park and the Zoo, with the pleasant residences of St. John's Wood. Paddington and Kensington contains various gardens and better-class residences. Here are the Albert Hall, and the great Natural History Museum. Hammer-

FIG. 189. Modern Boroughs within the 'County of London'. Note that this County has annexed portions of adjacent counties.

smith is the western borough; and next to it is Fulham where the Bishop of London lives, in what is said to be the oldest inhabited house in England. Chelsea is a low-lying suburb noted for the veterans' Hospital. It contains many apartment houses, and is popular with writers and artists. (Fig. 83.)

The next of the boroughs to be described is perhaps the most famous of all. It is Westminster, so named from the abbey built far to the west of the early city of London; though now Westminster is twelve miles from the western margin of built-in streets. The Courts of Justice, Trafalgar Square, the National Gallery, the great Administrative buildings of the empire, the Houses of Parliament, St. James Palace, and the Abbey itself, are all to be found in the southern part of the borough. The City has been described briefly already; with the Post Office and St. Paul's on the western of the two small hills, and Leadenhall Market on the other. Near the Tower and within the Post Office area are relics of the Roman Wall. Stepney and Poplar are two of the least attractive portions of London, owing to the slums and wharves which are so characteristic of the region along the Thames. Here are the West India Docks and to the east the great Albert Docks (Fig. 189).

Classification of Towns

Although the writer has already explained why he decided to devote most time to the more geographical features of town evolution, it will be time well spent to consider briefly several types of classification which consider sociological and political factors more than it is possible to explain fully in a volume of this size. One of the most comprehensive is that put forward by Marcel Aurousseau in the *Geographic Review* of October, 1921. He suggests no less than twenty-eight types of Active Towns, which he classifies under six heads. These are as follows:

1. *Administration.* a. Capital; b. Revenue.
2. *Defence.* a. Fort; b. Garrison; c. Naval.
3. *Culture.* a. Cathedral; b. University; c. Art; d. Pilgrimage; e. Religious.
4. *Production.* a. Manufactures; b. Crafts.
5. *Recreation.* a. Health; b. Tourist; c. Holiday.
6. *Communications,* which he subdivides again into three heads.
   A. *Collection.* a. Mining; b. Fishing; c. Forest; d. Depot.
   B. *Distribution.* a. Export; b. Import; c. Supply.
   C. *Transfer.* a. Market; b. Fall-line; c. Break of Bulk; d. Bridgehead; e. Tide Limit; f. Navigation head.

All of these categories are worthy of consideration, if we were specially concerned with *function.* But in many cases it would seem that there would be little difference in the siting or the distribution of the zones of folk in a town as between some of these classes. In the next classification —that of Chauncy Harris—the classes are much reduced, as will be seen. Howbeit the present writer prefers to make *site* and *evolution* the main features of his study.

Functions of American Cities

We owe Chauncy Harris a study which interprets the characters of most of the leading cities in the United States mainly in terms of their *commercial* interests.[1] These classes are based on the data of employment in 984 cities; and the groups are described in the following (abbreviated) definitions (Fig. 190).

Manufacturing cities	Manufacturing is over 60% of total.
Diversified cities	Manufacturing, Wholesale, and Retail, are respectively less than 60%, 20%, and 50%
Wholesaling cities	Wholesaling is at least 20% of total.
Retailing cities	Retailing is at least 50% of total
Transport cities	Transport employs at least 11% of workers.

The Manufacturing or *Industrial* cities naturally occur in the so-called Manufacturing belt of U.S.A. Philadelphia, Pittsburg, Buffalo, Detroit,

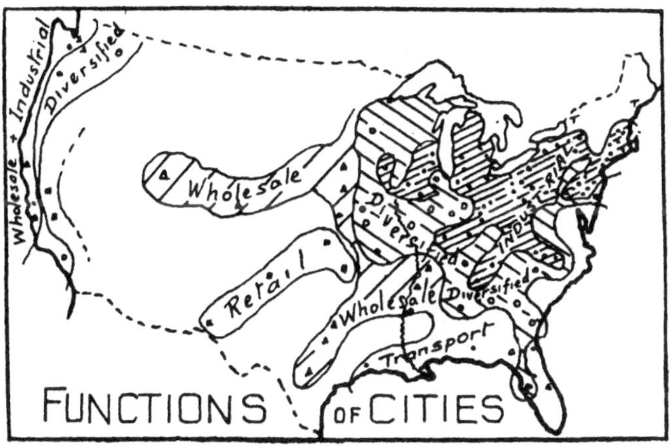

FIG. 190. A Classification of 984 cities of U.S.A. based on Employment data. (*Somewhat simplified from a map by Chauncy D. Harris.*)

Lowell, and Fall River are characteristic examples. This belt is found north of the Ohio Valley, with extensions to the south along the Piedmont and the Great Valley in the Appalachians. Here centres originally based on trading have been converted into industrial cities in the last generation or so. A few occur in the west, where they are usually lumber cities.[2]

[1] "Functional Classification of Cities of U.S.A.," *Geog. Rev.*, p. 92, 1943, New York.

[2] Based on the map in *American Journal of Sociology*, page 5, July, 1943, by Chauncy D. Harris.

In *Diversified* cities neither trade nor manufacturing are dominant, though both are important. Four of the five largest cities fall into this class; among them are Boston, New York and Baltimore in the east; Atlanta and Birmingham in the south-east; Chicago, Minneapolis, St. Louis and Indianapolis in the north centre; Houston and Fort Worth in Texas; and Spokane, Portland, Sacramento and Los Angeles in the west.

The chief *Wholesale* cities are New York and Chicago, but this is only one of their many occupations, and the typical 'Wholesale' cities are relatively small, such as Riverside (noted for oranges) or Wenatchee (apples). But many of the largest are engaged in distribution, and come under this class. Among such are San Francisco and Seattle in the west; Salt Lake and Denver in the centre; Omaha and Dallas further east.

Some geographers combine retail and wholesale towns, but Harris separates them. The *Retail* cities are smaller places on the border of the industrial belt; especially in the oil regions, such as Tulsa, Wichita and Shreveport. About half of the Retail cities lie near the eastern margin of the Great Plains; such as Fort Dodge, Lexington, Jackson, Lincoln, El Paso, Austin, Corpus Christi, Abilene, etc.

Some thirty-two cities out of the 900 are classed as *Transport* Towns. Of these eighteen are primarily railway centres and fourteen are ports. New Orleans is the largest transport city, but Galveston, Mobile and Savannah are other well-known examples. Altoona, Cumberland and Roanoke are sited where railways make use of passes through the mountains.

Harris also lists *University* Towns, such as Ann Arbor, Bloomington, Champaign-Urbana, Lawrence and Ithaca. *Mining* Towns are Hibbing, Butte, Wilkes-Barre, Connelsville, etc. *Resort* Towns are Phoenix, Tucson and San Diego in the west; Colorado Springs, Albuquerque, Santé Fe in the Rockies; Biloxi, Miami, Daytona in the south-east; Atlantic City and Saratoga Springs in the north-east.

The most striking feature of the distribution of these classes is that they seem to form zones about the north-east corner of the Republic (Fig. 190). This indicates probably that they represent *temporary stages* in city development, rather than permanent categories. In the early days industrial operations were almost confined to New England, and most of the territory to the west was in the Pioneer stage of development. Now the Industrial stage has spread almost to the Mississippi, and the Pioneer stage has moved west to the arid lands. But we may expect another generation to see the present Diversified zone become Industrialized, and the Diversified zone to move, in its turn, to the west.

Stages in the Evolution of a large City

In Chapter V the significant features in the growth of a village and town are described. The earliest type of all which I call the Sub-infantile stage is well illustrated by the numerous long-established but primitive settlements along the Mackenzie River. They are discussed in Chapter III. The classes of towns have been occupying my attention for nearly ten years; and it seems to me that we must introduce a definite break between the town of about 50,000 or so, and the growth of the agglomeration thereafter. At any rate when the town has reached a population of 50,000 it has passed through the stages labelled juvenile, adolescent, and mature, yet many new characteristics develop in the long period while the small city is growing into a metropolis or megalopolis. These we shall now examine.

Perhaps the main feature distinguishing the town from the city in this connection is the 'swallowing up' of the satellite villages in the vicinity of the growing city. This has already been described for New York (page 413); and a few notes on similar phenomena in Toronto and Los Angeles will be found of value in the discussion.

In Toronto (*fide* N. Deacon) the annexations of satellites took place in the following order.

Date	Toronto Population	Satellite	Date	Toronto Population	Satellite
1883	200,000	Yorkville (N.)	1908	300,000	East Toronto
1884		Brockton (W.)	1909		West Toronto
1884		Riverside (E.)	1909		Midway (E.)
1888		N.W. Toronto	1910		Dovercourt (N.W.)
1905	250,000	Rosedale (E.)	1912	400,000	North Toronto

A somewhat similar table for Los Angeles will be found in the volume *Planning of the Modern City*, by N. P. Lewis (New York, 1923).

Date	Satellite	Area	Date	Satellite	Area
1850	Original City	18,000 ac.	1915	San Fernando	109,000 ac.
1896	S. Addition	6,500 ,,	1916	Westgate	31,000 ,,
1906	Shoestring	12,000 ,,	1917	West Coast	8,000 ,,
1910	E. Hollywood	7,000 ,,	1918	Hansen Heights	5,000 ,,

Another major factor in the development of a city is the sub-division into City Wards. Unfortunately this is not always a regular or logical process, as we may gather from data kindly supplied by the City Clerk of Toronto. The city was incorporated in 1834 with five wards, and the later progress was as follows:

Date	Population	Wards	Date	Population	Wards
1834	9,000	5	1892	172,000	6
1847	21,000	6	1912	410,000	7
1856	42,000	7	1932	627,000	8
1874	69,000	8	1933	626,000	9
1876	67,000	9	1945	681,000	9
1883	105,000	10			
1884	105,000	12			
1889	170,000	13			

Thus we see that the city at first progressed in a steady fashion, with a new ward for each 30,000 increase of population. But in 1892 the number dropped from thirteen wards to six wards, and since then a new ward has been added for each 150,000 increase in population. For these and other reasons the author has not been able to use exact criteria in labelling the stages in the growth of a large city. Only tentative labels can be given; which will no doubt be much improved in later research.

Let us glance at various civic changes affecting other categories in Toronto, and link them with the population, much as appears in the graph (Fig. 188) for New York. It was about 1866 that Toronto changed from being a large *town* into the class of a *city* of 50,000. (Here I use the word 'city' in an evolutionary sense rather than a municipal sense.) But as early as 1834, it had been incorporated as a 'municipal city', though the population as stated earlier was only 9,000. In 1835 the police were appointed, and a general scheme of sewage was initiated. Lighting by gas came in 1841, and King's College lectures began in 1843. A Board of Trade was instituted in 1845. In 1851 Trinity College was started, and in 1852 Yorkville became a municipality; distinct from Toronto though only a mile or two to the north. All these features may be taken as characteristic of a large town rather than a true 'city'.

The transition years were marked by the erection of the University buildings in 1858, and by the first street rails being laid along Yonge Street. The *Daily Globe* newspaper started in 1853. About the same time was the first Provincial Exhibition, and Eaton's Department Stores started in a humble way about 1869. In 1866 Toronto had its first evening newspaper.

Later developments were the first Public Library in 1881, the installation of electric light in 1884, and the use of Lake Ontario waters on a large scale for the municipal water supply in 1893, though the municipality had been in control of the city's water supply since 1873. In 1907 Hamilton and Toronto shared the cost of an improved highway between the two cities. In 1912 Niagara hydro-electric power was brought into the city. In 1920 the street-car system was taken over by the Toronto Transport Commission, and about the same time some of the huge road viaducts, for which the city is noted (over the deep glacial valleys) were constructed. About 1920 Toronto's population reached the half million mark.

Classification of Towns and Cities

The early geographer, William Morris Davis, greatly clarified the study of topography by introducing the concept of the 'ages' of the various landscapes. He showed that an elevated area was slowly cut down to a rocky plain near sea level; and he gave anthropomorphic names (such as juvenile, senile, etc.) to these stages. The present writer in various recent papers has made an attempt to do the same thing in regard to the usual development of an occidental city.

In Chapter V the application of this principle to the growth of Toronto is described, and it will be useful to summarize the result in these closing paragraphs. The *sub-infantile* stage of a town—with a single ill-defined street—only lasted in Toronto for a short time about 1795, since a 'grid' of half a dozen streets was laid out at the start of the little town.

In the earliest stages of such a town there is no clear differentiation between industrial, commercial or residential areas. The first tendency is for the bigger houses to develop near the margins, where gardens and privacy are available. Factories of size have not yet been built. We see Toronto in the *Infantile* stage up to about 1818, when the population was about a thousand (Fig. 27).

In the next stage there is a fairly clear segregation of an extensive commercial quarter towards the centre of the town, including a number of streets without residences. But shops, offices and small industries are still rather mixed. A definite zone of better-class houses fringes the town. The poorer folk live near the centre, in many cases taking over the better houses of the former stage, but also building new small houses in any empty spaces. This may be called the *Juvenile* stage; and it is illustrated around 1842 in Toronto, which then had a population of about 20,000.

The beginning of adolescence is shown by a definite differentiation of the residences. The various types are displaced outwards as the years move on, though naturally examples of early houses survive in the expanding zone. Until the first-class houses are segregated we may call the stage *Adolescent*. The population of this stage varies from 30,000 to 50,000 according as the town is quick-growing or rather stagnant.

The next stage is the *Early Mature*, where there are definite areas of first-class residence, more or less approaching zones. Here the population is several thousand more than in the last stage. Maturity in Toronto may be said to be reached about 1885, but the population of any particular stage differs a good deal in various cities as suggested earlier (Fig. 27).

THE VARIOUS AGES OF TOWNS

Infantile Towns	Haphazard distribution of shops and houses, no factories.
Juvenile Towns	Differentiation of zones begins, shops are separated.
Adolescent Towns	Scattered factories, but no definite zone for first-class houses.
Early Mature Towns	A clear segregation of first-class houses.
Mature Towns	Separate commercial and industrial areas, four zones of houses, ranging from mansions to shacks.

TENTATIVE STAGES IN THE DEVELOPMENT OF A CITY OVER 50,000

First Stage	The first nearby villages become submerged by the growing city. University colleges are founded. Outlying water supplies are necessary for the large population.
Second Stage	Street railways spread, daily and evening newspapers are founded.
Third Stage	Satellite villages become separate municipalities.
Fourth Stage	Model Tenements spread through parts of the city.
Fifth Stage	Huge viaducts and bridges to improve traffic are erected.
Sixth Stage	Many outlying satellites are annexed to the city, greatly increasing the number of wards. Zoning laws develop.
Seventh Stage	Metropolitan Districts, County Councils, Commissions of management replace the early type of control.

In the table above an attempt is made to show the criteria which must be investigated in making a similar table for cities which have developed far beyond the simple patterns indicated for towns of 50,000 folk. The order of development will vary somewhat in most large cities, and the writer has not ventured to label the stages specifically at this early stage in his studies. But the table will be found of interest by all students who wish to obtain a clearer picture of the features which characterize the occidental city.

We have now investigated all the more important geographical aspects of urban communities, and have noted their gradual development from the irregular settlements of primitive peoples. Next have arisen the walled cities of classical times with their elaborate monuments and temples; to be followed by the even more closely defended trading cities of medieval times. (These latter belong to Mumford's *Eotechnic* Phase.) Most of our attention has been given to the modern industrial city (*Paleotechnic*, as Mumford terms it); and we have recently seen the rise of a few planned (or *Neotechnic*) cities. Without plan has developed the agglomeration of a million homes, which has been termed *Megalopolis*.

The final chapters have shown that modern civilization can easily improve on the haphazard paleotechnic city. The general conclusion of the experts is to establish a 'city unit' of about 50,000 inhabitants (i.e., the climax 'town'[1] of the preceding pages), and to make each additional unit of 50,000 more or less self-contained. It may develop near older units; but must be separated by considerable 'green belt' areas. No typical example of such a composite city has yet developed, though Canberra and other cities described in the preceding chapters are stages directed towards this ideal.

The writer believes that educated citizens realize that there are many disadvantages resulting from mere 'bigness'. He hopes that the urban communities of the future (which, in many nations, will soon vastly outnumber the rural populations) will give due attention to their surroundings; for the environment, even if it be one of bricks and mortar, is a major factor in the evolution of our civilization.

[1]After a city grows to forty or fifty thousand it fails to reproduce the full quota of its citizens (Mumford).

BIBLIOGRAPHY FOR FURTHER READING

BIBLIOGRAPHY FOR FURTHER READING

Chapter

3. EARLY SETTLEMENT

Adams, J. Q.: Settlements in N.E. Canadian Arctic. *Geog. Rev.*, pp. 112–127, Jan. 1941.
Photos., Pangnirtung, Lake Harbour.

Schwendeman, J. R.: Water Supply and Farms in Red R. *Econ. Geog.*, 17 maps and photos of control by water.

Wehrwein, G. S.: Rural-Urban Fringe (Indianapolis). *Econ. Geog.*, July 1942. 4 maps.

4. TOWNSHIPS

Anthony and Charles: Population Trends in Prince Edward Island. *Geog. Rev.*, October 1942. Two maps and photos.

Balk, H. H.: Rurbanisation of Worcester. *Econ. Geog.*, April 1945. 8 Satellite towns and links to city.

Beacon-Herald, Centenary Edition, 1932, Stratford. (Ont.) Re Stratford.

Beers, J. H.: *Country Atlas for Ontario*, 1877, Toronto.

Cannif, W.: *Settlement of Upper Canada*, 1869, Toronto.

Fraser, A.: *Land Settlement in Upper Canada*, 1921, Toronto

Gourlay, R.: *Upper Canada*, 1822, London.

Guillet, E. C.: *Early Life in Upper Canada*, 1933, Toronto

Harvey, D. C.: Charlottetown (*Can. Geog. Jnl.*), 1932, Montreal

Lauriston, V.: A Century of Goderich (*Can. Geog. Jnl.*), 1932, Montreal.

Schott, C.: *Landnahme und Kolonisation in Canada*, 1936, Kiel.

Taylor, Griffith: Evolution of a Capital (Canberra) (*Geog. Jnl.*), 1914, London.

Taylor, Griffith: Topographic Control near Toronto (*Can. Jnl. Econ.*), 1936, Toronto.

Taylor, Griffith: British Columbia (*Geog. Rev.*), 1942, New York.

Taylor, Griffith: Environment, Village and City (*Annals: Amer. Geog.*), 1942, New York.

Whitson, J. F.: Township Surveys in Toronto (*Ont. Land Survey*), 1905, Toronto.

6. PRIMITIVE AND ASIATIC SETTLEMENTS

Ackerman, E. A.: An Algerian Oasis Community. *Econ. Geog.*, July, 1936. One map, largely sociological.

Brass, L. J.: Stone Age Agriculture in New Guinea. *Geog. Rev.*: pp. 555–570, Oct. 1941. Grand Valley, walled villages.

Chapter

Waterman, T. T.: North American Indian Dwellings. *Geog. Rev.*, pp. 1-26, N.Y., Jan. 1924. 20 photos, Wigwams, village N. Carolina, Map of pit dwellings.

7. GREEK CITIES

Semple, E. C.: Templed Promontories of Mediterranean. *Geog. Rev.*, pp. 353-387, N.Y., July 1927. Old sketches, Bosporus, Sorrento, Halicarnassus, Syracuse.

8. ROMAN TOWNS

W. A. Bell, F. Cottrill, C. Spon: *London Wall.* London, 1937.

R. C. Bosanquet: *Roman Camp at Housesteads.* Newcastle. 1904.

F. Haverfield: *Roman Occupation of Britain.* 1904.

T. R. Holmes: *Ancient Britain.* 1907.

Middleton: *Remains of Ancient Rome.* 1892.

S. P. Platner: *Topography of Ancient Rome.* Boston. 1904.

9. EARLY MEDIEVAL CITIES

Ahlmann, H. W.: Study of Settlements. *Geog. Rev.*, pp. 93-129, N.Y., Jan. 1928. Many maps and photos Italian Towns, soils, Aversia, Perugia. Early Danish, Fardrup, Copenhagen, Myrdal.

Blache, Jules: Modes of Life in Moroccan Countryside. *Geog. Rev.*, pp. 478-502, New York, October 1921. Survival of medieval walls and forts, 22 Illustrations.

Dickinson, R. E.: Morphology of Medieval German Towns. *Geog. Rev.*, pp. 74-97, Jan. 1945. 34 good maps and plans.

Gradmann, R.: *Suddeutschland.* 2 vols. Stuttgart. 1931.

Hamm, E.: *Die Deutsche Stadt im Mittelalter.* 1935.

Hodgen, M. T.: Fairs of Elizabethan England. *Econ. Geog.* Oct. 1942. Seven maps.

Hoffbauer, T. J. H.: Paris, à travers les ages. Paris 1882.

Holtedahl, O.: Geography ot Fairs. *Geog. Rev.*, pp. 521-532. Oct. 1924. 11 maps of fairs and 5 photos. Describes types.

Klaiber, C.: *Die Grundrissgestaltung der deutschen Stadt im Mittelalter.* Berlin 1912.

10. LATE MEDIEVAL CITIES

Behrendt, W. C.: *Die Hollandische Stadt.* Berlin 1928.

Benson, E.: Life in a Medieval City (York). London 1920.

Boesch, H.: Basle, Switzerland. *Econ. Geog.* July 1936. No maps.

Darby, H. C.: Early Geography of the Fenland. *Geog. Jnl.* Nov. 1932. One map.

Chapter

Freeman and Hunt: *Historic Town Series.* 11 Vols. London 1889–93.

Gallois, L.: Origin of Paris. *Geog. Rev.*, pp. 345-368. New York. July 1923. 17 illustrations, good maps of growth in last century, old plans.

Geddes, P.: Cities in Evolution. *Geog. Rev.* London. 1915. Civic Survey of Edinburgh. Edinburgh. 1911.

Geisler: *Die Deutsche Stadt.* Stuttgart. 1924.

O'Dell, A. C.: Urbanisation of Shetlands. *Geog. Jnl.* June 1933. 10 illustrations.

Rees, H.: Growth of Bristol. *Econ. Geog.* October 1945. 4 Maps.

11. THE MODERN CITY

Aurousseau, M.: Review of Recent Urban Geography. *Geog. Rev.*, pp. 444-456. New York. July 1924.

Baugh, R. E.: Early Los Angeles. *Econ. Geog.* Jan. 1942. Three early maps.

Bromehead, C. E. N.: Geography and the Growth of London. *Geog. Jnl.* August 1922. Map showing river gravels. Deals chiefly with the central area.

Burgess, E. W.: *The Urban Community.* Chicago. 1926. A Study of the evolution of Urban Communities.

Carpenter, N.: *The Sociology of City Life.* (Longmans) 1931. Factors determining the location of cities.

Charlesworth, E.: Yorkshire Industry. *Geog. Jnl.* Apr. 1938. One map.

Cornish, Vaughan: *The Great Capitals.* London. 1923.

Dodge, S. D.: Bureau and Princeton Community. *Ann. Amer. Ass. Geog.*, pp. 159–209. Sept. 1932. Topography, circulation, limits, history and functions of two towns.

Douglass, H. P.: *The Suvurban Trend.* New York. 1925. An analysis of suburban growth in U.S.A.

Hartshorne, R.: Twin City District. *Geog. Rev.*, pp. 431-443. July 1932. 1 Map, 4 photos of Minneapolis and St. Paul.

Jefferson, M.: Great Cities of 1930. *Geog. Rev.*, pp. 90–101. Detroit. Jan. 1933. Graphs.

Jefferson, M.: Great Cities of U.S. *Geog. Rev.*, pp. 479–488. July 1941.

Lynd and Merrill: *Middletown.* New York. 1929. The resources of a small mid-western city in U.S.A.

Maxey, C. C.: Urban Democracy. *Geog. Rev.*. New York. 1927. Urban politics and government.

McClure, H. R.: Stryal in Cheshire. *Geog. Jnl.* June 1939. One map.

McClenehan, B. A.: *The Changing Urban Neighbourhood.* Los Angeles 1929. An ecological study of Los Angeles.

Moscheles, J.: Greater Prague. *Geog. Rev.*, pp. 414-430. July 1937. Sociological maps.

Mumford, Lewis: *Technics and Civilisation.* New York. 1934.

Chapter

Murphy, R. E.: Johnstown and York (Penn.). *Ann. Amer. Ass. Geog.*, pp. 175-196, Dec. 1935. Contrasts in two manufacturing towns.

Poete, Marcel: *Introduction a l'Urbanisme*. Paris. 1929.

Powell, L. P.: *Historic Towns of U.S.A.* New York. 1900. 3 volumes.

Schlesinger, A. M.: *The Rise of the City*. New York. 1933. American life in cities.

Smailes, A. E.: Population Changes near Newcastle. *Geog. Jnl.* March 1938. 10 illustrations.

Spate, O. H. K.: Industrial Revolution in London. *Geog. Jnl.* Nov. 1938. One map of Bermondsey.

Spate, O. H. K.: Development of Capitals. *Geog. Rev.*, pp. 622-631. Oct. 1942.

Spencer, J. E.: House types in S. Utah. *Geog. Rev.*, pp. 444-457. July 1945. 21 photos.

Thomas, L. F.: Sequence of Occupance in St. Louis. *Ann. Amer. Ass. Geog.*, pp. 75-90. June 1931. Five stages from 1785 to 1925.

Thompson, W. S.: *Population Problems*. New York. 1930.

Weber, A. F.: *Growth of Cities*. New York. 1899. A discussion of rural-urban migration.

Wilson, H. F.: Roads of Winsor. *Geog. Rev.*, pp. 379-398. July 1931. History of township. Maps 1853 and 1929. Farms, industrial changes.

Wright, J. K.: Diversity of New York City. *Geog. Rev.*, pp. 620-639. Oct. 1936. Time to get to work. Residences and business maps, negroes.

Zierer, C. M.: San Fernando, Calif. *Ann. Amer. Ass. Geog.*, pp. 1-28 March 1934. Site, vicinity, railways, functions, health.

12. CITIES OF THE PLAINS

Almagia, R.: Repopulation of the Roman Campagna. *Geog. Rev.*, pp. 529-556. N.Y. Oct. 1929. Maps and photos of estates.

Brown, R. H.: Roswell, New Mexico. *Geog. Rev.*, pp. 610-620. Oct. 1936. 9 illustrations. Artesian water and cotton in plains, block diagrams.

Davis, C. M.: Abandoning High Plains of Michigan. *Geog. Rev.*, pp. 664-673. Oct. 1938. Good maps. Only 14 villages left out of 100.

Finley and Scott: Great Lakes to Gulf Profile. *Geog. Rev.*, pp. 412-420. July 1940. Photos of types of houses.

Fleure, H. J.: Cities of Po Basin. *Geog. Rev.*, pp. 345-362. N.Y. July 1924. Various types of town. Photos.

Gerlach, A. C.: El Segundo (Cal.). *Econ. Geog.* April 1940. 2 maps. Oil.

Kallner and Rosenau: Palestine. *Geog. Rev.*, pp. 61-68. Jan. 1939. Note circular plan of Nahalal near Nazareth.

Miller, H. V.: New Albany, Indiana. *Econ. Geog.* Jan. 1938. Two maps.

Scofield, E.: Settlement patterns in Rural New England. *Geog. Rev.*, pp. 652-663. October 1938. Good patterns

Chapter

Spencer, J. E.: Changing Chungking. *Geog. Rev.*, pp. 46–61. Jan. 1939. With four maps.

Wright, A. J.: Ohio Town Patterns. *Geog. Rev.*, pp. 615–624. Oct. 1937. 8 illustrations. Reserves, canals, railways and roads effects.

13. CITIES ON RIVERS

Beynon, E. D.: Budapest. *Geog. Rev.*, pp. 256–276. April 1943. 6 good maps. Titled nobles, slums, etc. shown.

Brown, R. H.: Monte Vista, Colorado. *Geog. Rev.* N.Y., pp. 529–556. October 1928. Irrigated land.

Carls and Ristow: Seneca Falls. *Econ. Geog.* July 1936. Three maps.

Glendinning, R. M.: Population of Lake St. John. *Geog. Rev.*, pp. 278-291 April 1934. Maps showing population.

Henry, P. W.: Great Lakes Waterway. *Geog. Rev.*, N.Y., pp. 258-278 April 1927.

James, P. E.: Vicksburg. *Geog. Rev.*, N.Y., pp. 234-259. Oct. 1931. Changes in Mississippi. 7 Maps.

Renner, G. T.: Interpretation of the Fall Line. *Geog. Rev.* N.Y., pp. 278–287, April 1927.

Trewartha, G. T.: Prairie du Chien Terrace—a Confluence Site. *Ann. Amer. Ass. Geog.* June 1932., pp. 119–158. Site, soils, successive cultures. history, function.

14. PORTS

Adams, J. Q.: Prince Rupert, B.C., *Econ. Geog.* April 1938. Two maps.

Alcock, F. J.: Across Gaspé. *Geog. Rev.*, N.Y., pp. 197-215. Apr. 1924. 20 photos. Types of houses and scenery.

Buxton, L. H. D.: Malta. *Geog. Rev.*, N.Y., pp. 75-88. Jan. 1924. One map of towns.

De Geer, S.: Greater Stockholm. *Geog. Rev.*, N.Y., pp. 497-507. Oct. 1923. Functional maps.

Dobby, E. H. G.: Singapore. *Geog. Rev.*, pp. 84-110. Jan. 1940. Many photos and maps.

East, W. G.: Historical Geography of Whitby. *Geog. Jnl.* Dec. 1932. 4 maps.

Hitch, M. A.: Port of Tientsin. *Geog. Rev.*, pp. 367-382. July 1935. Silt and floods.

James, P. E.: Rio and Sao Paulo. *Geog. Rev.*, pp. 271-299, Apr. 1933. 12 good maps.

Levainville, J.: Port of Strasburg. *Geog. Rev.*, pp. 243-255, April 1923. 7 illustrations.

Mead, W. R.: Ports of Denmark. *Econ. Geog.* Jan. 1942. Good maps of Odensee, Aarhus and Aalborg.

Chapter

Muddle, I. E.: Port of Dover. *Geog. Jnl.* June 1934. Two maps.

McKean G. L.: Tacoma, Lumber Metropolis. *Econ. Geog.* July 1941.

Olsson, W. W.: Stockholm. *Geog. Rev.*, pp. 420–348, July 1940. Good historic maps.

Rees, H.: Mersey estuary settlements. *Econ. Geog.* April 1945. 2 maps.

Spate, O. H. K.: Rangoon. *Geog. Rev.* Jan. 1942., pp. 56–74. 9 maps.

Van Cleef, E.: Dantzig and Gdynia. *Geog. Rev.* Jan. 1933., pp. 101–108.

Van Cleef, E.: East Baltic Ports. *Geog. Rev.* Apr. 1945., pp. 257–272. Konigsberg and hinterland, Gdynia, etc.

Van Cleef, E.: *Trade Centers and Trade Routes.* New York. 1937.

Vaillaux, C.: Maritime and rural life in Norway. *Geog. Rev.* N.Y. Oct. 1924., pp. 505–519. 3 maps of population.

Zierer, C. M.: Melbourne as a Functional Center. *Ann. Amer. Ass. Geog.*, December 1941, pp. 251–28. Site, history, land use, functions, power, port.

Zierer, C. M.: Land Use Differentiation in Sydney, Australia. *Ann. Amer. Ass. Geog.* Sept. 1942, pp. 255–308. Origin, early maps, site, port, functions, factories.

Zierer, C. M.: Brisbane. *Econ. Geog.* Oct. 1941. Three maps and photos.

Whittlesey, D.: Dakar. *Geog. Rev.* Oct. 1941., pp. 609–639. Planned city for different cultures, also Goree. 18 illustrations.

15. MOUNTAIN CITIES

Garnett, A.: Insolation Topography and Settlement in Alps. *Geog. Rev.*, pp. 601–617. Oct. 1935. Study of noonday shadows, etc. Maps.

Leddihn, E. R. K.: Nauders. *Geog. Rev.*, pp. 239–257. April 1945. Cultural and photos.

Ogilvie, A. G.: Settlements in S. Macedonia. *Geog. Rev.* N.Y., pp. 172–198. April 1921. Block diagrams with position of settlement, Saloniki. 11 illustrations.

Unstead, J. F.: Lotschental. *Geog. Jnl.* Apr. 1932. Four maps.

16. MINING CITIES

Gleave, J. T.: Tees Steel Works. *Geog. Jnl.* May 1938. Four maps.

Harris, C.: Ruhr Coal-mining District. *Geog. Rev.*, pp. 194–222. April 1946. 12 maps.

Hoffmeister, H. A.: Central City Mining Area. *Econ. Geog.* Jan. 1940. Two maps and photos of mines in Colorado.

Kendall, O. D.: Steel Industry of Scunthorpe (Lincoln). *Econ. Geog.* July 1938. Three maps.

Platt, R. F.: Marquette, a Field Study. *Ann. Amer. Ass. Geog.*, pp. 52–73. March 1931. Site, pattern and function; relation to region.

Chapter

White, L.: Iron at Sparrows Point, Md. *Geog. Rev.*, pp. 244-259. A complete steel community near Baltimore. 3 maps.

Zierer, C. M.: Scranton. *Geog. Rev.*, N.Y., pp. 415-278. July 1927. Coal and functions.

Zierer, C. M.: Newcastle N.S.W. *Econ. Geog.* Jan. 1941. Three maps.

17. RESORTS

Davis, D. H.: The Finland Community, Minnesota. *Geog. Rev.*, pp. 382-394. July 1935. Mostly photos, some maps of two centres.

Toschi, U.: Vatican City State. *Geog. Rev.*, N.Y., pp. 529-539. Oct. 1931. 3 maps, 4 photos. Only 898 inhabitants.

18. PLANNED CITIES

Bressey, C.: Greater London Highways. *Geog. Jnl.* Nov. 1939. Two maps.

Brown, R. M.: The Business of Recreation. *Geog. Rev.* July 1935. Costs mainly.

Howard, E.: *Garden Cities of To-morrow.* London. 1902.

Le Corbusier: *The City of the Future.* New York. 1930.

McKenzie, R. D.: *Metropolitan Community.* New York. 1933. A Study of the Development of New York.

Purdon, C. B.: *The Garden City.* London. 1923.

Thompson, W. S.: *Population Problems.* New York. 1930. A discussion of the expansion of urban centres.

Various authors: *Regional Survey of New York and its Environs.* New York. 1927-29. Ten volumes. A survey of the growth and resources of the Metropolitan region.

19. REGIONAL SURVEYS

Abercrombie, P.: *Bristol and Bath.* Regional Plan. Liverpool 1930.

Abercrombie, P.: County of London Plan. *Geog. Jnl.* Nov. 1943. Map showing ring and radial roads, three photos.

Baker and Gilbert: Axial Belt of Industry in England. *Geog. Jnl.* Jan. 1944. Nine maps of England.

Barbour, G. B.: Tennessee Valley Project. *Geog. Jnl.* May 1937. Six Illustrations.

Bromehead, C. E. N.: Growth of London. *Geog. Jnl.* Aug. 1922. Geological map.

Dickinson, R. E.: Population in England. *Geog. Rev.*, pp. 279-286. N.Y. April 1932 (Conurbation).

Fawcett, C. B.: *Provinces of England.* London. 1919.

Fawcett, C. B.: Natural Divisions of England. *Geog. Jnl.* Feb. 1917. Twelve divisions including conurbations of Newcastle, W. Yorks., S. Lancs. and the Black Country. One large topographic map.

Chapter

Fawcett, C. B.: Urban Population in Great Britain. *Geog. Jnl.* Feb. 1932. Map of conurbations, also two maps showing changes in population.

Gilbert, E. W.: Practical Regionalism. *Geog. Jnl.* July 1939. 20 maps.

Hudson, G. D. et al: Knoxville-Chattanooga Area. *Econ. Geog.* July 1939. Eight maps and 20 photos.

Joerg, W. L. G.: Geography and National Land Planning. *Geog. Rev.*, pp. 177–208. April 1935. 13 good maps, erosion, forests, etc.

Stamp, L. D.: Land Utilisation in Britain. *Geog. Rev.* Oct. 1934. pp. 646–650.

Stamp, L. D.: The Scott Report. *Geog. Jnl.* Jan. 1943. No maps.

Stamp. L. D.: Land Utilisation Survey of Britain. *Geog. Jnl.* July 1931. One map.

Spate, O. H. K.: Industrial Evolution of London to 1850. *Geog. Jnl.* Nov. 1938. Map of Bermondsey 1799.

Taylor, Forrester, etc.: Regional Planning. *Geog. Jnl.* Feb. 1942. 4 maps.

Fagg, C. C. and Hutchins, C. E.: *Great Britain* (Weald). Cambridge. 1938.

20. CONTROL AND CLASSIFICATION

Bressey, C. *et. al.*: Greater London Highway Development. *Geog. Jnl.* Nov. 1939. Two maps of recommended roads.

Dickinson, R. E.: Metropolitan Regions of U.S.A. *Geog. Rev.*, pp. 278–291. April 1934. Wholesale trade, news, marketing, Chicago maps.

Duffus, R. L.: *Mastering a Metropolis.* New York. 1930. A summary of the nine-volume regional Survey of N.Y.

Kellogg, P. U.: *The Pittsburg Survey.* New York. 1914. 6 volumes.

Munro, W. B.: *Municipal Government* (Macmillans). 1923. The political organisation of a modern municiplaity.

Ogburn, W. F.: Social Characteristics of Cities. Chicago. 1937.

Parkins, A. E.: Profiles in Nashville. *Ann. Amer. Ass. Geog.*, pp. 164–175. Sept. 1930. Factors in the localisation of retail districts.

Purdom, C. B.: *The Building of Satellite Towns.* New York. 1925.

Taylor, G. R.: *Satellite Cities.* New York. 1915.

Thompson, J. G.: *Urbanization.* New York. 1927.

INDEX

(Geographical Terms and Foreign Words in *Italics*.)

For Product Safety Concerns and Information please contact our EU representative GPSR@taylorandfrancis.com
Taylor & Francis Verlag GmbH, Kaufingerstraße 24, 80331 München, Germany

www.ingramcontent.com/pod-product-compliance
Lightning Source LLC
LaVergne TN
LVHW011104300926
849160LV00008B/273
* 9 7 8 0 4 1 5 4 8 9 5 8 4 *